The European Landscape Convention

Volume 13

Landscape Series

Volume 13

Aims and Scope

Springer's innovative Landscape Series is committed to publishing high-quality manuscripts that approach the concept of landscape from a broad range of perspectives. Encouraging contributions on theory development, as well as more applied studies, the series attracts outstanding research from the natural and social sciences, and from the humanities and the arts. It also provides a leading forum for publications from interdisciplinary and transdisciplinary teams.

Drawing on, and synthesising, this integrative approach the Springer Landscape Series aims to add new and innovative insights into the multidimensional nature of landscapes. Landscapes provide homes and livelihoods to diverse peoples; they house historic – and prehistoric – artefacts; and they comprise complex physical, chemical and biological systems. They are also shaped and governed by human societies who base their existence on the use of the natural resources; people enjoy the aesthetic qualities and recreational facilities of landscapes, and people design new landscapes.

As interested in identifying best practice as it is in progressing landscape theory, the Landscape Series particularly welcomes problem-solving approaches and contributions to landscape management and planning. The ultimate goal is to facilitate both the application of landscape research to practice, and the feedback from practice into research.

For further volumes:
http://www.springer.com/series/6211

Michael Jones · Marie Stenseke
Editors

The European Landscape Convention

Challenges of Participation

Springer

Editors
Prof. Michael Jones
Norwegian University of Science
and Technology
Department of Geography
Trondheim
Norway
michael.jones@svt.ntnu.no

Dr. Marie Stenseke
Gothenburg of University
School of Business, Economics
and Law
Department of Human
and Economic Geography
PO Box 630
Göteborg
Sweden
Marie.Stenseke@geography.gu.se

ISSN 1572-7742
e-ISSN 1875-1210
ISBN 978-90-481-9931-0
e-ISBN 978-90-481-9932-7
DOI 10.1007/978-90-481-9932-7
Springer Dordrecht Heidelberg London New York

Printed on acid-free paper

Springer is part of Springer Science+Business Media (www.springer.com)

Preface

The present book provides the first extensive presentation of the challenges of public participation in relation to the European Landscape Convention. The idea for the book arose out of a series of special sessions, organized by the editors, on 'The European Landscape Convention and Participatory Development Planning', held during the 23rd Session of the Permanent European Conference for the Study of the Rural Landscape (PECSRL) at Lisbon and Óbidos, Portugal, 1–5 September 2008. A total of 21 papers were presented in five sessions. Twelve chapters of this book have developed out of 14 of the papers presented in Portugal. The studies include cases from 11 countries in northern, southern, western, and Eastern Europe. While the planned chapters on Italy, Germany and Russia unfortunately did not materialize, an additional invited chapter provides a case study from Britain. Examples from both signatories and non-signatories of the European Landscape Convention are included.

The chapters are arranged in two main sections. Part I deals with implementation of public participation in relation to the European Landscape Convention, both theoretically and through case studies. Part II is concerned with participatory methods in practice, again through selected cases. The case studies presented here provide illustrations of both successful and less successful applications of participatory approaches to landscape protection, management, and planning. Some lessons that may be drawn from these studies are presented in the concluding chapter.

Each of the main chapters in this book has been subject to peer review by two reviewers, one external and one within the group of contributors. We would like to thank these 26 anonymous reviewers for their constructive suggestions, as well as the seven anonymous reviewers of the original book proposal. We also thank Radmil Popovic of the Department of Geography, Norwegian University of Science and Technology, for technical assistance with a number of the maps and diagrams in this volume, and Linda Clark for compiling the index.

Trondheim, Norway
Gothenburg, Sweden
10 June 2010

Michael Jones
Marie Stenseke

Contents

Contributors

Henk Baas Cultural Heritage Agency, Ministry of Education, Culture and Science, NL-3800 BP Amersfoort, The Netherlands, h.baas@cultureelerfgoed.nl

Elinor Bjärnborg Alvesta kommun, Centralplan 1, SE–342 80 Alvesta, Sweden, elinor.bjarnborg@alvesta.se

Morten Clemetsen Department of Landscape Architecture and Spatial Planning, Norwegian University of Life Sciences, NO-1432 Ås, Norway, morten.clemetsen@umb.no

Karoline Daugstad Department of Geography, Norwegian University of Science and Technology, NO-7491 Trondheim, Norway, karoline.daugstad@svt.ntnu.no

Berezi Elorrieta Faculty of Geography and History, University of Barcelona, 08001 Barcelona, Spain, berezielorrieta@gmail.com

Bert Groenewoudt Cultural Heritage Agency, Ministry of Education, Culture and Science, NL-3800 BP Amersfoort, The Netherlands, b.groenewoudt@cultureelerfgoed.nl

Mats Gyllin Department of Work Science, Business Economics, and Environmental Psychology, Swedish University of Agricultural Sciences, P.O. Box 88 SE–230 53 Alnarp, Sweden, mats.gyllin@ltj.slu.se

Christine Haaland Department of Landscape Management, Design and Construction, Swedish University of Agricultural Sciences, P.O. Box 66 SE–230 53 Alnarp, Sweden, christine.haaland@ltj.slu.se

Thierry Joliveau University of Lyon & CNRS UMR EVS, F-42023 Saint-Etienne, France, thierry.joliveau@univ-st-etienne.fr

Michael Jones Department of Geography, Norwegian University of Science and Technology, NO-7491 Trondheim, Norway, michael.jones@svt.ntnu.no

Erling Krogh Department of Mathematical Sciences and Technology, Norwegian University of Life Sciences, NO-1432 Ås, Norway, erling.krogh@umb.no

Mart Külvik Institute of Agricultural and Environmental Sciences, Estonian University of Life Sciences, EE-51014, Tartu Estonia, mart.kylvik@emu.ee

Anders Larsson Department of Landscape Architecture, Swedish University of Agricultural Sciences, P.O. Box 58 SE–230 53 Alnarp, Sweden, anders.larsson@ltj.slu.se

Anna Majchrowska Department of Physical Geography, University of Lodz, 91-139 Lodz, Poland, majchrow@uni.lodz.pl

Yves Michelin Clermont University & UMR Metafort, VetAgro Sup, F-63370 Lempdes, France, y.michelin@vetagro-sup.fr

Anna Peterson Department of Landscape Architecture, Swedish University of Agricultural Sciences, P.O. Box 58 SE–230 53, Alnarp, Sweden, anna.peterson@ltj.slu.se

Claire Planchat-Héry Clermont University and UMR Metafort Agro Paris Tech, BP50085, F-63172, Aubière, France, metafort.planchat@gmail.com

Edwin Raap Landschapsbeheer Nederland, NGO, NL-3506 GT Utrecht, The Netherlands, e.raap@landschapsbeheer.nl

Isabel Loupa Ramos Centre for Studies of Regional and Urban Systems (CESUR), Technical University of Lisbon, 1049-001 Lisbon, Portugal, isa.ramos@civil.ist.utl.pt

Dolores Sánchez-Aguilera Faculty of Geography and History, University of Barcelona, 08001 Barcelona, Spain, dsanchez.aguilera@gmail.com

Neil Spencer Sustainable Futures, Ashreigney Chulmleigh, Hayes, EX18 7NG Devon, UK, neil.spencer@sustainablefutures.co.uk

Marie Stenseke Department of Human and Economic Geography, University of Gothenburg, SE–405 30 Göteborg, Sweden, Marie.Stenseke@geography.gu.se

Monika Suškevičs Institute of Agricultural and Environmental Sciences, Estonian University of Life Sciences, EE-51014 Tartu, Estonia, monika.suskevics@emu.ee

Theano S. Terkenli Department of Geography, University Hill, University of the Aegean, Lesvos 81100, Greece, t.terkenli@aegean.gr

Kine Halvorsen Thorén Department of Landscape Architecture and Spatial Planning, Norwegian University of Life Sciences, NO-1432 Ås, Norway, kine.thoren@umb.no

Chapter 1
The Issue of Public Participation in the European Landscape Convention

Michael Jones and Marie Stenseke

Abstract The chapter introduces the European Landscape Convention (ELC) and its innovative features compared to earlier approaches to landscape. The Convention provides a new definition of landscape. It applies to all landscapes, not just selected ones, and underlines the diversity of landscapes as a value. It emphasizes that landscape is not an exclusive field for scientific and technical specialists but the concern of everybody, and advocates an enhanced role for public participation in landscape issues. Further, it highlights the principle of subsidiarity, requiring that

M. Jones (✉)
Department of Geography, Norwegian University of Science and Technology, NO-7491 Trondheim, Norway
e-mail: michael.jones@svt.ntnu.no

M. Jones, M. Stenseke (eds.), *The European Landscape Convention*,
Landscape Series 13, DOI 10.1007/978-90-481-9932-7_1,

landscape matters should be dealt with as closely to the affected population as possible. Next, the chapter provides a brief discussion of landscape concepts. Three prevailing notions of landscape are presented – landscape as morphology, landscape as scenery, and landscape as polity – and then the ELC's definition of landscape as an 'area as perceived by people'. Following this, the chapter discusses the diversity of landscapes as an important common value. Respect for and promotion of cultural diversity is part of the Council of Europe's objective of promoting a democratic culture based on respect for law while actively involving civil society and citizens. Participation as provided for by the Aarhus Convention and followed up by the European Landscape Convention is then presented, followed by a discussion of the provisions in the ELC for implementation. The chapter concludes with a section on the ELC and participation in practice, briefly introducing the individual chapters of the book.

Keywords Landscape definition · Landscape concepts · Diversity of landscapes · Public participation · Implementation of European Landscape Convention (ELC)

1.1 The European Landscape Convention

The European Landscape Convention (ELC) is a step in the project of the Council of Europe to protect human rights, pluralistic democracy, and the rule of law. Founded by the Treaty of London in 1949, the Council of Europe aims to achieve greater unity among the nations of Europe in order to safeguard their common heritage of values and ideals that form 'the source of individual freedom, political liberty and the rule of law' (Council of Europe, 1949: Preamble and Article 1). The political agenda of the Council of Europe (since 2007 with 47 member countries) is expressed in its Action Plan adopted in 2005. This includes the commitment to 'promoting common fundamental values: human rights, rule of law and democracy', under which the goal is specified of strengthening democracy and good governance nationally, regionally and locally through, among other things, citizens' participation (Council of Europe, 2005: I).

The European Landscape Convention (ELC) celebrates its tenth anniversary in 2010. The Convention was opened for signature at Florence on 20 October 2000 and hence is often referred to as the Florence Convention (Council of Europe, 2000a). It entered into force on 1 March 2004 after the required first ten ratifications had been obtained. By September 2010, 32 Parties had signed and ratified it while a further seven had signed but not yet ratified it. Of the 27 member countries of the European Union (EU), all but three (Austria, Estonia and Germany) have signed the Convention, although two of the signatories (Malta and Sweden) have not yet ratified it. However, the Swedish government announced in November 2010 its decision to ratify the Convention. The EU as such has not so far acceded to the ELC.

The Convention notes that landscape is 'an important public interest' and 'an important part of the quality of life for people everywhere'. Landscape 'contributes to the formation of local cultures' and 'is a basic component of the European natural and cultural heritage, contributing to human well-being and consolidation of the European identity'. Noting that 'changes in the world economy are in many cases accelerating the transformation of landscapes', the ELC expresses a response 'to the public's wish to enjoy high quality landscapes and to play an active part in the development of landscapes'. Landscape is seen as 'a key element of individual and social well-being' and 'its protection, management and planning entail rights and responsibilities for everyone' (Council of Europe, 2000a: Preamble).

The stated aims of the ELC 'are to promote landscape protection, management and planning, and to organise European cooperation on landscape issues' (Council of Europe, 2000a: Article 3). Parties to the ELC undertake to implement four General Measures (Council of Europe, 2000a: Article 5):

a. to recognise landscapes in law as an essential component of people's surroundings, an expression of the diversity of their shared cultural and natural heritage, and a foundation of their identity;
b. to establish and implement landscape policies aimed at landscape protection, management and planning...
c. to establish procedures for the participation of the general public, local and regional authorities, and other parties with an interest in the definition and implementation of the landscape policies...
d. to integrate landscape into its regional and town planning policies and in its cultural, environmental, agricultural, social and economic policies, as well as any other policies with possible direct or indirect impact on the landscape.

Specific Measures that Parties undertake to follow up are: raising awareness among civil society, private organizations, and public authorities of the value of landscapes; training and education related to landscape matters; identification and assessment of landscapes; definition of Landscape Quality Objectives; and implementation of landscape policies (Council of Europe, 2000a: Article 6).

To facilitate the application of the ELC, an *Explanatory Report* was published together with the Convention, discussing in more detail its provisions (Council of Europe, 2000b). It was drawn up by experts under the authorization of the Council of Europe's Committee of Ministers, although it is not meant as an authoritative interpretation of the treaty's provisions (Olwig, 2007: 588).

In 2008, *Guidelines for the Implementation of the European Landscape Convention* were recommended by the Council of Europe's Committee of Ministers (Council of Europe, 2008). They contain 'a series of theoretical, methodological and practical guidelines' for the implementation of the ELC and set out a number of general principles. It is stated that: 'The identification, description and assessment of landscapes constitute the preliminary phase of any landscape policy'. Landscape strategies are to be drawn up at each administrative level. The landscape dimension is to be integrated into territorial polices (spatial management plans) and sectorial policies. Every planning action or project should comply with Landscape Quality Objectives, aiming to improve landscape quality or at least prevent a decline (Council of Europe, 2008: §I.1). Landscape policies are not to be considered as additional to other policy themes but as an integral part of them. Operationally, a transition is presupposed from policy based only on protecting outstanding landscape features to policy based on the 'quality of all living surroundings, whether outstanding, everyday or degraded'. New forms of collaboration between various bodies and levels of administration are advocated. Territories should be viewed as a whole rather than places to be protected simply being identified. Several approaches should be combined, linking ecological, archaeological, historical, cultural, perceptual, and economic perspectives, and both social and economic aspects should be considered (Council of Europe, 2008: §I.4).

The ELC is the first international treaty specifically devoted to landscape as a unity. The history of the ELC's origins is summarized in the Convention's *Explanatory Report* (Council of Europe, 2000b: I). The Convention was drawn up on the initiative of the Congress of Local and Regional Authorities of the Council of Europe (CLRAE), a political assembly representing c.200,000 local and regional authorities. Hence it has administratively a bottom-up rather than a top-down impetus. Inspiration was provided by the Mediterranean Landscape Charter, adopted in Seville in 1993, and aimed at developing landscape conservation and management policy to meet the threats that uncontrolled development posed for ecological and historical landscape values (Sarlöv Herlin, 2007). The predecessor of the CLRAE, the Standing Conference of Local and Regional Authorities of Europe, called in 1994 for a framework European convention on the management and protection of cultural landscapes based on the Mediterranean Landscape Charter. The following year, the European Environment Agency of the EU published *Europe's Environment: The Dobříš Assessment* (Stanners and Bourdeau, 1995), in which the Council of Europe was encouraged to take the lead in drawing up a European convention on rural landscapes. The same year the World Conservation Union (IUCN) advocated an international convention on protection of rural landscape in Europe in its report *Parks for Life* (IUCN, 1995). The work of drafting the European Landscape Convention, beginning in 1994, involved extensive consultation with interested international organizations, national governments, and regional authorities, as well as scientific bodies and non-governmental organizations.

Maguelonne Dejéant-Pons, Head of the Spatial Planning and Landscape Division of the Council of Europe, in describing the scope and originality of the Convention (Dejéant-Pons, 2006: 365–367), states that it 'represents an important contribution to the implementation of the Council of Europe's objectives'. She points out that the ELC is 'the first international treaty to be exclusively concerned with all dimensions of European landscape'. Unlike UNESCO's Convention Concerning the Protection of the World Cultural and Natural Heritage of 1972, the ELC covers all landscapes, not just those of 'outstanding universal value'. Its objective is not to list landscape assets of exceptional value, but to promote rules and principles for landscape protection, management, and planning relevant for all types of landscape. In the wording of the ELC's *Explanatory Report* (Council of Europe, 2000b: §45):

> The Convention's original feature is that it applies to ordinary landscapes no less than outstanding ones, since all decisively influence the quality of Europeans' surroundings. Thus everyday, outstanding and damaged landscapes all come within its scope. This comprehensive coverage is justified for the following reasons: every landscape forms the setting for the lives of the population concerned; urban and rural landscapes interlock in complex ways; most Europeans live in towns and cities (large or small), the quality of whose landscapes greatly affects their lives; and rural landscapes occupy an important place in the European consciousness. It is also justified by the profound changes which European landscapes, particularly peri-urban ones are now undergoing.

All in all, the ELC contains a number of innovative features compared to earlier approaches to landscape. It provides a new definition of landscape. It applies to all landscapes, not just selected ones, and underlines the diversity of landscapes as a value. It emphasizes that landscape is not an exclusive field for scientific and technical specialists but the concern of everybody, and advocates an enhanced role for public participation in landscape issues. It highlights the principle of subsidiarity, requiring that landscape matters should be dealt with as closely to the affected population as possible.

1.2 The Landscape Concept

The advent of the ELC has led to discussion of how understandings of landscape in legislation, policy, planning, and management may be affected (e.g. Ermischer, 2004; Howard, 2004; Sarlöv Herlin, 2004; Scazzosi, 2004; Groening, 2007; Jones, 2007; Jones et al., 2007; Olwig, 2007). Not least, the enhanced role for public participation is likely to influence the conceptualization and use of the term.

In the Convention's approach, the landscape is not simply 'a given assemblage of physical objects, which can be objectively analysed by the natural or social scientist', but a product of 'changeable cultural perceptions and identities' (Olwig, 2007: 581). The meaning of the term 'landscape' is broader than that of a view or scenic panorama, which characterized many environmental and historical heritage policies earlier, and broader than 'nature' or 'environment' (Scazzosi, 2004: 337). The Convention challenges perceptions by some scientists, technicians, and planners of landscape as a form of scenery (Olwig, 2007: 582). Nor is the landscape an objective scenic location but 'a place constituted through the tangible and intangible social and cultural practices that shape the land'; further, 'it is not primarily the experts who are to plan and develop this landscape, but rather, the people whose daily practices and perceptions shape the social and physical landscape' (Olwig, 2007: 581). Olwig (2007: 584–585) points out that through the role of the CLRAE the genesis of the European Landscape Convention lies in a convening of local and regional authorities. In Olwig's words (2007: 579–580):

> . . .the 'conventional' meaning of landscape does not lie in the establishment of a fixed, theoretically founded, definition from which planning is to proceed (as in classic top down planning). Rather this meaning must be found in the process that sets in motion a plethora of gatherings involving members of various interest groups, polities and communities, in

which the common perception of landscape that emerges provides a basis for subsequent practice. This perception of landscape is therefore largely the outcome of public discursive practice rather than scientific reasoning.

The term 'landscape' has a diversity of contemporary meanings as well as historical layers of meaning. Discussions of the landscape concept have produced a vast literature (recent examples include Cosgrove, 2000; Ingold, 2000; Mitchell, 2000: 89–144; Mitchell, 2002; Olwig, 2002; Rose, 2002; Jones, 2003; Winchester et al., 2003; Widgren, 2004; Wylie, 2007; Jones and Olwig, 2008; Gray, 2009; Morin, 2009; Wylie, 2009). In the following (based on Jones 2010), three prevailing notions of 'landscape' are presented – landscape as morphology, landscape as scenery, and landscape as polity – and then the definition of 'landscape' in the European Landscape Convention.

1.2.1 Landscape as Morphology

The conception of landscape as morphology focuses on the material forms of our physical surroundings. Landscape in this sense is studied by scientists, ostensibly in an objective manner, as an areal unit of distinctive physical character, associated forms or interrelated features. A distinction is often made between natural forms of the landscape, studied from a natural science perspective, and cultural forms, studied from a humanities or social science perspective, although what is natural and what is cultural is subject to discussion. The landscape is variously depicted in maps, photographs, or perspective drawings, as well as being presented in descriptive texts and, for quantified information, in tables and graphs.

These presentations appear objective but nonetheless express a particular view. When addressing landscape change, this approach focuses on changing material forms such as land cover (especially vegetation), buildings, settlements, and other artefacts. The choice of what landscape elements and landscape changes are specifically examined is bound up with ideas of what is important or significant. Although dealing with objectively perceivable phenomena, these ideas of significance often paradoxically contain implicit or explicit judgements of what is 'beautiful' or 'ugly', 'good' or 'bad', 'right' or 'wrong', and 'desirable' or 'undesirable'. Such value judgements may be hidden in the terminology that is used. When we speak of the 'impacts' of humans on nature, they are frequently seen as harmful and therefore regarded in a negative light (e.g. carbon emissions, pollution, habitat fragmentation, or technical installations), and similarly in the case of impacts of nature on humans (e.g. volcanic eruptions, earthquakes, or tsunamis). Globalization is frequently presented as having negative impacts on the landscape. When a distinction is made between 'deliberate' and 'unintended' landscape changes, the latter are regarded as more problematical than the former because their consequences are less easy to foresee. Landscapes that show visible signs of social deprivation and poverty are frequently judged negatively. Physical planning, nature conservation, and cultural heritage management are activities that typically involve description, registration, and inventory of the landscape's morphology (among other things) before making recommendations concerning which landscape forms are 'good' or 'desirable' and hence worthy of preservation.

1.2.2 Landscape as Scenery

The conception of landscape as scenery relates to the visual content of an area observed from a particular viewpoint. This frequently refers to the aesthetic experience of the landscape. Landscape is here studied as an expression of subjective human experiences, feelings, and emotions. The human experience of the physical surroundings varies not only according to the season, weather or time of day, but can also be affected by the mood or fantasy of the observer. This meaning of landscape developed from the Renaissance onwards, and was constituted through theatre, art, and literature. Landscape as 'a way of seeing' (Cosgrove, 1984) initially expressed the view of property owners, which was made to seem natural through the use of perspective drawing. Gillian Rose (1992) has argued that, as the landowner was generally a man, this was also the landscape of the male gaze. However, although it was an elite view, it resonated among a wider population, especially in the period of national romanticism in nineteenth century, when landscape paintings were a means of evoking strong feelings of national sentiment. Such representations of landscape expressed the experiences of artists and writers, but when they were reproduced and disseminated they contributed to expectations concerning the landscape among a wider public. In this way, ideals of landscape were 'socially constructed'.

When the landscape changes, these ideals provide a measure against which the changes can be assessed (frequently negatively). Such ideals have a strong influence on physical planners and conservationists regarding acceptable change and visions of future landscapes, which when implemented can in turn lead to changes in the physical landscape. Representations of landscape, too, vary over time as a result of changing interpretations and ideologies, changing artistic ideals, and changing media (e.g. photography and film). A significant feature in our times is the tourist industry, and its effects on perceptions of landscape beauty as well as on the shaping of physical landscapes (Urry, 2002).

1.2.3 Landscape as Polity

The conception of landscape as polity is the earliest use of the term 'landscape' and is closely related to law. It referred to historical administrative-territorial units in which the land was literally shaped according to the customs and laws of the people, including specific systems of land rights. Kenneth R. Olwig (2002) has demonstrated that the medieval notion of 'landscape' incorporated the characteristics and conditions of a land, including its customs, institutions, and law-making bodies. Historically, the territorial *landskap* or *landskab* of Scandinavia was a politically organized unit or polity within which the shaping of the land expressed the practices of the area's legal system and culture. In the German-speaking areas of Schleswig-Holstein, once under the Danish crown, the last political *Landschaften*, as they were called, disappeared in the mid-nineteenth century. In Sweden, although no longer existing as formal administrative areas, *landskap* are remembered and remain important for people's feelings of regional identity. The internally autonomous *Landskap* of Åland in Finland is an example of a modern self-governing landscape polity.

The role of custom in the landscape polity has helped inspire newer ideas of landscape as a reflection of habitus, practice, and performance. Custom changes according to need and circumstance, yet in a manner that is seen to be in accordance with precedence. Changing customary usages and practices lead to changes in the landscape in ways that are considered acceptable and which do not represent a radical break with the past (Olwig, 2001).

1.2.4 Landscape as 'An Area as Perceived by People'

All of the previously mentioned prevailing notions of landscape – as morphology, scenery, and polity – are subsumed in the European Landscape Convention's concept of landscape, and at the same time given the widest possible interpretation (Jones 2010). As morphology, the landscape includes all types of physical landscape as well as waterscape. As scenery, landscape is perceived not primarily by an elite but by people in general. As polity, landscape is the responsibility of elected authorities together with a participating population.

The Convention defines 'landscape' as 'an area, as perceived by people, whose character is the result of the action and interaction of natural and/or human factors' (Council of Europe, 2000a: Article 1a). Hence landscape is not something objective; it is more than just an area, as 'it also expresses the perceptions of an area that people share, value and use' (Olwig, 2007: 581).

'As perceived by people' implies that the views of all groups should be included, not just the views of a scientific or political elite. The *Explanatory Report* states that: 'Official landscape activities can no longer be allowed to be an exclusive field of study of action monopolised by specialist scientific and technical bodies' (Council of Europe, 2000b: §22). It further specifies that landscape defined in accordance with the Convention may be perceived by 'local inhabitants or visitors' (Council of Europe, 2000b: §38). Landscape protection, management, and planning hence concern the characteristics of the landscape that the involved population wishes to give recognition to in their surroundings (Jones, 2007: 615). The Specific Measures required of the Parties to the Convention include identification of landscapes and analysis of 'their characteristics and the forces and pressures transforming them', and assessment of landscapes 'taking into account the particular values assigned to them by the interested parties and the population concerned' (Council of

Europe, 2000a: Article 6.C.2). According to the *Explanatory Report*, the quality of landscapes should be assessed taking into account the particular values of different kinds assigned to them by the general public and interested parties such as landowners and land users or land managers. The point of this evaluation is to provide a basis for judging what landscape features of an area are so valuable that they should be protected; what features need management in order to maintain the quality of the landscape; and what features or areas should be considered for enhancement (Council of Europe, 2000b: §57).

The *Explanatory Report* specifies that the definition of landscape reflects the idea that landscapes evolve through time, acted upon by both natural forces and human beings, and that the landscape's natural and cultural components form a whole and should not be taken separately (Council of Europe, 2000b: §38). The Convention applies to all types of landscape: natural, rural, urban and peri-urban areas; inland waters and marine waters; and landscapes considered to be 'outstanding', as well as 'everyday' and 'degraded' landscapes. It applies to the entire territory specified by the Parties (Council of Europe, 2000a: Article 2).

In the *Guidelines*, it is stated that the identification, description, and assessment of landscapes involve 'an analysis of morphological, archaeological, historical, cultural and natural characteristics and their interrelations, as well as an analysis of change. The perception of landscape by the public should also be analysed from the viewpoint of both its historical development and its recent significance' (Council of Europe, 2008: §I.1.B). However, going on to discuss the Convention's concept of landscape, the *Guidelines* make the point that it differs from concepts that regard landscape as an 'asset' (heritage concept of landscape) and that assess it as 'cultural' or 'natural' landscape considered as part of physical space. The 'new concept' of the Convention focuses on 'the theme of the quality of the surroundings where people live; this is recognised as a precondition for individual and social well-being (understood in the physical, physiological, psychological and intellectual sense) and for sustainable development, as well as a resource conducive to economic activity' (Council of Europe, 2008: §1.2).

The *Explanatory Report* observes that the quality of the surroundings of the European population 'to some extent has to do with the feelings aroused in them by contemplating the landscape. They have come to realise that the quality and diversity of many landscapes are deteriorating as a result of a wide variety of factors and that this is having an adverse effect on the quality of their everyday lives' (Council of Europe, 2000b: §21). The *Guidelines* similarly refer to subjective experience when they discuss the landscape concept (Council of Europe, 2008: §I.2):

> Sensory (visual, auditory, olfactory, tactile, taste) and emotional perception which a population has of its environment and recognition of the latter's diversity and special historical and cultural features are essential for the respect and safeguarding of the identity of the population itself and for individual enrichment and that of society as a whole.

Taking into account the 'social perception of landscape and popular aspirations in choices regarding landscape protection, management and planning' is an important part of the 'concept of participation'. The *Guidelines* explicitly relate this approach to landscape to a particular democratic agenda: In this sense, the concept of landscape proposed by the Convention implies an exercise in democracy whereby differences are accepted, common characteristics found and operational compromises eventually reached; these represent an alternative to the drawing up by experts of hierarchical classifications of landscape qualities (Council of Europe, 2008: §II.2.3.A).

Alongside public involvement is the role of local and regional authorities, in which the principle of subsidiarity is upheld. Protection, management, and planning of landscapes are considered to be most effective if responsibility is entrusted to the competent authorities closest to the communities concerned (Council of Europe, 2000b: §§48–49). The Convention (Council of Europe, 2000a: Article 4) refers to the principle of subsidiarity in relation to the Council of Europe's European Charter of Local Self-Government (signed in Strasbourg 1985) (Council of Europe, 1985a). This Charter was also an initiative in its time of the Standing Committee of Local and Regional Authorities of Europe, asserting the principles of representative democracy and local autonomy (Council of Europe, 1985b).

1.2.5 Diversity of Landscapes

The emphasis on public participation in the European Landscape Convention is closely related to maintenance of the diversity of European landscapes as an important common value and to recognition of the usefulness of diverse approaches to landscape protection, management, and planning rather than a single universal approach. This is in line with the Council of Europe's Action Plan of 2005, in which protecting and promoting cultural diversity is one of the means of building a more humane and inclusive Europe. Respect for and promotion of cultural diversity is part of the Council of Europe's objective of promoting a model of democratic culture based on respect for law while actively involving civil society and citizens (Council of Europe, 2005: III). UNESCO's Convention on the Protection of Cultural Diversity the same year, affirming that cultural diversity is a defining characteristic of humanity, similarly encourages the active participation of civil society to achieve the convention's objectives of promoting and protecting the diversity of cultural expressions (UNESCO, 2005).

While taking account of the particular values assigned to landscapes by interested parties and the population concerned, the ELC also expresses some general landscape values. First, it is specified that landscape constitutes a resource for economic activity, and its protection, management, and planning can contribute to job creation. Second, landscape contributes to local cultures as well as to European natural and cultural heritage. Third, landscape is part of people's quality of life. Fourth, 'the quality and diversity of European landscapes constitute a common resource' (Council of Europe, 2000a: Preamble). The *Explanatory Report* states that, besides having local significance, Europe's landscapes are of value to all Europeans, and are cherished outside the locality and beyond national borders. 'In their diversity and quality, the cultural and natural values linked to European landscapes are part of Europe's common heritage'; hence landscape is a collective European concern (Council of Europe, 2000b: §§29–30).

The diversity of European landscapes is a value that runs throughout the Convention. Landscape as an expression of shared cultural and natural heritage and as a foundation of identity is the justification for the ELC's first General Measure, the recognition of landscapes in law (Council of Europe, 2000a: Article 5a). The deteriorating quality and diversity of many landscapes is considered to have an adverse effect on people's everyday lives (Council of Europe, 2000b: §21). The Convention is not confined to either cultural or natural components of landscapes but is concerned with all forms of landscape (Council of Europe, 2000b: §26). Recognition of the landscape's diversity is regarded as essential for people's collective and individual identity and enrichment (Council of Europe, 2008: §I.2).

The value of maintaining diversity is reflected in the approach to landscape measures and policies. The *Explanatory Report* states that extending landscape action to the whole of national territories 'does not imply that the same measures and policies must be applied to all landscapes' but they 'should be adaptable to particular types of landscape, which, depending on their specific characteristics, will need various forms of treatment at local level, ranging from the strictest conservation via protection, management and planning to actual creation' (Council of Europe, 2000b: §27). Parties to the Convention are left with 'the choice of means to be used within their internal legal arrangements to fulfil their obligations. The legal, administrative, fiscal and financial arrangements made in each country should fit in as comfortably as possible within that country's traditions' (Council of Europe, 2000b: §34). The *Guidelines* take 'account of advances and developments in the concept of landscape in Europe and of the diverse existing and practical experience in applying the convention' and pay 'due regard to the freedom, and particularly the creativity, of the authorities of each state to draw up legal, operational, administrative and technical landscape-related instruments' (Council of Europe 2008: Introduction). Further, each state decides on its own institutional organization in landscape matters according to its administrative and cultural traditions and existing structures, whether centralized, decentralized or federal (Council of Europe, 2008: §II.1).

The various experimental practices being developed or already in operation in different European countries 'show a diversity of approach to knowledge production that also reflects the diversity of cultural concepts'. Approaches should, however, be cross-disciplinary to avoid disciplinary compartmentalization of knowledge. Nonetheless, measures 'should not be too interventionist' regarding methods and stakeholders involved in the process of knowledge production (Council of Europe, 2008: §II.2.1).

Finally, the definition of Landscape Quality Objectives should 'link the social requirements and values attached to the landscape by the public to the choice of policy decisions' and 'particular importance should be devoted to the range of social perceptions, which reflect the population's diversity' (Council of Europe 2008: §II.2.2).

1.3 Participation

The European Landscape Convention recognizes that landscape is political and advocates principles of landscape governance that actively involve the broad population. The ELC refers in its Preamble to the United Nation's Economic Commission for Europe's Aarhus Convention of 1998 (in force 2001) on Access to Information, Public Participation in Decision-making and Access to Justice in Environmental Matters (UNECE, 1998).

The term 'public' can be taken to mean civil society in the broad sense, according to Michel Prieur and Sylvie Durousseau (2006: 165), experts to the Council of Europe. Public participation complements official decision-making by involving individuals and groups who are otherwise outside the formal decision-making process. It can be compared with co-management in resource utilization, defined as 'the sharing of power and responsibility between government and local resource users' (Zachrisson, 2004: 12). The ELC states that participation is to include the general public, local and regional authorities, and other parties with an interest in the implementation of landscape policies (Council of Europe, 2000a: Article 5c). According to Prieur and Durousseau (2006: 205), 'the interpretation of "public" ought to be extended to its meaning in the broadest sense, including individuals regardless of their place of residence'. In other words, participation is intended to be non-discriminatory, which is in accordance with the provision of the Aarhus Convention that it applies 'without discrimination as to citizenship, nationality or domicile' (UNECE, 1998: Article 3.9). This implies that everyone is entitled to a say: administrators, professionals and ordinary people; women, men and children; residents and visitors; citizens and immigrants; and different ethnic groups (see Jones, 2007: 620–622 on the challenges that this provides).

Public participation is implicit in the ELC's definition of landscape as an area 'as perceived by people', and in the definition of Landscape Quality Objectives, meaning for a specific landscape 'the formulation by the competent public authorities of the aspirations of the public with regard to the landscape features of their surroundings' (Council of Europe, 2000a: Article 1 a and c). Under Specific Measures, each Party is explicitly to undertake identification and assessment of landscapes 'with the active participation of the interested parties ... and with a view to improving knowledge of its landscapes'. Once landscapes have been identified, their characteristics, and the forces and pressures transforming them, are to be analysed, the changes are to be taken note of, and the landscapes thus identified are to be assessed 'taking into account the particular values assigned to them by the interested parties and the population concerned' (Council of Europe, 2000a: Article 6.C.1). Further, each Party undertakes to define Landscape Quality Objectives for the landscapes identified and assessed 'after public consultation' (Council of Europe, 2000a: Article 6.D).

The *Explanatory Report* provides the following justifications for participation (Council of Europe, 2000b: §§23, 24 and 36):

> 23. Landscape must become a mainstream political concern, since it plays an important role in the well-being of Europeans who are no longer prepared to tolerate the alteration of their surroundings by technical and economic developments in which they have had no say. Landscape is the concern of all and lends itself to democratic treatment, particularly at local and regional level.
>
> 24. If people are given an active role in decision-making on landscape, they are more likely to identify with the areas and towns where they spend their working and leisure time. If they have more influence on their surroundings, they will be able to reinforce local and regional identity and distinctiveness and this will bring rewards in terms of individual, social and cultural fulfilment. This in turn may help to promote the sustainable development of the area concerned, as the quality of landscape has an important bearing on the success of economic and social initiatives, whether public or private.
>
> 36. . . .The landscape is important as a component of the environment and of people's surroundings in both town and country and whether it is ordinary landscape or outstanding landscape. The public is accordingly encouraged to take an active part in landscape management and planning, and to feel it has responsibility for what happens in the landscape. . . .

Regarding identification and assessment of landscapes, the *Explanatory Report* states that it is vital that professional fieldwork 'involves the local community, the general public and the various other stakeholders by means of surveys and information meetings' (Council of Europe, 2000b: §56). However, the *Explanatory Report* then addresses the problems that assessment raises in a revealing manner (Council of Europe, 2000b: §57):

> . . . This process must take account of the concerned people's opinion and the interests linked to sectoral policies, and here views may well be highly subjective and differ considerably. It may well be worth performing the evaluation according to objective criteria first, then comparing the findings with the various assessments of the landscape by people concerned and other interest groups. If necessary, this comparison could be carried out by public enquiry, with the interested parties having the right to express their opinion. Public participation in this type of procedure could be fostered by providing the public with information, consulting all representative bodies, using the media and conducting awareness-raising campaigns at all levels.

There appears to be a mismatch here between the Convention and the *Explanatory Report* regarding the relationship between experts and the public. Public perceptions of landscape are inevitably subjective and variable, but recommending an objective evaluation – as if any criteria can be objective – is questionable. Although public participation in the form of a public enquiry is referred to, the procedure described is very top-down (Jones, 2007: 619–620). This has been pointed out by Olwig (2007: 591) in a critical analysis of the 'discursive tension' between the Convention and its *Explanatory Report*, when he states:

> The kind of 'public enquiry' envisioned in the explanatory report treats landscape as something that is known to experts and inculcated into the populace through information campaigns before that population is then allowed to exercise the right bestowed upon them to express an opinion.

Olwig (2007: 588–591) observes that where the Convention takes a broad cultural approach to landscape, some parts of the *Explanatory Report*, especially relating to the Specific Measures, are often quite technical and instrumental. It can be added that its approach here very much views the landscape as morphology. For example, regarding awareness-raising, the *Explanatory Report* states that, in the 'crucial question' of public awareness, 'every citizen has a share in the landscape and in the duty of looking after it, and the well-being of landscapes is closely linked to the level of public awareness'; however, the means proposed for awareness-raising are 'campaigns for informing and educating the public, elected representatives and associations about the value of present and future landscapes' (Council of Europe, 2000b: §52). Regarding training and education in landscape matters, these are supposed to be multidisciplinary, yet the *Explanatory Report* seems satisfied with specialist training, improvement of technical expertise, and development of school and university courses related to landscape 'so that young people become aware if the issues concerning the environment in which they live' (Council of Europe, 2000b: §53). Regarding identification and assessment of landscapes, the *Explanatory Report* advocates the use of geographical information systems and computerized mapping to study physical features of the landscape (Council of Europe, 2000b: §55). Surveys, information meetings, awareness-raising campaigns, and use of the media are a very one-sided approach to public involvement.

Olwig (2007: 591) concludes that, although technical expertise is useful and should play a role, what may be more needed is 'the cultural expertise necessary to interpret and make conscious the daily landscape practices that are often taken for granted, and which can only be sustained if their value is recognized and their continuation encouraged.'

The *Guidelines*, issued 5 years after the *Explanatory Report*, are considerably less one-sided regarding involvement of the public. With regard to public participation, the *Guidelines* state (Council of Europe, 2008: §I.1.G):

> All action taken to define, implement and monitor landscape policies should be preceded and accompanied by procedures for participation by members of the public and other relevant stakeholders, with the aim of enabling them to play an active role in formulating, implementing and monitoring landscape quality objectives.

Under the definition of landscape, the *Guidelines* emphasize 'the rights and responsibilities of populations to play an active role in the processes of acquiring knowledge, taking decisions and managing the quality of the places where they live. Public involvement in decisions ... is regarded not as a formal act but as an integral part of management, protection and planning procedures' (Council of Europe, 2008: §I.2). Further: 'Participation, consultation, pooling of ideas and approval (between institutions and the population, horizontal and vertical) should be organised at all stages in this process' (Council of Europe, 2008: §II:2). The development of landscape knowledge should, among other things, include 'recognition of characteristics and value systems based on analysis by experts or knowledge of the social perceptions of landscape ... gained through various forms of public involvement in the process of landscape policy definition...' (Council of Europe, 2008: §II.2.1). Regarded as a means of strengthening the identities of populations through recognizing themselves in their surroundings, participation is presented in the following words (Council of Europe, 2008: §II.2.3.A):

> Public participation, which may entail contradictions resulting from the diversity of the value systems espoused by the various social groups, should be regarded as enriching and as an opportunity to validate knowledge and the definition of objectives and action.
>
> Participation implies two-way communication from experts and scientists to the population and vice versa. The population possesses empirical knowledge (local and naturalistic knowledge) that may be useful in completing and contextualising specialist knowledge.
>
> This also has an influence on 'assessment' activity, understood as a dialectical comparison between analyses by experts and the values attached by the population to landscape, in the knowledge that different systems of 'value' and 'non-value' exist that may be well-entrenched or still in the process of definition; these value systems (universal, specific to national cultures, to local cultures, to each individual's culture) belong to both scholarly culture and to popular culture: they are qualitative and not quantifiable and some of them are sometimes mutually opposed.

Further, the *Guidelines* propose a wide range of awareness-raising methods, where the emphasis is on exchanges between local people affected by planning on the one hand, and scientists and experts possessing technical knowledge on the other (Council of Europe, 2008: §II.2.3.B).

Two levels of participation in relation to landscape have been identified by Prieur and Duousseau (2006). The first is in the definition of landscape policy. The second is in the implementation of landscape policy. They note that 'the public, as a rule, is more sensitive to visible operations than to plans' and that the public takes most notice 'during the actual implementation of projects in the field' when 'decisions are made to build or carry out works, the often irreversible character of which will have an impact on the environment, whether on landscape, soil or biological diversity' (Prieur and Duousseau 2006: 203–204). A major challenge is to get the public involved early, before implementation has gone so far that public participation is too late to be effective (Jones, 2007: 619).

The Convention is concerned with problems caused by globalization ('changes in the world economy') and the need to achieve sustainable development, but is little concrete on the challenges for landscape protection, management, and planning resulting from other issues of major importance in the early twenty-first century, such as climate warming, loss of biodiversity, economic recession, increasingly multicultural societies, and terrorism. These are issues in which the instincts of governments are often to act in a rather authoritarian, top-down manner rather than to approach them through broad public participation at an early stage. The solutions adopted for many of these problems will in themselves affect landscapes in multiple ways. The ways in which participation is practised in relation to the ELC, and the importance given to it by governments and other administrative authorities, will indicate how far the ideals of the Council of Europe will be followed in tackling the major issues facing European society in the near future.

1.4 Implementation

With regard to participation, the *Explanatory Report* confirms that: 'Landscape is an issue which affects the whole population and care for the landscape requires collaboration between a wide range of individuals and organisations' (Council of Europe, 2000b: §50.c). Implementation is to take place at all levels from the European level to the local level. The *Explanatory Report* states (Council of Europe, 2000b: §25):

> The general purpose of the Convention is to encourage public authorities to adopt policies and measures at local, regional, national and international level for protecting, managing and planning landscapes throughout Europe so as to maintain and improve landscape quality and bring the public, institutions and local and regional authorities to recognize the value and importance of landscape and to take part in related public decisions.

At the European level, Parties undertake to cooperate regarding the landscape dimension of international policies and programmes, to render each other mutual assistance and exchange information, to encourage cooperation on transfrontier landscapes, and to monitor the implementation of the Convention (Council of Europe, 2000a: Articles 7–10). The Convention has established the Landscape Award of the Council of Europe as a distinction for lastingly effective landscape policies or measures that can serve as an example (Council of Europe, 2000a: Article 11). The Council of Europe organizes at intervals Conferences of Member States of the European Landscape Convention as well as regular Workshops for the Implementation of the European Landscape Convention (Council of Europe n.d.). The work of implementing the ELC is supported by a number of networks: the European Network of Local and Regional Authorities for the Implementation of the European Landscape Convention (RECEP-ENELC, 2009) was founded in 2006; the European Network of Universities for the Implementation of the ELC (UNISCAPE, 2009) and Non-Governmental Organisations for the ELC (CIVILSCAPE, 2010) were established in 2008. Another, more recent, network is LANDSCAPE EUROPE (2010), an interdisciplinary network of national research institutions with expertise in landscape assessment, planning, and management. These networks arrange their own conferences on European landscapes. Other conferences with European participation are organized by national bodies.

Each Party to the Convention is to implement it according to its own division of powers (Council of Europe, 2000a: Article 4). In accordance with the principle of subsidiarity, Parties should implement the Convention at the 'most appropriate level'. Each country should set out the tasks and measures for which each level is responsible – national, regional, or local – and lay down rules for inter-level coordination (Council of Europe, 2000b: §§48–49). Application should be adaptable, allowing a choice of means, accepting that there is no universally acknowledged method for studying, identifying, and evaluating landscapes (Council of Europe, 2000b: §§27, 34 and 58).

In order to put landscape policies into effect, Parties to the ELC undertake 'to introduce instruments aimed at protecting, managing and/or planning the landscape' (Council of Europe, 2000a: Article 6.E). Protection, management, and planning are three terms frequently used in the Convention and are clearly defined (Council of Europe, 2000a: Article 1):

- *Landscape protection* means 'actions to conserve and maintain the significant or characteristic features of a landscape, justified by its heritage value'. It consists according to the *Explanatory Report* of 'measures to preserve the present character and quality of a landscape which is greatly valued on account of its distinctive natural or cultural configuration' and involves 'upkeep measures to preserve significant features of a landscape'.
- *Landscape management* means actions in accordance with the principle of sustainable development 'to ensure the regular upkeep of a landscape, so as to guide and harmonise changes which are brought about by social, economic and environmental processes'. The explanation underlines that the management approach must be dynamic and 'seek to improve landscape quality on the basis of the population's expectations'.
- *Landscape planning* means 'strong forward-looking action to enhance, restore or create landscapes'. The *Explanatory Report* elaborates this as 'the formal process of study, design and construction by which new landscapes are created to meet the aspirations of the people concerned', including in particular the areas most affected by change and badly damaged areas which need to be radically reshaped (Council of Europe, 2000b: §40).

The *Guidelines* elaborate on the concepts of landscape protection, management, and planning, emphasizing their dynamic character as landscape actions. Protection 'includes the idea that landscape is subject to changes which, within certain limits, have to be accepted'. Protective measures 'should not be designed to stop time or to restore natural or human-influenced characteristics that no longer exist' but 'guide changes in sites in order to pass on their specific, material and immaterial features to future generations'. Management is 'a continuing action aimed at influencing activities liable to modify landscape' and 'can be seen as a form of adaptive planning, which itself evolves as societies transform their way of life, their development and surroundings'. Landscape planning 'can anticipate new social needs by taking account of ongoing developments.' It also covers 'the rehabilitation of degraded land (mines, quarries, landfills, wasteland, etc.) so that they meet the stipulated landscape quality objectives' (Council of Europe, 2008: §I.5).

The *Guidelines* include a long discussion of criteria and instruments for landscape policies (Council of Europe, 2008: §§II.2 and II.3.3, and Appendix 1). On methods of implementation, a distinction is drawn between regulatory and voluntary implementation. The former is contained in legislation or policy documents, while the latter is based on agreements between the authorities and stakeholders, such as landscape management agreements (Council of Europe, 2008: §II.3).

Periodically, reports on the status of implementation of the ELC have been sent in by European states to the Council of Europe. The first time this was done was in 2003, before the Convention had entered into force (Council of Europe, 2003). Information was received from 27 states and 2 autonomous regions (in Belgium) –

including not only signatories – and provided answers to a series of questions. The questions concerned definitions of the term 'landscape' in each language, legal organization, and administrative organization. The information was presented in summary and tabular form. In 2007 and 2009, the questionnaires returned by each state were presented (Council of Europe, 2007, 2009). In addition to the questions answered in 2003, there were answers to questions on implementation of the ELC's General and Specific Measures, European cooperation, and landscape awards. In 2007, there were 13 reports (including two from Belgium); all but one were from countries that had answered in 2003. In 2009, there were 10 reports (including two non-signatories); two countries had also answered in both 2003 and 2007, six had previously answered only in 2003, and two were new. It is difficult to say whether the declining number of reports indicates declining interest in the ELC or whether it reflects a judgement that it is not necessary to send in reports every time. The amount of information and thoroughness of the answers varies considerably from country to country. This may reflect varying enthusiasm for the ELC, or it may reflect varying administrative capacity. The United Kingdom and Wallonia in Belgium have reported on all three occasions and are among the countries providing the most detailed reports.

The reports reveal a diversity of administrative and legal arrangements concerning landscape, which is in accordance with the terms of the Convention. For example, ministerial responsibility may be concentrated in a single ministry (most often environment, or environment combined with planning, agriculture or cultural heritage), or split among several ministries. In federal states or states with regional devolution, landscape may be primarily the responsibility of the regions. This is made clear in the reports from Austria, and may be a contributory reason why (like Germany) it has not signed the Convention. Belgium has ratified the Convention both nationally and in each of its regions. In Spain landscape is primarily the responsibility of the regional autonomous communities, while in the United Kingdom it is both a national responsibility and a devolved responsibility in Scotland, Wales, and Northern Ireland. As far as resources earmarked for landscape in the different countries are concerned, the answers are fairly evenly divided between those countries that have specifically allocated funding or personnel to landscape matters, and those that deal with landscape within existing budgets.

As many respondents have legal definitions of landscape as do not, some have several definitions, but only three (Croatia, Cyprus and Wallonia in Belgium) have formally adopted the definition of the ELC. Some states refer to landscape in their constitution or basic law, a few have a specific law dealing with landscape, while in most cases landscape is included in a variety of laws concerning environment, cultural heritage, planning and/or agriculture.

With regard to public participation, the reports vary considerably in the amount of detail provided. Most detail is provided by the United Kingdom, which lists consultation, public inquiries, stakeholder partnerships, and involvement of community groups. A number of countries refer to legal provisions for public participation, most commonly in planning legislation. Where the type of participation is specified, meetings, hearings, and inquiries are the most common forms. A few countries report that participatory procedures in accordance with the Convention have yet to be implemented. Taken as a whole, the reports give only a very small glimpse into the reality of public participation in landscape issues in Europe, and the status and significance given to it.

1.5 The European Landscape Convention and Participation in Practice

The present book illustrates through a number of case studies the workings and experiences of public participation in relation to landscape in selected European countries. The objective is to contribute towards an understanding of the state of public participation in European landscapes. One aim is to explore the manner in which the European Landscape Convention has been implemented regarding the obligations to recognize landscape in law and to establish and implement procedures for public participation in landscape matters. A second aim is to provide a basis for comparing experiences in different countries. The benefits, difficulties, and limits of the participatory approach are examined though examples from countries that have both ratified and not ratified the Convention. The case studies include a country that ratified the ELC at an early stage (Norway in 2001), countries that ratified it more recently (Belgium and Poland in 2004, the Netherlands and Portugal in 2005, France and the United Kingdom in 2006, and Spain in 2007) or have only just ratified it (Greece in 2010), a country that has signed but not yet ratified it (Sweden signed in 2001, while it announced its decision to ratify it in November 2010), and a country that has not signed it (Estonia).

The approaches in the different chapters to the theme of participation are in part theoretical, in part methodological, and in part empirical. The first part of the book deals with the implementation of participation theoretically and through case studies. Michael Jones provides initially a theoretical analysis of participatory procedures in which lessons are drawn from the literature on participation, including a critique of prevailing orthodoxy regarding participatory approaches in Third World development projects. The extent to which participation has been implemented varies in Europe from country to country. Henk Baas, Bert Groenewoudt, and Edwin Raap examine how implementation of the ELC has gained a fair degree of success in the Netherlands through well-considered efforts to involve the general public, scientists, and local authorities in a process of working together. Karoline Daugstad discusses how ideology and practice affect the implementation of participatory approaches in nature conservation in Norway. Anna Majchrowska examines how lack of strong commitment at ministerial level has hindered the drawing up of a national landscape policy in Poland and provided an obstacle to the introduction of effective public participation. The chapter by Berezi Elorrieta and Dolores Sánchez-Aguilera shows that delegation of landscape regulatory powers to Spain's regional autonomous communities respects regional differences but results in varying fulfilment of the objectives of the ELC. Theano S. Terkenli explores the absence of a well-developed landscape conscience in Greece, resulting in a lacking concern for landscape issues in both public and private life.

The second part of the book presents examples of participatory methods in practice. The chapter by Yves Michelin, Thierry Joliveau, and Claire Planchat-Héry discusses the advantages and limitations of different tools in participatory processes concerning landscape projects in France and presents a typology of techniques for landscape mediation. Claire Planchat-Héry also presents the Prospective Vision as a participatory method applied in two communities in respectively France and Belgium. Isabel Loupa Ramos addresses experts' and stakeholders' perspectives regarding the future of the landscape in a remote part of Portugal through the use of landscape scenarios in relation to the formulation of Landscape Quality Objectives. The chapter by Morten Clemetsen, Erling Krogh, and Kine Halvorsen Thorén examines a methodology involving 'sense of place' as a means of bringing in local people's perceptions of landscape in Norway. Neil Spencer provides a case study of participatory management of a river catchment landscape in England. The example from Sweden, by Anders Larsson, Anna Peterson, Elinor Bjärnborg, Christine Haaland, and Mats Gyllin, is a pilot study for a Regional Landscape Strategy, focusing on methods of public participation involving equestrians and landowners. The case study from Estonia, by Monika Suškevičs and Mart Külvik, provides lessons from landowner participation in Natura 2000 designations. The case studies illustrate both successful and less successful applications of participatory approaches to landscape protection, management, and planning. Some lessons that may be drawn from these studies are presented in the concluding chapter.

The presentation of case studies from a range of different countries reveals the way in which Europe's social and cultural diversity is reflected in varying approaches to landscape, law, and public participation. The European Landscape Convention allows a large degree of freedom regarding how the Convention and the requirement of public participation are implemented. This is necessary in order to take into consideration the large variety of administrative arrangements in different European countries as well as to take into account the aspirations of the many different types of stakeholders and the large number of regional and local authorities involved. Effective public participation faces the challenges of meeting the costs involved, arguing for the benefits, creating the trust necessary for a successful process, combating apathy or passive and even active opposition, and overcoming powerful vested interests. Participatory approaches are not limited to local participation but include participation at all levels. It is vital to combine expert scientific and technical knowledge with the empirical knowledge and experiences of the general public.

The present book does not provide a recipe for successful public participation, but presents examples of participation in practice. Different methods of participation may be suitable in different situations. Participation is a process as much as a method or set of methods. The main purpose of the methods is to provide tools for communication in order to make conflicting interests specific and to create the conditions for dialogue.

We hope that the examples and lessons presented in this book will lead to cross-fertilization of ideas and approaches, and provide inspiration for practitioners in landscape protection, management, and planning. We hope, too, that it can contribute to an assessment of how far the Council of Europe's objective of enhanced democracy through citizen's participation has been achieved, and that it can stimulate further work to meet the challenges of participation that are revealed.

References

CIVILSCAPE (2010) Non-Governmental Organisations for the European Landscape Convention. http://www.civilscape.eu/civilscape/content/en/index.html. Accessed 17 Jun 2010

Cosgrove D (1984) Social formation and symbolic landscape. Croom Helm, London

Cosgrove D (2000) Cultural landscape. In: Johnston RJ, Gregory D, Pratt G, Watts M (eds) The dictionary of human geography, 4th edn. Blackwell, Malden, MA, pp 138–141

Council of Europe (1949) Statute of the Council of Europe, London, 5.V.1949. European Treaty Series Nos. 1/6/7/8/11. http://conventions.coe.int/Treaty/en/Treaties/Word/001.doc. Accessed 8 June 2010

Council of Europe (1985a) European Charter of Local Self-Government, Strasbourg, 15.X.1985. ETS No. 122. http://conventions.coe.int/treaty/en/treaties/html/122.htm. Accessed 16 Jun 2010

Council of Europe (1985b) European Charter of Local Self-Government (ETS No. 122): explanatory report. http://conventions.coe.int/Treaty/EN/Reports/HTML/122.htm. Accessed 16 Jun 2010

Council of Europe (2000a) European Landscape Convention, Florence, 20.X.2000. ETS No. 176. http://conventions.coe.int/Treaty/en/Treaties/Html/176.htm. Accessed 6 Mar 2007

Council of Europe (2000b) European Landscape Convention (ETS No. 176): explanatory report. http://conventions.coe.int/Treaty/EN/Reports/Html/176.htm. Accessed 6 Mar 2007

Council of Europe (2003) European Landscape Convention – Florence Convention – synthesis of the received information concerning summary descriptive note on the landscape policies pursued in the Council of Europe Member States. T-FLOR 2 (2002) 11 of 15 May 2003, T-FLOR 3 (2003) 11 rev. Council of Europe, Strasbourg. http://www.coe.int/t/dg4/cultureheritage/heritage/landscape/tflor-2003-11rev-synthese_en.pdf. Accessed 17 Jun 2010

Council of Europe (2005) CM(2005)80 Final 17 May 2005 Action Plan. http://www.coe.int/t/dcr/summit/20050517_plan_action_en.asp. Accessed 8 Jun 2010

Council of Europe (2007) Synoptic presentation of the status of landscape policies pursued by the Member States of the Council of Europe 2006–2007. T-FLOR 3 (2007) 7 rev. – Bil. Strasbourg: Council of Europe. http://www.coe.int/t/dg4/cultureheritage/heritage/Landscape/TFLOR-2007-7rev-Synoptic_bil.pdf. Accessed 17 Jun 2010

Council of Europe (2008) Recommendation CM/Rec(2008)3 of the Committee of Minister to Member States on the guidelines for the implementation of the European Landscape Convention. https://wcd.coe.int/ViewDoc.jsp?id=1246005. Accessed 24 Sept 2009

Council of Europe (2009) European Landscape Convention: 5th Conference of the Council of Europe on the European Landscape Convention: presentation of the landscape policies in the Member States of the Council of Europe. CEP-CDPATEP (2009) 3Bil. Council of Europe, Strasbourg. http://www.coe.int/t/dg4/cultureheritage/heritage/Landscape/ReunionConf/5eConference/CEP-CDPATEP-2009-3-PresentationPolitiques_bil.pdf. Accessed 17 June 2010

Council of Europe (n.d.) The European Landscape Convention. http://www.coe.int/t/dg4/cultureheritage/heritage/landscape/default_EN.asp. Accessed 17 Jun 2010

Dejéant-Pons. M (2006) The European Landscape Convention. Landsc Res 31:363–384

Ermischer G (2004) Mental landscape: landscape as idea and concept. Landsc Res 29:371–383

Gray M (2009) Landscape: the physical layer. In: Clifford NJ, Holloway SL, Rice SP, Valentine G (eds) Key concepts in geography, 2nd edn. Sage, London, pp 265–285

Groening G (2007) The 'landscape must become law' – or should it?. Landsc Res 32:595–612
Howard P (2004) Spatial planning for landscape: mapping the pitfalls. Landsc Res 29:423–434
IUCN (1995) Parks for life: actions for protected areas in Europe. IUCN, Cambridge
Ingold T (2000) The perception of the environment: essays in livelihood, dwelling and skill. Routledge, London
Jones M (2003) The concept of cultural landscape: discourse and narratives. In: Palang H, Fry G (eds) Landscape interfaces: cultural heritage in changing landscapes. Kluwer, Dordrecht, pp 21–51
Jones M (2007) The European Landscape Convention and the question of public participation. Landsc Res 32:613–633
Jones M (2010). The heritage of landscape – Driving force or counterforce? In: Proceedings, 8th meeting of the Council of Europe workshops for the implementation of the European Landscape Convention: Landscape and Driving Forces, Malmö, Sweden, 8–9 Oct 2009. European spatial planning and landscape no. 93. Council of Europe Publishing, Strasbourg, pp 111–127
Jones M, Howard P, Olwig KR, Primdahl J, Sarlöv Herlin I (2007) Multiple interfaces of the European Landscape Convention. Nor Geogr Tidsskr–Norw J Geogr 61:207–215
Jones M, Olwig KR (Eds) (2008) Nordic landscapes: region and belonging on the northern edge of Europe. University of Minnesota Press, Minneapolis
LANDSCAPE EUROPE (2010) http://www.landscape-europe.net/. Accessed 17 June 2010
Mitchell D (2000) Cultural geography: a critical understanding. Blackwell, Oxford
Mitchell D (2002) Cultural landscape: the dialectical landscape. Progr Hum Geogr 26:381–389
Morin KM (2009) Landscape: representing and interpreting the world. In: Clifford NJ, Holloway SL, Rice SP, Valentine G (eds) Key concepts in geography, 2nd edn. Sage, London, pp 286–299
Olwig KR (2001) 'Time out of mind' – 'mind out of time': custom vs. tradition in environmental heritage research and interpretation. Int J Herit Stud 7:339–354
Olwig KR (2002) Landscape, nature and the body politic: from Britain's Renaissance to America's New World. University of Wisconsin Press, Madison
Olwig KR (2007) The practice of landscape 'conventions' and the just landscape: the case of the European Landscape Convention. Landsc Res 32:579–594
Prieur M, Durousseau S (2006) Landscape and public participation. In: Landscape and sustainable development: challenges of the European Landscape Convention. Council of Europe Publishing, Strasbourg, pp 165–207
RECEP-ENELC (2009). European Network of Local and Regional Authorities for the Implementation of the European Landscape Convention. http://www.recep-enelc.net/. Accessed 17 June 2010
Rose G (1992) Geography as a science of observation: the landscape, the gaze and masculinity. In: Driver F, Rose G (eds) Nature and society: essays in the history of geographical knowledge. Historical Geography Research Series no. 28. Historical Geography Research Group, Cheltenham, pp 8–18
Rose G (2002) Landscapes and labyrinths. Geoforum 33:455–467
Sarlöv Herlin. I (2004) New challenges in the field of spatial planning: landscapes. Landsc Res 29:399–411
Sarlöv Herlin I (2007) The European Landscape Convention – a brief presentation. Nor Geogr Tidsskr–Norw J Geogr 61:207–208
Scazzosi L (2004) Reading and assessing landscape as cultural and historical heritage. Landsc Res 29:299–320
Stanners D, Bourdeau P (1995) Europe's environment: the Dobříš assessment. European Environmental Agency, Copenhagen
UNECE (1998) Convention on Access to Information, Public Participation in Decision-Making and Access to Justice in Environmental Matters done at Aarhus, Denmark on 25 June 1998. United Nations Economic Commission for Europe. http://www.unece.org/env/pp/treatytext.htm. Accessed 11 Mar 2010

UNESCO (1972) Convention Concerning the Protection of the World Cultural and Natural Heritage adopted by the General Conference at its Seventeenth Session, Paris, 16 November 1972. United Nations Educational, Scientific and Cultural Organisation. http://whc.unesco.org/archive/convention-en.pdf. Accessed 9 June 2010

UNESCO (2005) Convention on the Protection and Promotion of the Diversity of Cultural Expressions, Paris, 20 October 2005. United Nations Educational, Scientific and Cultural Organisation. http://unesdoc.unesco.org/images/0014/001429/142919e.pdf. Accessed 16 June 2010

UNISCAPE (2009) European Network of Universities for the Implementation of the European Landscape Convention. http://www.uniscape.eu/. Accessed 17 June 2010

Urry J (2002) The tourist gaze, 2nd edn. Sage, London

Widgren M (2004) Can landscapes be read?. In: Palang H, Sooväli H, Antrop M, Setten G (eds) European rural landscapes: persistence and change in a globalising environment. Kluwer, Dordrecht, pp 455–465

Winchester HPM, Kong L, Dunn K (2003) Landscapes: ways of imagining the world. Pearson, Harlow

Wylie J (2007) Landscape. Routledge, London

Wylie J (2009) Landscape. In: Gregory D, Johnston R, Pratt G, Watts MJ, Whatmore S (eds) The dictionary of human geography, 5th edn. Wiley/Blackwell, Chichester, pp 409–411

Zachrisson A (2004) Co-management of natural resources: paradigm shifts, key concepts and cases. Mountain Mistra Programme report, Report no. 1. Mountain Mistra Programme, Umeå

Part I
Implementing Participation

Chapter 2
European Landscape and Participation – Rhetoric or Reality?

Michael Jones

Abstract The European Landscape Convention (ELC) obliges parties to establish procedures for the participation of the general public, local and regional authorities, and other interested parties in landscape matters. This indicates that the views of

M. Jones (✉)
Department of Geography, Norwegian University of Science and Technology, NO-7491 Trondheim, Norway
e-mail: michael.jones@svt.ntnu.no

M. Jones, M. Stenseke (eds.), *The European Landscape Convention*, Landscape Series 13, DOI 10.1007/978-90-481-9932-7_2,

all interested groups should be considered, not just scientific or political elites. Participatory, dialogue-based approaches mean that values and meanings attached to landscapes by different groups need to be negotiated between competing interests. Justifications for participation include reinforcing local identity, democratization, legitimacy, information exchange, tackling conflicts, and social justice. Introducing effective public participation in landscape protection, management, and planning has wide-ranging and radical implications for policy-makers and administrators. Successful participation involves sharing knowledge and negotiating power relations, and can challenge oppression and injustice. However, participation has been criticized as time-consuming and costly. Participation rhetoric may conceal inequalities in bargaining power and divergent motivations of participating stakeholders, and allow manipulation by powerful interests. Participatory projects may mask power structures in local communities, conceal the oppressions of daily life (e.g. gender) and override legitimate decision-making bodies. Drawing lessons from literature on participation, in particular literature critiquing the prevailing orthodoxy regarding participatory approaches in Third World development projects, a theoretical analysis is provided of participation rhetoric, attitudes to participation, and advantages, disadvantages and effectiveness of participation related to landscape issues in the European context.

Keywords Participation theory · Participatory approaches · Justifications for public participation · Challenges to public participation · Democracy

2.1 Introduction

The European Landscape Convention (ELC) obliges Parties 'to establish procedures for the participation of the general public, local and regional authorities, and other parties with an interest in the definition and implementation of . . . landscape policies' (Council of Europe, 2000a: Article 5c). The ELC's *Explanatory Report* specifies in relation to this Article that: 'Landscape is an issue which affects the whole population and care for landscape requires collaboration between a wide range of individuals and organisations' (Council of Europe, 2000b: §50c). It is made clear that the views of all interested groups should be considered, not just scientific and technical or political elites (Council of Europe, 2000b: §§22–23) – and, by implication, also not the general public alone. According to the *Guidelines for the Implementation of the European Landscape Convention*, recommended by the Committee of Ministers in 2008, participation involves a two-way process of communication between experts and scientists on the one hand and the general population on the other. Further, the ELC's concept of landscape 'implies an exercise in democracy whereby differences are accepted, common characteristics found and operational compromises eventually reached; these represent an alternative to the drawing up by experts of hierarchical classifications of landscape qualities' (Council of Europe, 2008: §II.2.3.A).

Discussions of justifications for and challenges of participation have generated a wide range of literature in recent years, partly in relation to landscape but more widely in relation to planning in general. A pioneer work, edited by W.R. Derrick Sewell and J.T. Coppock, was *Public Participation in Planning* (1977a). There is a growing body of literature on participation in environmental issues. In relation to Third World development projects, considerable experience of participatory planning has been gained during the last 30 years. Since the advent of the ELC in 2000, a range of studies have dealt with participation specifically in relation to European landscapes. These have included discussions of: participatory landscape ecology in Germany (Luz, 2000); social barriers to participatory landscape development in Switzerland (Buchecker et al., 2003; Höppner et al., 2007); use of scenarios in participatory landscape planning in Denmark (Tress and Tress, 2003); community participation in planning and management of cultural landscapes in Europe generally (Selman, 2004); community participation in landscape planning in an Irish national park (O'Rourke, 2005); participation in Landscape Character Assessment in England (James and Gittins, 2007) and Denmark (Caspersen, 2009); participation in rural landscape conservation in Italy (Borsotto et al., 2008); and local participation in cultural landscape maintenance in Sweden (Stenseke, 2009). However, despite the celebration of the ELC's tenth anniversary in 2010, there is still relatively little literature that specifically examines participation in relation to the provisions of the Convention (Prieur and Durousseau, 2006; Jones, 2007; Olwig, 2007).

The present chapter poses the question of whether the obligation of public participation in the ELC is rhetoric or reality. This is of course a rhetorical question. Rhetoric is the art of effective or persuasive speaking and writing. The term is not infrequently used and interpreted in a negative sense as referring to persuasive or impressive language that is lacking in sincerity or meaningful content. In this sense, the question could be reformulated as whether participation in the ELC is 'all talk and no action'. In a positive sense, however, rhetoric can be understood as a means of effective communication in the exercise of democracy and thus does not need to be opposed to broad public participation. In this sense, the object of rhetoric is to 'let the best argument win'. The intention of participation in the ELC is to bring landscape issues into the public domain by reaching decisions through discursive and dialogic processes rather than leaving landscape character to be something determined by purportedly 'objective' technocratic approaches. The purpose of this chapter is to examine arguments for and against public participation in landscape and related issues, and in particular to bring into the discussion an awareness of the danger that persuasive arguments for public participation may mask practices that in reality reduce genuine participation.

The chapter is intended as a theoretical contribution to the question of participation in landscape issues. First, justifications for participation, conditions for successful participation and challenges facing participation will be presented. Next, recent literature critiquing the prevailing orthodoxy regarding participatory approaches in Third World development projects will be focused upon with a view to drawing lessons that might be useful for the implementation of participation in accordance with the ELC. A recent study of participation in the field of social policy in Britain will also be referred to. The chapter will conclude with some theoretical observations concerning issues that need to be addressed in participation procedures in the European context.

2.2 Justifications for Participation

The ELC (Council of Europe, 2000a) refers in its Preamble to a number of European conventions concerning matters such as protection and management of natural and cultural heritage, regional and spatial planning, and local self-government. Last (but not least), the list refers to the Convention on Access to Information, Public Participation in Decision-Making and Access to Justice in Environmental Matters of 1998, generally referred to as the Aarhus Convention after the Danish city where it was signed. This is a convention of the United Nations Economic Commission for Europe and entered into force in 2001. It recalls in its preamble the Rio Declaration on Environment and Development of 1992, one of the principles of which is that environmental issues are to be handled with the participation of all concerned citizens at the relevant level, and that each individual is to have access to information on their environment held by public authorities as well as the opportunity to participate in decision-making processes (UN, 1992: Principle 10). The Aarhus Convention specifically applies 'without discrimination as to citizenship, nationality or domicile...' (UNECE, 1998: §9). By 2 November 2009, 44 Parties had ratified, accepted, approved or acceded to the Convention, including 26 of the 27 members of the European Union (EU) as well as Norway. It has been signed but not ratified by five states, including Ireland within the EU, and Switzerland and Iceland outside it (although neither Russia nor Turkey is listed).

The EU is a Party to the Aarhus Convention in its own right. A directive providing for public participation in environmental matters became part of EU law in 2003. According to this, effective public participation in environmental decision-making enables the public to express its opinions and concerns, and the decision-maker to take them into account (EC, 2003: §3). The directive also states that sufficient time should be allowed for informing the public, and for the public concerned to prepare and participate effectively in environmental decision-making procedures (EC, 2003: §4).

Discussing landscape and public participation, Council of Europe experts Michel Prieur and Sylvie Durousseau (2006: 165) state that 'the term "public" should be taken to mean civil society in the broad sense'. Public participation means involving individuals and groups who are outside the formal decision-making processes of the government and local authorities. However, it is not the same as the complete delegation of powers to local communities. It should not be seen as a substitute for official decision-making but as a complement to it (O'Riordan and Stoll-Kleemann, 2002: 90; Barnes et al., 2007: 186). Its objective is to draw into decision-making the views of all concerned groups in environmental matters – hence implicitly including local communities, residents, visitors, landholders, specialists, and deprived groups – before representative, democratically elected bodies make the final decisions.

Several justifications for the desirability of public participation are found in the literature (Jones, 2007: 616–618). *Reinforcement of local and regional identity* is emphasized in the ELC's *Explanatory Report* (Council of Europe, 2000b: §24):

> If people are given an active role in decision-making on landscape, they are more likely to identify with the areas and towns where they spend their working and leisure time. If they have more influence on their surroundings, they will be able to reinforce local and regional identity and distinctiveness and this will bring rewards in terms of individual, social and cultural fulfilment. This may in turn help promote the sustainable development of the area concerned, as the quality of landscape has an important bearing on the success of economic and social initiatives, whether public or private.

Democratization is stressed by Prieur and Durousseau (2006: 166–167):

> A landscape policy which involved only experts and administrators, who themselves are often specialists, would result in landscapes that were imposed on the public, just as in the days when landscape was produced by and for an elite. Democratization of the landscape ... is also reflected in this collective and individual appropriation of all landscapes, through the requirement that there be direct participation for all in all phases of decision-making regarding landscape alteration, supervision of landscape evolution and prevention of reckless landscape destruction.

This recognizes that landscape is not just for an elite, and not just the concern of experts and administrators, but the concern of everybody. Public participation in decision-making enhances the democratic system. Bringing people into the management process is a 'democratic necessity' and recognizes people's self-worth, vital role and citizenship credentials (O'Riordan and Stoll-Kleemann, 2002: 106).

Further, the *legitimacy* of decisions is improved with public participation. The Aarhus Convention aims 'to further the accountability of and transparency in decision-making and to strengthen public support for decisions on the environment' (UNECE, 1998: preamble). Effective and efficient management requires the support of those affected (O'Riordan and Stoll-Kleemann, 2002: 106). Swedish political scientist Anna Zachrisson (2004: 24) argues that:

> When people are listened to, paid attention to, treated politely and with respect, the legitimacy for the final decisions is increased. ... Decentralised management regimes can make use of local people's place-specific knowledge and their social norms to design well-adapted rules and sanctions. The capacity to respond to environmental and social feedbacks is also suggested to be developed faster by social systems. ... Central organs can provide ... coordination as well as other forms of support such as economic incentives and technical assistance.

Interaction can facilitate cooperation by creating trust. Decisions gain greater legitimacy if the public has participated in the formulation of visions, objectives and decision-making criteria. Thereby decisions can become better anchored among interest groups (Jonsson and Lundqvist, 2006).

Part of this process involves *information exchange*. The Aarhus Convention (UNECE, 1998: preamble) recognizes that:

> ...improved access to information and public participation in decision-making enhance the quality and the implementation of decisions, contribute to public awareness of environmental issues, give the public the opportunity to express its concerns and enable public authorities to take direct account of such concerns.

Information exchange should hence be a two-way process, in which the authorities both inform the public and listen to the public. On the one hand, knowledge and awareness of landscape are enhanced when the authorities share relevant information with the general public. On the other hand, public participation gives the authorities a better overview of problems as perceived by the public and allows the incorporation of the public's knowledge, values, viewpoints, and behaviour in the decision-making process, thus facilitating the finding of appropriate solutions (Jonsson and Lundqvist, 2006). Sharing knowledge is important because there is no single, 'correct' understanding. Individuals have different frameworks of knowledge and assumptions depending on their social and occupational settings. Local communities are rarely politically cohesive or homogenous entities. Information exchange allows multiple perspectives to be brought on a problem (O'Riordan and Stoll-Kleemann, 2002: 106).

In extension of this, participation may help the *tackling of conflicts* by giving the different parties and stakeholders a better idea of each other's viewpoints. Conflicts may be reduced through actors meeting to discuss problems (Zachrisson, 2004; Jonsson and Lundqvist, 2006).

Finally, public participation without discrimination is a form of *social justice* that involves the acceptance of heterogeneity. The ELC (Council of Europe, 2000a: Article 5a) specifically recognizes landscapes 'as an essential component of people's surroundings, an expression of the diversity of their shared cultural and natural heritage', indicating that landscape heterogeneity is regarded as a value.

2.3 Conditions for and Challenges to Successful Participation

Three reasons for the increasing interest for participation in environmental management are advanced by environmental researchers Tim O'Riordan and Susanne Stoll-Kleemann (2002: 89). First, it is impossible to govern effectively without some degree of involvement of interested parties and groups. Second, new electronic media have opened an array of sources and means of communicating information. Third, the general public have increasing expectations of consultation.

To be successful, participation requires the sharing of knowledge and the negotiation of power (Cooke and Kothari, 2001: 13). There are a number of prerequisites for such a pluralistic power relationship. One is a free press. Another is the capacity of different stakeholders to articulate their stake. Third is their ability to mobilize so as to gain attention and exert pressure. Fourth is their ability to be informed. A fifth prerequisite is an open political system and open bureaucracy. Sixth is non-manipulation of information by governors and governed alike. Seventh is willingness to share power and to show respect. Eighth is the adaptability and responsiveness of management. Further, successful participation requires critical assessments of performance (O'Riordan and Stoll-Kleemann, 2002: 90).

However, successful participation faces serious challenges, discussed in a growing body of literature that has critically analysed participation theory and practice. It is argued that a disadvantage of participation is that it is time-consuming and costly (O'Riordan and Stoll-Kleemann, 2002: 105, 107). A recurring problem is public apathy because of social barriers to public involvement (Diduck and Sinclair, 2002; Buchecker et al., 2003). In cases where compromises are reached, sub-optimal solutions may be the result (O'Riordan and Stoll-Kleemann, 2002: 105, 107). Some conflicts are intractable, precluding solutions acceptable to all parties due to incompatible aims among different interests.

In a study of public participation in Environmental Impact Assessment, environmental researchers Nicola Hartley and Christopher Wood (2005) found that, although the Aarhus Convention advocates 'early' and 'effective' participation (UNECE, 1998: Article 6, §§2–4), these terms remain largely undefined and there remain questions concerning how to implement effectively the Aarhus principles. They developed ten evaluation criteria for assessing the degree of achievement of 'early' and 'effective' participation in concrete cases: communication; fairness; timing; accessibility to information; information provision; influence on decision-making; competence of the public; interaction; compromise; and trust. They identified certain barriers impeding 'early' and 'effective' participation in environmental impact analyses of proposed waste disposal sites in the United Kingdom: the participation time provisions of current legislation; technical complexity of project proposals; varying developer and stakeholder attitudes concerning what constitutes 'effective' participation; poor provision of legal and procedural information; and financial constraints. They concluded that the degree to which participation procedures according to the Aarhus Convention will be strengthened will depend on how its ideals are interpreted and incorporated into legislation and practice.

Sewell and Coppock (1977b: 7–11) call attention to six crucial questions, the answers to which will affect the level and form of public participation in practice:

- Who should participate?
- Who is likely to participate?
- How much participation is possible and desirable?
- On what issues and at what stages in decision-making is public participation desirable?
- What weight should be attached to the views of well-organized, articulate interest groups as against the views of the unorganized public?
- How can meaningful views on regional and national issues be obtained?

The power dimension is critical for how genuine participation is in practice. Politicians, bureaucrats, and experts may be reluctant to concede influence or may remain sceptical to the efficiency of participation (O'Riordan and Stoll-Kleemann, 2002: 105; Thorell, 2008: 34). Lip-service may be paid to participation, but it is open to manipulation. Elite structures of power tend to be self-generating and are often stable over long periods of time. The power of the few with the requisite resources and skills is maintained through networks of influence and mutual protection (O'Riordan and Stoll-Kleemann, 2002: 92). Further, the micro-politics of participatory forums should not be ignored. Conflicts and uneven power relations exist in all social groups; people may have multiple roles, and there may be domination by forceful leaders leading to the marginalization of certain voices and weaker groups (Barnes et al., 2007: 189). The role of local communities is frequently romanticized, but they are also characterized by unequal constellations of power and influence; local communities are not socially homogenous (O'Riordan and Stoll-Kleemann, 2002: 98; Cooke and Kothari, 2001: 6). Local elites, local customs or tacit social conventions may lead to the exclusion of certain groups (Buchy and Hoverman, 2000: 21–22; O'Riordan and Stoll-Kleemann, 2002: 107; Daugstad et al., 2006: 132–133; Svarstad et al., 2006) – women may be disregarded, or children silenced, or old people neglected, or the physically and mentally impaired forgotten, or the homeless ignored, or immigrants discriminated against, or outsiders excluded.

The extent to which nominal participation means genuine participation varies in the relationship between local communities and governing bodies. This has been described in various typologies of participation, all broadly similar but with certain variations. The seminal work is health policy specialist Sherry Arnstein's 'Ladder of citizen participation' (1969), exemplified by federal social programmes in the USA. Her typology contains eight levels of participation with each rung of the ladder corresponding to the extent of citizen's power in determining the end product. At the bottom are (1) *manipulation*, aiming in the name of participation to 'educate' citizens in what is essentially a public relations vehicle for the power-holders, and (2) *therapy*, aiming to 'cure' the participants through activity resembling group therapy, aiming to adjust their values and attitudes to those of society at large. Manipulation and therapy represent according to Arnstein 'non-participation' rather than genuine participation. The next two rungs are (3) *informing* citizens of their rights, responsibilities and options, frequently through one-way information via news media, pamphlets, posters, and responses to inquiries, and (4) *consultation* by means of attitude surveys, neighbourhood meetings, and public hearings. Informing and consultation allow citizens to hear and be heard, but do not guarantee that their views will be heeded, and are therefore termed 'tokenism' by Arnstein. Rung (5) is *placation*, in which citizens begin to have some influence by being invited to offer advice, although tokenism is still apparent in that power-holders retain the right to decide after judging the legitimacy or feasibility of the advice. Further up the ladder, citizens can enter into (6) *partnership*, enabling them to negotiate with power-holders through participation in joint boards and committees. Partnership works most effectively, says Arnstein, when citizen leaders are accountable to the community, and have access to financial resources, giving them real bargaining power. The topmost rungs are (7) *delegated power*, where citizens have the dominant decision-making authority over a particular plan or programme, and (8) *citizen control*, in which participants or residents govern an institution or neighbourhood, with full control over policy and management without outside interference. Arnstein points out that her typology has its limitations. Neither deprived groups, citizens nor power-holders are homogenous groups but encompass divergent views, competing vested interests, and various sub-groups. The achievement of genuine participation can be hindered by racism, paternalism, and resistance to power redistribution on the part of power-holders. Or it can be hindered by inadequate political or socioeconomic infrastructure and knowledge base on the part of citizens, including difficulties of organizing representative and accountable citizen groups, particularly in deprived communities characterized by alienation and distrust.

Arnstein's work has been adapted in Canada in connection with co-management between indigenous groups and authorities (Zachrisson, 2004: 12–13). Zachrisson (2004) applied it in Sweden to the co-management of natural resources, particularly in relation to reindeer-herding, fish and game, large carnivores, and tourism in protected areas. In her typology of citizen participation (Zachrisson, 2004: 13), participatory approaches range from tokenism and placation through genuine partnership to more or less autonomous community control. *Tokenism* means that the governing power sets the agenda and retains all the decision-making power, while getting some input from the local level. In the worst case, communication is one-way, with the community simply being informed about decisions already made. Slightly better is consultation in which local community input is heard but not necessarily heeded; it often comes late in the decision-making process, and involvement is limited by the governing agency's agenda. Somewhat better again is a degree of cooperation in which local knowledge is solicited and local community members are involved as assistants or guides, but still limited by the governing agency's agenda. *Placation* involves local communities to a greater degree, but their role is still only advisory. Through information exchange, local concerns begin to enter management plans, but without joint jurisdiction being established. Partnership in decision-making may begin to develop beyond this through the establishment of local advisory committees, but they have advisory powers only and their decisions are non-binding. Under *genuine participation*, the community is given the opportunity to participate in developing and implementing plans, with local input playing more than an advisory role. In a partnership of equals, joint decision-making is institutionalized and formally recognized, with control being delegated to the community where feasible. Full *community control* means that most or all of the management power is delegated to the local community, which can make decisions independently of government at higher level or with very limited government control. In this case, participatory involvement by higher-level authorities is limited. Zachrisson's variant of Arnstein's typology has been adapted to studies of landscape management in Sweden by geographers Marie Stenseke (2006a, b, 2009) and Kristina Thorell (2008).

There is often an in-built tension between deliberative democracy and representative democracy. While deliberative democracy is dependent on participatory approaches, representative democracy gains its legitimacy from elections. However, philosopher Finn Arler (2008) argues that the aim stated in the ELC's *Explanatory*

Report (Council of Europe, 2000b: §64) to create a 'true landscape democracy' brings into play more dimensions than these. He focuses on three different sets of values that are all associated with democracy. The first is *personal freedom and self-determination*. This emphasizes respect for private property and consumer sovereignty, although public authorities might intervene in cases of market failure or impose restrictions on the free use of private property if necessary for the common good. Privatization and market mechanisms (including the creation of virtual markets through willingness-to-pay surveys for common goods) are seen as the most democratic arrangement for landscape. The second set of values is *co-determination and participation* in common affairs. This emphasizes voting rights, the right to be heard and taken seriously in public negotiations, and to have one's interests taken into consideration. Democracy is ensured on the one hand through elections and preference surveys reflecting the popular vote and on the other hand by public engagement to ensure that the solutions adopted for landscape gain wide support. The third set of values is *objectivity and impartiality*. This emphasizes that decisions are to be made with due respect for valid arguments. These should address impartially the public as a whole and not just one privileged section of it, although giving experts and landscape specialists a role in providing good arguments. Democracy is safeguarded by open debate and fair decision procedures. However, Arler stresses that landscape is not formed simply by landscape policy. Commodity markets, globalization, and political decisions that are not concerned with landscape all influence the way in which landscape develops.

2.4 Lessons from Participation in Third World Development

The effectiveness of participatory approaches is dependent on the degree to which government authorities or others with power allow real involvement by the public and different interest groups. Some useful lessons can be drawn from literature critiquing the prevailing orthodoxy regarding participatory approaches in Third World development projects (Selman, 2004: 368–371). The idea of participation has long been applied to development programmes and projects in Third World countries, especially in resource management. Participatory approaches in the management of forests, watersheds, water, wildlife, and biodiversity are variously termed community management, joint management and co-management. Although landscape is not generally specified as an issue in such projects, resource management issues are in a sense landscape issues.

The agricultural historian Jules N. Pretty (1995) has studied participation in connection with the promotion of sustainable agriculture in development programmes. He identifies two overlapping schools of thought regarding participation in agricultural development. The first views participation as a means of increasing efficiency – 'if people are involved, they are more likely to agree with and support the new development'. The second regards participation as a fundamental right, the aim of which 'is to initiate mobilization for collective action, empowerment and institution building' (Pretty, 1995: 1251). Pretty has developed a much-cited typology of participation, which appears to derive in part from Arnstein's. He identifies seven types of participation, ranging from 'manipulative and passive participation, where people are told what is to happen and act out predetermined roles, to self-mobilization, where people take initiatives largely independent of external institutions' (1995: 1253). *Manipulative participation* is simply a pretence, in which local people are ostensibly represented but have no power. *Passive manipulation* involves sharing information without listening to people's responses. In *participation by consultation* people answer questions but the problems are defined and information-gathering determined by external agents. *Participation for material incentives* is where people participate by providing labour in return for financial or other incentives, but they have no stake in prolonging their involvement when the incentives some to an end (landscape management agreements provide a European example). These first four types of participation provide stakeholders with little or no real influence and are unlikely to have positive lasting effects on people's lives. *Functional participation* involves affected groups as a means of achieving project goals and reducing costs; there may be some shared decision-making, but only after the major decisions have been made by external agents. In *interactive participation* people take part in the analysis of problems and development of plans; participation is seen as a right, and people have responsibility for local decisions, hence having a stake in maintaining new structures or practices. Finally, under *self-mobilization* people take initiatives independently of external institutions.

Pretty emphasizes that the term 'participation' must be used and interpreted with great care. Studies of development projects have shown that participation is a critical component of success. However, there is a danger that lip-service is paid to participation in name rather than in reality. Pretty (1995: 1251–1252) sums up the pros and cons:

> It has been associated with increased mobilization of stakeholder ownership of polices and projects; greater efficiency, understanding and social cohesion; more cost-effective services; greater transparency and accountability; increased empowering of the poor and disadvantaged; and strengthened capacity of people to learn and act.... The dilemma for many authorities is they both need and fear people's participation. They need people's agreement and support, but they fear that this wider involvement is less controllable, less precise and so likely to slow down planning processes. But if this fear permits only stage-managed forms of participation, then distrust and greater alienation are the most likely outcomes. This makes it all the more crucial that judgements can be made on the type of participation in use.

An important contribution to critical research on participatory approaches in Third World development projects is contained in the collection of articles edited by development researchers Bill Cooke and Uma Kothari under the title *Participation: The New Tyranny?* (2001). Their book contains a critique of participatory development orthodoxies, in particular Participatory Rural Appraisal as advocated by Robert Chambers in 1983. It is argued that a potential consequence of participatory approaches can be tyranny, in the sense of illegal or unjust exercise of power, despite the rhetoric of empowerment. Three types of tyranny are identified. First is the tyranny of decision-making and control: participation may override legitimate decision-making processes. Second is the tyranny of the group, within which group dynamics reinforce the interest of the powerful. Third is the tyranny of method: alternatives are neglected even though established participatory processes may not always give the intended result.

Cooke and Kothari's book brings into focus the power dimension of participation. The complexity of power relations is stressed. It is pointed out that stakeholders have varying motivations for participating or not as the case may be. They also have varying bargaining power. Participation may provide opportunities for powerful groups to get their agenda accepted. There are dangers of manipulation by powerful interests, for instance through political cooption or bribery. It is claimed that by manipulation participation may mask centralization in the name of decentralization. It is further argued that questionable assumptions are often made about what the 'local community' is and what its role is in public participation. In the micro-politics of local knowledge production, local elites may define what is regarded as local knowledge. An over-emphasis on micro-level decisions may paradoxically obscure local inequalities and injustices. It is often easier to relate to formal rather than informal organizations – but many people may not have organizations that speak for them. Participation may in practice conceal the daily oppressions that rule people's lives, relating to for example gender or social class. Hence the power structures of local communities may be masked.

The editors maintain that they are not against participation as such. Genuine participation means sharing knowledge and negotiating power relations. They note that political activism and engagement in social movements can provide means to challenge oppression and injustice, while recognizing that some forms of what is termed participation may paradoxically reinforce oppression and injustice. They conclude that a more reflexive understanding of power requires the study of participation rhetoric and practice.

In response to what can be seen as a backlash against participatory approaches produced by Cooke and Kothari's book, development researchers Samuel Hickey and Giles Mohan edited a collection articles exploring new approaches to participation in development under the title *Participation: From Tyranny to Transformation?* (2004; summarized in Hickey and Mohan, 2005). Here a number of prerequisites for successful participatory approaches in the Third World context are set out. They argue that participation is most likely to succeed where it is part of a radical political project focusing on power relations and not just on technical solutions. Participation needs to be conceived as a means of specifically including marginal and subordinate groups. It must engage with underlying processes of development rather than being limited to specific interventions; in other words its needs to be part of a broader social movement. The conditions of success involve examining the political economy of participation – the distribution of power, wealth and patronage. There is a need for criteria to evaluate forms of participation that seek to transform the practices of development agencies and professionals. There is a need for engagement with the politics of difference, respecting group differences without oppression. However, there are two cautionary warnings: it is important not to romanticize the capacity of the poor and marginalized, and it is important to avoid treating all local knowledge as incontrovertible.

2.5 Participation in Social Policy in English Cities

The debate over participation in Third World development projects provides issues for consideration when implementing the obligation of public participation in the European Landscape Convention. If participation is to be meaningful, it must be genuine. Further lessons can be drawn from a recent study of the role of participation in social policy in Britain, based on 17 case studies in two unidentified English cities, undertaken by policy researchers Marian Barnes, Janet Newman and Helen Sullivan (2007). This study investigated participation in forums such as citizen's juries, area committees, neighbourhood forums, tenant groups, and user groups. Some were established by public bodies for the purpose of dialogue, while others were independent voluntary, charitable or political action groups. Among the findings is that a general shift in social policy towards a more open, collaborative and innovative policy system has led to more differentiated state–citizen interactions. Further, participation complements rather than challenges representative democracy, although the relationship between participatory forums and decision-makers remains uncertain. Plural voices are recognized – but some still struggle to be heard. New partnerships between social movements and the authorities have developed, leading to new forms of citizenship. However, these new partnerships also contain the danger that social movements can be 'captured' by the authorities and lose their autonomy. A difficult issue concerns the representation and representativeness of active participants – how should they be represented and who do they represent. This issue requires negotiation.

Barnes et al. (2007) find that there are varying institutional dynamics of participation. At one end of the spectrum is symbolic conformance to the idea of participation by reluctant public bodies. This leaves existing practices largely unchallenged. In the middle are new partnerships established by government policy. However, those in authority generally speaking retain the power to constitute what is considered 'the public' through discourses defining 'the community' and by naming participant groups. The authorities set the rules and agendas, decide the legitimacy of different voices, and can choose to take account of the views expressed or not. At the other end of the spectrum are voluntary or community organizations that are separate from the authorities. These maintain their independence and serve as 'safe places' from which to criticise the authorities.

Barnes et al., observe that new forms of public participation lead to improved dialogue and communication, respect for the position of others, and new means of social agency. However, they question the way in which 'the public' is constituted and point out the danger that the terminology used by both officials and citizens may be exclusive, with certain groups being marginalized, or even stigmatized or demonized. Groups such as beggars, gypsies and asylum-seekers come to mind here. Finally, Barnes et al. note the paradox that governments appear to want to promote active citizenship, yet at the same time seem uncomfortable with community activism and new social movements.

2.6 Tentative Conclusions

The critical literature on participatory approaches in development projects and social policy raise issues that also need to be addressed as procedures are established for participation in the definition and implementation of landscape policies.

The *institutional dynamics* of participation need to be considered, focusing on the legitimacy of different types of participation and the power relations involved (Barnes et al., 2007). An important question is who sets the premises for participation and who exercises the power of definition. Another question is how to safeguard the process of participation from the danger of manipulation – a neutral control instance may be necessary, but what form should it take? A third question is how to assess the representativeness of actively participating stakeholders in relation to the broad range of those with an interest in a landscape. Related to this are questions of who are included and who tend to be excluded in the participatory process – how can the views of marginalized groups best be brought in? How can respect for difference and social justice be ensured?

A second issue concerns the *cost-benefits* of participation (Buchy and Hoverman, 2000: 19). To what extent does it represent a cost and to what extent does it represent an investment? Landscapes are generally complex and it takes time to debate and research all aspects, frequently involving many interests. Hence participation is often seen by the authorities as causing delay in planning and involving costly expenditure. On the other hand, the cost of not ensuring effective participation may in the long term be higher if the results are long ongoing protest actions or the alienation of sectors of the public.

A third issue is that of *local versus non-local stakeholders* (Jones, 2007: 622–623). Discussions of public participation frequently focus on local participation. It is argued that local people, 'insiders', have a greater right to be heard on matters relating to 'their' landscapes than 'outsiders'. Individuals or groups who have stable, long-term attachments to particular places, sometimes inherited over several generations, frequently (although not always) have deep-rooted and often tacit knowledge about the history and distinctive characteristics of a landscape. However, there may be a danger of romanticizing local knowledge. Local communities are important but not the only legitimate stakeholders. Many outsiders also have legitimate interests in a landscape. Further, the distinction between 'outsiders' and 'insiders' is not clear-cut. New residents moving into an area may be regarded as 'outsiders' by long-established residents yet express strong attachments to the landscape of their adopted home. More contentious are the views of immigrants from other countries, introducing unfamiliar religious or ethnic elements to the landscape. Conversely, people who have moved away from an area may continue to visit it because of family ties, property or continuing emotional bonds, and may express strong feelings about landscape changes. Visitors have varying degrees of attachment. They may be tourists who visit a place once or only a few times, holiday-makers who return regularly to the same place, or second-home owners who spend time in an area on a weekly or seasonal basis, sometimes becoming semi-permanent residents. Landscape specialists in academia or government administration are frequently 'outsiders' who tend to have especially strong views on landscape change based on their particular disciplinary training. The perceptions of all these groups are relevant in terms of the European Landscape Convention. The challenge is to find a modus vivendi – or in a participatory process a modus operandi – between the differing interests and values of different groups.

A fourth issue is the challenge of *combining deliberative democracy and representative democracy* (Barnes et al., 2007: 2). Questions arise concerning how the deliberations of extensive public participation can be made compatible with the decision-making responsibility of elected representatives. How are participatory processes ultimately reflected in decision-making? To what degree do elected councils and officials take heed of different viewpoints in making final decisions? How do they reconcile or arbitrate between conflicting viewpoints? Two levels of public participation have been identified: the first is in the definition of landscape policy, and the second is in its implementation (Prieur and Durousseau, 2006: 165). A challenge is to get the public involved in discussions of landscape policy at an early stage, giving decision-makers a broad input of views before decisions are made and before implementation has gone so far that public participation is too late to be effective.

A fifth issue concerns *dispute-resolution procedures*. Many disputes cannot be settled by discourse and debate alone, even when all views have been heard. Often some form of arbitration is necessary (O'Riordan and Stoll-Kleemann, 2002: 100). A role needs to be discussed for mediators who have broad respect and are perceived as being without any interest of their own, and who can weigh evidence and recommend solutions.

Participation is not easy but at the same time cannot be ignored (O'Riordan and Stoll-Kleemann, 2002: 109). Government authorities both need and fear participation – they need support but fear loss of control (Pretty, 1995: 1252). Participatory processes are one of several ways in which civic society may affect policy formulation and implementation. Participation lies on a spectrum between exercising purchasing power in markets or voting in elections at one end, and demonstrating dissatisfaction through peaceful social protest, covert resistance or in the worst case violent protest at the other end (O'Riordan and Stoll-Kleemann, 2002: 102–103). The concrete results in the physical landscape will ultimately reflect the interaction between practices based on the ideal of communicative rationality (Habermas, 1990) on the one hand and the reality of power relations (Foucault, 1980) on the other hand.

While the European Landscape Convention has established the principle of participation in landscape protection, management, and planning, participatory approaches to landscape are still in their infancy in much of Europe. They are applied in a piecemeal fashion and are often considered experimental. It is necessary to learn from both the successes and failures of these examples in order to make a reflective, collective assessment of performance. Examples of best practice in the implementation of participatory procedures need to be sought, but it is also necessary to focus on problems and critical aspects in order to bring out hindrances to and the limitations of participation. The study and assessment of what actually happens in practice includes the role of rhetoric. The effectiveness of participation needs to be assessed both in terms of efficiency – working productively with a minimum of wasted effort and expense – and in terms of the exercise of the fundamental right of all interested parties to have a say without discrimination.

Acknowledgments I would like to thank Tor A. Benjaminsen, Cathrine Brun and Marie Stenseke for drawing my attention to central literature references used in this chapter. The ideas expressed here have crystallized in part as a result of discussions following my presentations on the European Landscape Convention and participation at meetings of the Nordic Landscape Research Network at Oscarsborg, Norway, 29–30 August 2007, the Nordic Forum for Ethnography at the Department of Social Anthropology, University of Oslo, Norway, 4 December 2007, and the Permanent Conference for the Study of the Rural Landscape (PECSRL) in Lisbon and Óbidos, Portugal, 1–5 September 2008, where a first draft of the present paper was presented.

References

Arler F (2008) A true landscape democracy. In: Arntzen S, Brady E (eds) Humans in the land: the ethics and aesthetics of the cultural landscape. Unipub, Oslo, pp 77–101

Arnstein SR (1969) A ladder of citizen participation. AIP J 35:216–224

Barnes M, Newman J, Sullivan H (2007) Power, participation and political renewal: case studies in public participation. Policy Press, Bristol

Borsotto P, Henke R, Macri MC, Salvioni C (2008) Participation in rural landscape conservation schemes in Italy. Landsc Res 33:311–330

Buchecker M, Hunziker M, Kienest M (2003) Participatory landscape development: overcoming social barriers to public involvement. Landsc Urban Plann 64:29–46

Buchy M, Hoverman S (2000) Understanding public participation in forest planning: a review. Forest Policy Econ 1:15–25

Caspersen OH (2009) Public participation in strengthening cultural heritage: the role of landscape character assessment in Denmark. Geogr Tidsskr 109:33–45

Chambers R (1983) Rural development: putting the last first. Longman, London

Cooke B, Kothari U (eds) (2001) Participation: the new tyranny? Zed Books, London

Council of Europe (2000a) European Landscape Convention, Florence, 20.X.2000. ETS No. 176. http://conventions.coe.int/Treaty/en/Treaties/Html/176.htm. Accessed 6 Mar 2007

Council of Europe (2000b) European Landscape Convention (ETS No. 176): explanatory report. http://conventions.coe.int/Treaty/EN/Reports/Html/176.htm. Accessed 6 Mar 2007

Council of Europe (2008) Recommendation CM/Rec (2008)3 of the Committee of Ministers to Member States on the guidelines for the implementation of the European Landscape Convention. https://wcd.coe.int/ViewDoc.jsp?id=1246005. Accessed 24 Sept 2009

Daugstad K, Svarstad H, Vistad OI (2006) Community-based nature protection: lessons from Norway. Int J Environ Cul Econ Soc Sustainability 2:127–135

EC (2003) Directive 2003/35/EC of the European Parliament and of the Council of 26 May 2003 providing for public participation in respect of the drawing up of certain plans and programmes relating to the environment and amending with regard to public participation and access to justice Council Directives 85/337/EEC and 96/61/EC.

Official Journal of the European Union, L 156, 25/06/2003, 17–25. EUR-Lex. http://eur-lex.europa.eu/LexUriServ/LexUriServ.do?uri=OJ:L:2003:156:0017:0024:EN:PDF. Accessed 6 May 2010

Foucault M (1980) Power/knowledge: selected interviews and other writings 1972–1977. Gordon C (ed) (trans: Gordon C, Marshall L, Mepham J, Soper K). The Harvester Press, Brighton

Habermas J (1990) Moral consciousness and communicative action. (trans: Lenhardt C, Nicholson SW). Polity Press, Cambridge, UK – MIT Press, Cambridge, Mass.

Hartley N, Wood C (2005) Public participation in environmental impact assessment – implementing the Aarhus convention. Environ Impact Assess Rev 25:319–340

Hickey S, Mohan G (eds) (2004) Participation: from tyranny to transformation? Exploring new approaches to participation. Zed Books, London

Hickey S, Mohan G (2005) Relocating participation within a radical politics of development. Dev Change 36:237–262

Höppner C, Frick J, Buchecker M (2007) Assessing psycho-social effects of participatory landscape planning. Landsc Urban Plan 83:196–207

James P, Gittins JW (2007) Local landscape character assessment: an evaluation of community-led schemes in Cheshire. Landsc Res 32:423–442

Jones M (2007) The European Landscape Convention and the question of public participation. Landsc Res 32:613–633

Jonsson A, Lundqvist LJ (2006) Engagera sig i vattenfrågor – varför, hur mycket och var? In: Jöborn A, Danielsson I, Oscarsson H (eds) På tal om vatten: Om vägen mot en hållbar vattenförvaltning. Vastra rapport 6. Vastra, Göteborg, pp 93–122

Luz F (2000) Participatory landscape ecology – a basis for acceptance and implementation. Landsc Urban Plann 50:157–166

Olwig KR (2007) The practice of landscape 'conventions' and the just landscape: the case of the European Landscape Convention. Landsc Res 32:579–594

O'Riordan T, Stoll-Kleemann S (2002) Deliberative democracy and participatory biodiversity. In: O'Riordan T, Stoll-Kleemann S (eds) Biodiversity, sustainability and human communities: protecting beyond the protected. Cambridge University Press, Cambridge, pp 87–112

O'Rourke E (2005) Landscape planning and community participation: local lessons from Mullaghmore, the Burren National Park. Landsc Res 30:483–500

Pretty JN (1995) Participatory learning for sustainable agriculture. World Dev 23(8):1247–1263

Prieur M, Durousseau S (2006) Landscape and public participation. In: Landscape and sustainable development: challenges of the European Landscape Convention. Council of Europe Publishing, Strasbourg, pp 165–207

Selman P (2004) Community participation in the planning and management of cultural landscape. J Environ Plann Manage 47:365–392

Sewell WRD, Coppock JT (eds) (1977a) Public participation in planning. Wiley, London

Sewell WRD, Coppock JT (1977b) A perspective on public participation in planning. In: Sewell WRD, Coppock JT (eds) Public participation in planning. Wiley, London, pp 1–14

Stenseke M (2006a) Lokalt inflytande i landskapsvården: En forskningsöversikt med bäring på skötseln av naturbetesmarker. Occasional Papers 2006:1. Kulturgeografiska institutionen, Handelshögskolan vid Göteborgs Universitet, Göteborg

Stenseke M (2006b) Vad kan vi lära av "Ölandsmodellen?": Om förutsättningar för lokal delaktighet i landskapsvården. Occasional Papers 2006:5. Kulturgeografiska institutionen, Handelshögskolan vid Göteborgs Universitet, Göteborg

Stenseke M (2009) Local participation in cultural landscape maintenance: lessons from Sweden. Land Use Policy 26:214–223

Svarstad H, Daugstad K, Vistad OI, Guldvik I (2006) New protected areas in Norway: local participation without gender equality. Mt Res Dev 26:48–54

Thorell K (2008) Naturvårdsplanering med förankring i det lokala: Villkor för delaktighet och underifrånperspektiv i vården av värden i odlingslandskapet. Meddelanden från Göteborgs universitets geografiska institutioner, serie B, nr 114. Handelshögskolan vid Göteborgs universitet, Göteborg

Tress B, Tress G (2003) Scenario visualisation for participatory landscape planning – a study from Denmark. Landsc Urban Plan 64:161–178

UN (1992) Report of the United Nations Conference on Environment and Development (Rio De Janeiro, 3–14 June 1992) Annex I: Rio Declaration on Environment and Development, A/CONF.151/26 (Vol. I). United Nations General Assembly. http://www.un.org/documents/ga/conf151/aconf15126-1annex1.htm. Accessed 24 Jun 2010

UNECE (1998) Convention on Access to Information, Public Participation in Decision-making and Access to Justice in Environmental Matters done at Aarhus, Denmark on 25 June 1998. United Nations Economic Commission for Europe. http://www.unece.org/env/pp/treatytext.htm. Accessed 11 Mar 2010

Zachrisson A (2004) Co-management of natural resources: paradigm shifts, key concepts and cases. Mountain Mistra Programme report, Report no. 1. Mountain Mistra Programme, Umeå

Chapter 3
The Dutch Approach

Public Participation and the Role of NGOs and Local Authorities in the Protection, Management and Development of Cultural Landscapes in the Netherlands

Henk Baas, Bert Groenewoudt, and Edwin Raap

Abstract The preservation of the cultural landscape of the Netherlands is a complex matter, partly because responsibilities are divided between several ministries and administrative levels. Guided by the motto 'conservation through development', the Belvedere Programme was launched in order to combine efforts and to

H. Baas (✉)
Cultural Heritage Agency, Ministry of Education, Culture and Science, NL-3800 BP Amersfoort, The Netherlands
e-mail: h.baas@cultureelerfgoed.nl

M. Jones, M. Stenseke (eds.), *The European Landscape Convention*, Landscape Series 13, DOI 10.1007/978-90-481-9932-7_3,

develop and put into practice projects aiming at actively taking into account historical values in the process of spatial planning. In accordance with the European Landscape Convention (ELC), the care of the cultural landscape has increasingly become the responsibility of local authorities. Consequently landscape protection is to an increasing degree dependent on arousing public support. In shaping local policy towards landscape, all parties involved are faced with the challenge of combining scientific knowledge with local demands and initiatives. To achieve this, Landscape Development Plans and research guided by the 'landscape biography' concept have proven to be an effective strategy. Especially on a local level, the long-term history of humans and landscape can be used to inspire future developments.

Keywords Conservation through development · Landscape Development Plans · Landscape biography · Landscape Manifesto

3.1 Introduction

As a result of a new Spatial Planning Act (*Wet Ruimtelijke Ordening, WRO*) in 2008 (Ministerie van VROM, 2008), responsibility for landscape management in the Netherlands is increasingly being delegated to the municipalities. This development is in accordance with the European Landscape Convention (ELC). Nonetheless, there is also opposition against making municipalities responsible for the landscape. Their ability to make sound decisions with regard to landscapes is being questioned. There is, moreover, a potential conflict between, on the one hand, local economic targets, which find their expression in housing developments and business parks, and, on the other, the desire to improve landscape quality.

We believe that a local policy that respects the character of a historical landscape is perfectly feasible, provided it meets certain conditions. These include maximum use of existing policy instruments, as well as complete integration and application of all available expertise on landscape and culture history. During the past 10 years, the Belvedere Programme in particular has generated much experience in this field.

This chapter gives an overview of the 'Dutch approach', which focuses on the local scale and juxtaposes different approaches. We examine ways in which care for the historical landscape can be developed at the level of the local community. Central to this is the Landscape Development Plan (*Landschapsontwikkelingsplan*), a municipal policy instrument which is particularly suitable for combining historical and modern dimensions. This plan can be supplemented by Village Surroundings Plans (*Dorpsomgevingsplannen*), which assign a significant role to the inhabitants of an area. One of the questions we address is to what extent the wishes and the expertise of the local population can be incorporated into the planning process. We believe that the concept of *landscape biography* provides excellent opportunities for this. We conclude the chapter by offering some suggestions for policy and research, which need to inform each other now perhaps more than ever before.

3.2 Conservation Through Development

Landscape is the object of much attention in the Netherlands. Increasing interest is reflected in several policy documents, such as the Land Use Planning Memorandum or *Nota Ruimte* (Ministries of VROM, LNV, V&W and EZ, 2005) and the recent Agenda Landscape (*Agenda Landschap*) (Ministries of LNV & VROM, 2008). Twenty National Landscapes have been designated thus far (Fig. 3.1), each of them being an area that:

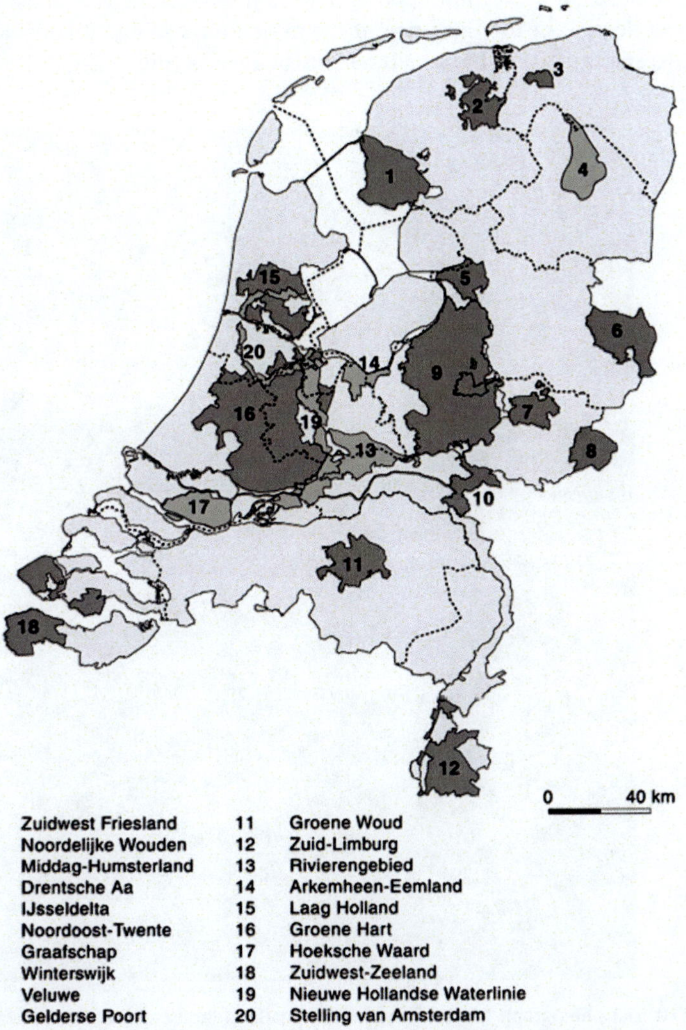

Fig. 3.1 Location of the 20 National Landscapes in the Netherlands

> ...possesses a unique combination of cultural-historical and natural elements. ... The National Landscapes are characterized as areas with specific coherence between nature, surface, land use and occupation. They are not museums, but areas where people live, work, venture, and recreate (Nationale Landschappen n.d.).

Although some of the National Landscapes are also on the UNESCO World Heritage List, not all of them are necessarily the most valuable areas from a cultural heritage point of view.

The fact that the Netherlands now possesses twenty National Landscapes is remarkable, for all earlier attempts since the 1970s had failed because of the political influence of the agricultural sector, which feared potentially adverse effects of any form of landscape protection upon agricultural development (Renes, 2008). After the recent decline of the influence of the agricultural lobby, it became possible to put landscape protection back on the political agenda (Fig. 3.2).

Fig. 3.2 Aerial photograph of the characteristic peatland landscape of the western Netherlands (National Landscape Groene Hart), 2007. In the centre of the photo is a *geriefhoutbosje*, a coppice for private use by farmers (Photo: Courtesy of Paul Minkjan)

As a result of the designation of the twenty National Landscapes, 20% of the total land surface of the Netherlands is more or less protected – 'more or less', since the current regime does not provide much actual protection. The guiding principle is that development is allowed provided it takes certain aspects into account, such as the preservation of cultural heritage. This is a typically Dutch solution: some areas are protected, but then again, they are not really protected. Whether this approach will result in effective protection of complete cultural landscapes is doubtful, for it requires constant attention as well as continuity in policy (Janssen et al., 2007). According to Dirkx (2009), the state has set a bad example in this respect, leaving the protection of National Landscapes almost completely in the hands of provincial authorities and providing hardly any guidelines or recommendations.

Interest in landscape and cultural heritage culminated in 1999 with the Belvedere Memorandum on cultural heritage and spatial planning (*Nota Belvedere*) (Ministries of OCW, LNV, VROM and V&W, 1999). This memorandum, which was endorsed by no less than four ministries, introduced an approach named 'conservation through development', which involves the incorporation of cultural-historical values (archaeology, built heritage, and historical landscape) into new developments (e.g. building, infrastructure, and nature development). This strategy has proved to be very successful (Evaluatie Belvedere, 2008). The Belvedere memorandum also contained a map of national cultural-historical values, which has had great impact on national and provincial policy. Moreover, the project agency has taken several initiatives. Thanks to project subsidies, over 300 projects (local, regional, national, and knowledge-based) could be executed, all of them variations on the central concept of 'conservation through development'. Since then, attention to the cultural landscape as part of the cultural heritage has increased, and historical geographers and archaeologists no longer have to beg for it. Quite the reverse; they have found themselves cooperating with landscape architects and spatial planners. This has proved to be a fruitful relationship.

3.3 Landscape Biography

The national government in the Netherlands is responsible for the protection of what are termed the 'core qualities' of the twenty National Landscapes (such as maintenance of characteristic field patterns or different kinds of small-scale landscape elements). The state, in close cooperation with the provincial authorities, provides National Landscapes with a certain degree of protection through the application of spatial planning instruments. These 'core qualities' are described in the Land Use Planning Memorandum, but only imprecisely and in general terms. One may also wonder whether the inhabitants of the National Landscapes recognize these 'core qualities' as the most important qualities of the landscape they live in. Research has shown that people have more appreciation for local and regional history, and for the stories that relate to the history of their own landscape, than they have for larger, more abstract issues (Koedoot, 2004).

Research into historical cultural landscapes in the Netherlands increasingly involves the concept of 'landscape biography'. Dutch archaeologist Jan Kolen introduced this concept in the Netherlands. He considered it to be a helpful instrument in the continuous struggle between, on the one hand, the material landscape and, on the other, the world of ideas, meanings, representations, and memories (Kolen, 2005). Landscapes change constantly, just like people and even objects (see e.g. Kopytoff, 1986). These changes involve their material shape, their meaning, and their interpretation (Kolen, 1993; Hidding et al., 2001). Hence landscapes can be read like a book, albeit a book without a clear beginning and definitely without an end. The concept of landscape biography is particularly useful when studying and describing long-term developments in the relation between people and their environment, as well as the processes related to this. It does, however, demand an interdisciplinary approach involving the disciplines such as physical geography, archaeology, historical geography, building history, and historical ecology (Groenewoudt, 2006). The most important goals and points of consideration (Hidding et al., 2001) are:

- The effects of processes of change, dynamics, and 'breaks' in the history of landscapes on the dimensions of continuity and sustainability
- The perception of landscape history in the past
- Landscape development considered in the (very) long term, including both the earliest (pre-5000BP) and the youngest (post-120BP) habitation history
- The relation between the historical and the present dimensions of a landscape; this last point implies an active role for the present inhabitants and users of the landscape.

A landscape-biographical approach also makes it possible to describe a landscape not merely as a physical but also as a social and a mental reality. Studies of the physical landscape (e.g. soil, vegetation, reclamation patterns, settlement patterns, and infrastructure) can incorporate cognitive and cultural perspectives, such as past and present appropriation of the landscape and mental ordering of space (Roymans, 1995; Rooijakkers, 1999). Folk tales and place-names, among other things, can be used as sources of information for this.

The awareness that landscape itself is a history book is important for closing the gap between national and local or regional interests. Applying the concept of 'landscape biography' in daily practice, however, is far from easy because of its multiple perspectives. It aims to produce scientific knowledge about a landscape and its history, but at the same time also focuses on the knowledge and perceptions of the inhabitants and users of that landscape (Abrahamse et al., 2008).

Recently, the Cultural Heritage Agency, in cooperation with several Dutch universities and a large number of other organizations, explored the implementation of the 'biography concept' in the context of the research programme 'Protecting and Developing the Dutch Archaeological-Historical Landscape (2001–2007)' (Knaap and Valk, 2006). Within this programme, four regional research projects were designed. In each of these interdisciplinary projects, dealing respectively with parts of the northern, southern, western and eastern Netherlands, attempts were made to establish useful connections between science, policy, and daily practice. The chosen research methods and results displayed a considerable degree of variation, depending on the specific topics addressed, the local socio-political circumstances, and the backgrounds and viewpoints of the researchers involved. For instance, within the project dealing with the Drentsche Aa region, situated in the north-eastern part of the Netherlands, a wide variety of quite successful 'applied' and interlinked projects were given shape in close collaboration with the inhabitants. Examples are an interactive digital cultural atlas and a project dealing with field names, resulting in an easy accessible book (Elerie and Spek, 2009). What all these activities have in common is that they aim at using heritage to inspire future developments.

Cooperation with local inhabitants is especially important for organizations involved in the protection of historical landscapes. Such cooperation offers opportunities to link the biography approach to other instruments, such as the Landscape Development Plan (LDP). It is thus able to close the gap between 'official' cultural heritage and 'popular' cultural heritage with regard to how each group values a landscape or its elements. According to Duineveld (2006), this gap can be a problem, especially for archaeology.

The biography approach also makes it possible to combine landscape history with landscape architecture and landscape planning. Studying the forces that shaped the landscape as we know it today can help us to tackle current problems.

3.4 Landscape Development Plans

The Landscape Development Plan is the main instrument available to local governments for landscape management and landscape development (Fig. 3.3). The first LDP (called Landscape Policy Plan until 2001) was completed in 1988. It was essentially a 'green plan', dealing mainly with greenery around farms and alongside roads in agricultural landscapes. This was to be expected in the circumstances of the time, when natural values received little attention. During the next 15 years, two thirds of all municipalities in the Netherlands drafted a Landscape Policy Plan (Baas and Herwaarden, 2001). The special subsidy from the Ministry of Agriculture proved to be very successful. A municipality is entitled to a refund of up to 50% of the costs, and if several municipalities combine their application this may reach 75%. This is conditional on the outsourcing of the drafting of the plan to a bureau that employs at least one registered landscape architect. Towards the end of the twentieth century, when the Belvedere Memorandum was published, the traditional conservation policies were replaced by the 'conservation through development' strategy. The Landscape Policy Plan became the Landscape Development Plan, and the role of cultural heritage was emphasized (Baas and Herwaarden, 2001). On the occasion of the formal introduction of the LDP, Geke Faber, then undersecretary of the Ministry of Agriculture, Nature and Food Quality, clearly formulated the goals of the LDPs:

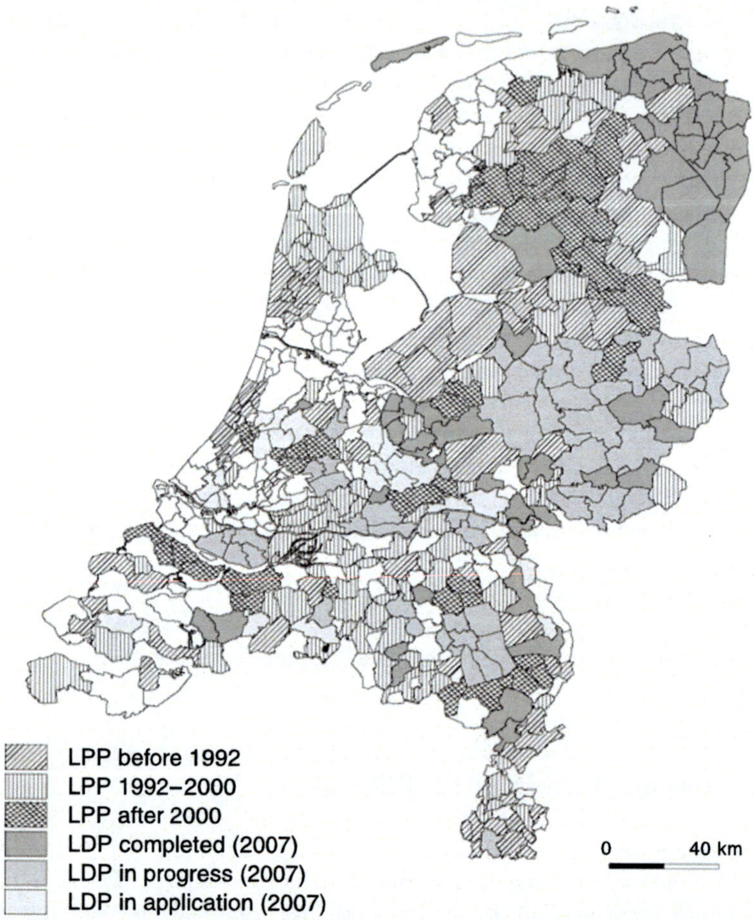

Fig. 3.3 Location of current Landscape Policy Plans (LPP) and Landscape Development Plans (LDP) in the Netherlands. There are notably fewer such plans in the densely populated, western part of the country

they were to contribute to the dynamics and by extension to the quality of the landscape. Furthermore, they were expected to support local and regional initiatives in such a way that these would contribute to the preservation of the specific character and variety of the Dutch landscape (Fig. 3.4). The role of the general public in these processes was important, and the general public were to be involved to a greater extent than had hitherto been the case (Woestenburg, 2008).

Another aspect of LDPs is their emphasis on implementation. Evaluation of previous policy instruments has brought to light their lack of attention to this particular issue (Verhoeven, 2000). The new policy and its associated subsidies actively stimulate the implementation of plans, for example by offering compensation for the appointment of a landscape coordinator. Evaluation has demonstrated

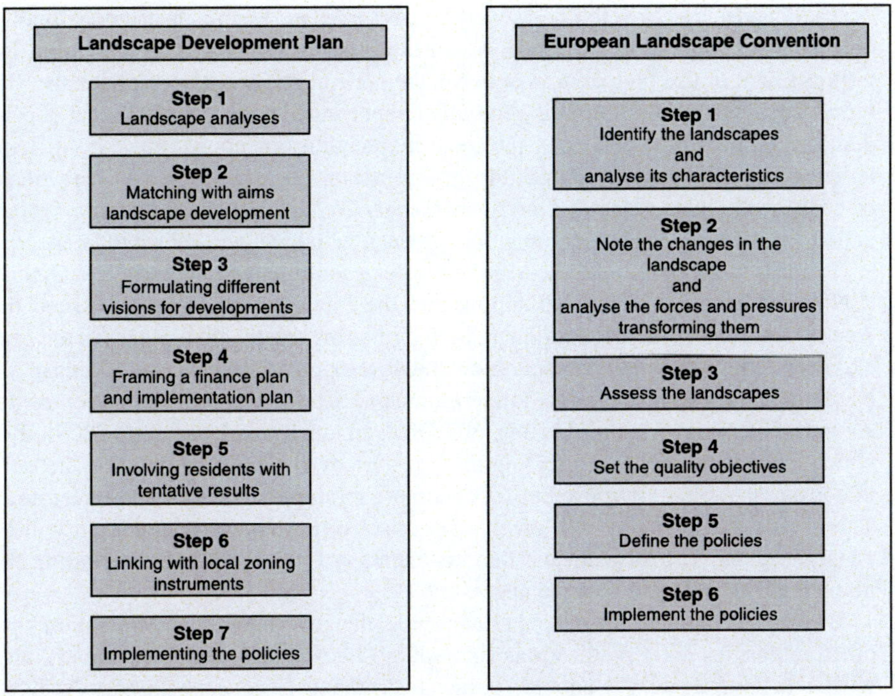

Fig. 3.4 The phased plan for the implementation of Landscape Development Plans, largely proceeding in accordance with the policy process proposed by the European Landscape Convention

that the presence of such a coordinator greatly contributes to the success of an LDP (Landschapsbeheer Utrecht, 2000).

An important element of an LDP is analysis of landscape genesis, which in turn generates information important to the formulation of a landscape vision. However, the specific methods for such an analysis are not prescribed, and each company that is involved is free to make its own choices. The provisions of the subsidy do not specify this point. Nonetheless, the municipality can make other demands, such as for the involvement of specialists.

The aspect of participation by the general public makes LDPs very attractive to politicians and policy-makers. Research into the perception of landscape by the general public demonstrates that there is a growing interest in landscape and landscape history. Although scientific research on this is scarce, it is estimated that a quarter of the Dutch population (i.e. roughly 4 million people) are actively involved with landscape, for example by helping with maintenance, by buying local products, or by participating in local landscape policies (Overbeek and Vader, 2008). Local people can and wish to take care of their own affairs, particularly when the landscape they live in is involved. More and more people can be expected to take the initiative, and local authorities should encourage this. This is what is meant by 'local ownership' (Vos et al., 2007).

The Landscape Development Plan is an easily accessible instrument that is designed in cooperation with local organizations. However, there is still room for improvement of this cooperation between the general public and professionals. The influence of non-professionals is generally rather limited and to a great extent dependent on the help of others (e.g. the local municipality or companies). In order to increase this influence, a Dutch non-governmental organization (NGO) involved in countryside management, *Landschapsbeheer Nederland*, has initiated a project called *Thuis in Groen* ('At Home in the Green'), which we will return to later.

Towards the end of 2008, the Ministries of Agriculture, Nature and Food Quality (LNV) and Housing, Spatial Planning and the Environment (VROM) issued the Agenda Landscape. In this document, both ministries present their plans for the next 12 years with regard to the preservation and development of landscape values in the Netherlands. It explicitly points to the municipalities for the implementation of the Agenda. Subsidies for the drafting of LDPs remained available until 2009, after which the government is redirecting these resources towards a stimulation measure that is comparable in purpose but focuses more on implementation and integration. This means that planning will receive fewer subsidies but implementation will be encouraged more. It also means that co-financing by other parties, including the municipality itself, will become necessary.

Besides introducing the Agenda Landscape, the government, in cooperation with other authorities and organizations (including *Landschapsbeheer Nederland*), also wishes to investigate yet another policy instrument: Landscape Impact Analysis (LIA), which investigates expected changes and developments and their effect on the landscape.

Nonetheless the LDPs will not disappear, for every individual municipality can still decide to draft one. It is to be expected, however, that with the disappearance of some of the existing subsidies the growth in the number of new LDPs will slow down compared to previous years. As long as financial resources are redirected towards the implementation of the plans, the landscape could still benefit. The future will tell if this will be the case. What has already become apparent is that the landscape coordinators appointed by some municipalities are doing excellent work, trying to get the policy targets formulated in the LDPs implemented.

Reformulating their conditions with regard to landscape analysis (which are at present rather vague) in more specific and detailed terms, in combination with explicit references to culture history, would turn LDPs into even more powerful instruments.

Box 3.1 An example: the past landscape of *Berlewalde*

The forgotten medieval *Berlewalde* wilderness (Fig. 3.5) may serve as a powerful source of inspiration to shape new local developments within the context of a Landscape Development Plan. *Berlewalde* (the name is *pars pro toto*) once covered much of the low-lying and formerly marshy centre of the Achterhoek region (Groenewoudt and Keunen, 2008; Van Beek, 2009). For many centuries this was a border zone. Nowadays low-lying areas are predominantly pasture and used intensively for cattle-breeding and milk production. The landscape is flat, parcelled out in large sections and open. It was reshaped after the vast, virtually treeless commons of the seventeenth and eighteenth centuries were partitioned in the nineteenth century and subsequently reallotted in the twentieth century.

Between the Iron Age and the fifteenth to sixteenth centuries AD, however, the area looked radically different. *Berlewalde* was a sparsely populated landscape covered by a mosaic of wood pasture (*Hudewald*), dense woodland (predominantly alder carr), coppice, shrubs, marshland, and raised bog. Habitation was for a long time restricted to a few isolated sandy ridges and the banks of the river Berkel. *Berlewalde* initially had a spectacular fauna: brown bear, elk, red deer, roe deer, wild boar, wolf, beaver, common crane, etc. Medieval sources up until c.1500 AD also mention the herding of 'forest horses' and 'wild horses'.

The area was reclaimed from the thirteenth century onwards. The last woodlands disappeared largely in the seventeenth and eighteenth centuries, although some patches may have survived longer (Groenewoudt et al., 2007).

Berlewalde could potentially become an attractive historical reference point, and a source of inspiration for the realization of policies concerning water management and nature conservation, certainly for the development of recreation and tourism, and for the establishment of the 'new country estates'. There are several possibilities:

1. Spatial planning and landscape architecture could try to mark a deliberate contrast between the highly domesticated and planned modern landscape and the older, more chaotic and mysterious *Berlewalde* wilderness.
2. *Berlewalde* could become an authentic and inspiring frame of reference for nature conservation, nature development, and water management.
3. *Berlewalde* could contribute to regional branding (and encourage a sense of regional identity).
4. *Berlewalde* definitely would make an ideal breeding ground for more adventurous recreational activities.
5. The reintroduction of semi-wild horses would fit in well with the ongoing 'horsification' of the Dutch countryside. *Wildpferde* (wild horses) like those that formerly roamed *Berlewalde* have survived in the Merfelder Bruch nature reserve near Dülmen in Germany, where they are a major tourist attraction (Die Dülmener Wildpferde... n.d.) (Fig. 3.6).
6. The reintroduction of charcoal-burning (terminated in 1906) would give tourists and local people the opportunity to use *Berlewalde* charcoal for their barbecues. It could also help to make traditional coppicing

profitable again. The charcoal-burning itself could also be a tourist attraction.

7. Resuming coppicing would have a positive effect on the (now declining) biodiversity and recreational value of woodland and hedges.
8. Investigating, explaining, and using landscapes of the past may help to communicate the reality that landscape change is normal.

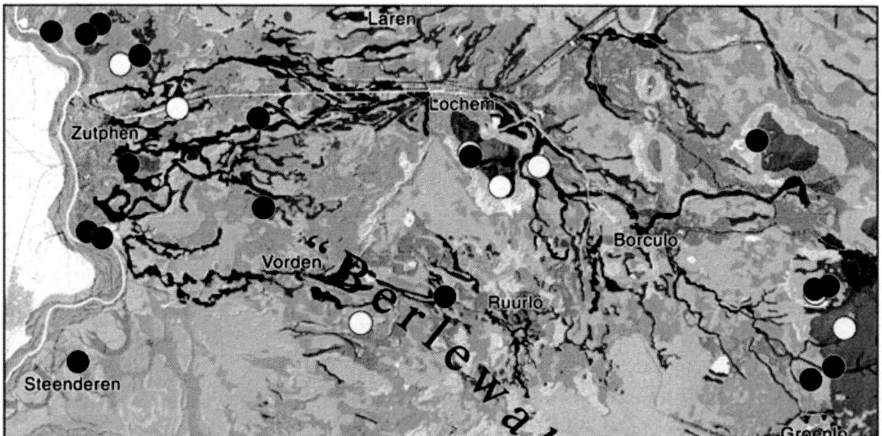

Fig. 3.5 The medieval *Berlewalde* wilderness in the Achterhoek region of eastern Netherlands. In general, the area contains very little archaeology predating the late Middle Ages. This is illustrated by the distribution of urnfields (1000–500 BC)

Fig. 3.6 Wild horses are a major tourist attraction in the German Merfelder Bruch nature reserve near Dülmen (2008). Reintroducing wild horses to *Berlewalde* would fit in with the ongoing 'horsification' of the modern landscape and stimulate tourism (Photo: Courtesy of Gitta Gesing)

3.5 Biography Approach and Landscape Development Plan

An interdisciplinary approach with a biographical perspective can be an important stimulus for local interest and involvement (see Box 3.1 for an example of the biography approach). This makes the biography concept a particularly useful tool for the study and description of landscapes, for instance when a Landscape Development Plan is to be drafted.

We will illustrate this point using a Landscape Development Plan for a part of the Dutch Achterhoek region in the east of the Netherlands, involving the municipalities of Bronckhorst, Lochem and Zutphen. The eastern Netherlands is rapidly being transformed from a predominantly agricultural area into the multifunctional eastern rim of the Dutch 'Delta Metropolis' (*Deltametropool*). The area is also increasingly becoming a transit zone between Rotterdam harbour and the growing markets in the east, especially in eastern Europe. These developments are accompanied by major changes that in turn have important effects upon the landscape. There is growing demand for recreational and housing facilities, water management offers multiple challenges, and agriculture is undergoing rapid changes. The implementation of the National Ecological Network (*Ecologische Hoofdstructuur*) involves nature development as well as the designation of several National Landscapes. Furthermore, there are plans for the development of a number of 'new country estates' (*nieuwe landgoederen*). A new country estate is a publicly accessible area, including a residential building with up to two or three units and a minimum size of at least 5 ha of forest. The house itself creates an architectural unity with the surrounding greenery.

The interdisciplinary Eastern Netherlands Project (*Oost-Nederland Project*) of Wageningen University and the Cultural Heritage Agency was initiated to study in-depth, and for the first time, the landscape history of the area, and to generate ideas and tools to keep its past alive in a rapidly changing environment. The participants in this project included municipalities, counties, water boards, and nature conservation organizations (Beek and Keunen, 2006).

The new expertise generated by the Eastern Netherlands Project will be applied to the drafting of new Landscape Development Plans. For the LDP Bronckhorst-Lochem-Zutphen, this process has already been initiated through workshops (Ziel and Baarslag, 2008). In these workshops several organizations contributed: the municipalities of Bronkhorst, Lochem and Zutphen; the Regional Tourism Board; the Federation of Private Landowners; several interest groups in the fields of agriculture and horticulture, liveability of the countryside and small villages, and nature and landscape preservation; the District Water Board; and the provincial authorities of Gelderland. The workshops indicated that there was much local interest, particularly in the unexpected degree to which the landscape appeared to have been dynamic, and in the previously unknown 'chronological layers' in the history of the landscape. Knowledge of these forgotten landscapes offers unexpected opportunities to shape the landscape of the future. Raising awareness of and using past landscapes may also help to communicate the incontestable fact that landscapes always and inevitably change.

3.6 Landscape Manifesto and the European Landscape Convention

The European Landscape Convention was ratified by the Netherlands in 2005. All ratifying countries must implement a 'systematic landscape policy', and they must also guarantee the involvement of local communities. With the publication of two national memorandums and the designation of the twenty National Landscapes, the Netherlands fulfilled the first requirement. At the lower levels of administration, the policy has either not yet been worked out in sufficient detail, or focuses too much on nature conservation, and there is still much work to do in this respect.

The ELC was the inspiration for a unique cooperation of thirty-four organizations in the Netherlands, resulting in the formulation of the Landscape Manifesto (Landschapsbeheer Nederland, 2006). The main purpose of this manifesto is to emphasize to politicians, decision-makers, and the general public the importance of landscape. Its ultimate purpose is to create a more attractive landscape. Besides organizations involved with nature and landscape, participants also include organizations with other goals such as the Dutch Society of Property Developers. Cooperation between these different types of organization is essential, because landscape is everywhere and always present, and is owned by everyone. As stated in the ELC, the landscape is integral, regional and cross-border in scope.

For the past 3 years, several study groups of the Manifesto have worked hard to convert its goals into actual projects. One of the results is the establishment of Civilscape, a platform of NGOs from all over Europe that support the European Landscape Convention (Civilscape n.d.). In 2009 the platform started a campaign to raise awareness of landscape, in cooperation with the Ministry of Agriculture, Nature and Food Quality.

The wide interest in landscape is demonstrated by the existence of the twenty National Landscapes, the European Landscape Convention, and the Landscape Manifesto. There are, moreover, numerous local initiatives as well as several instruments that aim to create and improve a durable landscape. Many of these organizations, projects, and initiatives involve to a greater or lesser extent local inhabitants. These initiatives are based on the principle of 'local ownership', which is concerned with small-scale participation by an area's inhabitants in local projects (Vos et al., 2007).

3.7 Dutch Approaches to Participatory Planning

Besides instruments designed to make information on the historical cultural landscape available to policy-makers, it is equally important – in the spirit of the European Landscape Convention – to involve non-professionals in all the planning stages. Several recent initiatives attempt to do just that, for example the project called 'At Home in the Green' started by *Landschapsbeheer Nederland* in cooperation with the National Association of Small Communities (*Landelijke Vereniging Kleine Kernen*). The project aims to test the ECOVAST (European Council for the Village and Small Town) method (Fig. 3.7) (ECOVAST, 2006). ECOVAST resembles the Landscape Character Assessment (LCA), a method used in the UK by which local residents are stimulated to become involved in their environment (Swanwick, 2002). Where historical and archaeological values are involved, a variant of this method is the Historic Landscape Character Assessment (HLC) (Fairclough n.d.). ECOVAST operates on a smaller scale than LCA, and places more emphasis on the role of local residents. 'At Home in the Green' attempts in a structured manner to create a sense of place within the landscape by closely cooperating with the residents. The method can be used to allow spatial development to fit in better in the local environment, to accentuate important landscape types, and potentially to stimulate people to take care of the landscape themselves. At present it is still too early for conclusions, but the first results from the province of Friesland look promising.

The ECOVAST method assists municipal councils both with the protection of existing landscapes and with attempts to achieve desired situations. The guiding

Fig. 3.7 The ECOVAST method helps local residents assess their landscape (2006), and can at local level facilitate the planning, protection, maintenance, and improvement of landscapes (Photo: Courtesy of Landschapsbeheer Nederland)

principle of the method is to determine which aspects of the landscape are important to the residents of the village, district, or neighbourhood. On the basis of information about what is already valuable to them and what perhaps need extra attention, plans will be drawn up to realize the ideas that came up, e.g. forming 'green teams', which will enthusiastically take care of the landscape and its nature. Crucial to the success of ECOVAST (and indeed any LCA) is that the results receive formal status by being incorporated in the spatial plans of the local government, for this will prove to the residents that their efforts have been successful.

This approach is an example of what is variously called participatory planning, communicative planning, interactive planning, and consensus planning (Sager, 1994; Roo et al., 2001). As elsewhere in Europe, this type of planning process has rapidly gained ground in the Netherlands, in reaction to the classic type of planning process in which the authorities and property developers would launch major plans without first properly consulting residents and other stakeholders. In the new type of planning, these groups are involved at a very early stage in the process, allowing them to present any alternative ideas they might have. The extra time invested early in the planning process will be regained at a later stage.

A fine example of participatory planning took place in Den Hoorn, a small village on the largest Dutch Wadden Sea island, Texel (Vos et al., 2007). A group of farmers and local entrepreneurs took the initiative for a Village Surroundings Plan. The main goal of this plan was to develop the potential for sustainable tourism and recreation. The study group and the municipality did not want an abstract document but instead a straightforward list of things to do that could give new impulses to existing ideas. The Village Surroundings Plan was drafted during an interactive process involving residents and other interested people, with a professional landscape architect assisting. A group of representatives from the municipality, the province, and the international cross-border project 'Landscape and Cultural Heritage of the Wadden Sea Region' (LanceWad n.d.) provided feedback to the project. The residents were, and still are, themselves responsible for the execution of the plan, during which a typically agrarian landscape such as Texel's will gradually be transformed into a recreational one. This transformation involves a variety of actions:

- New small landscape elements will be planted or created (as the case may be) in cooperation with local farmers (Fig. 3.8)
- Old elements such as the characteristic sheepfolds of the area will be converted into hikers' cabins
- New landscape elements will guide visitors through the history of Den Hoorn and the surrounding landscape
- Pilot beacons will refer to the maritime history of Texel and Den Hoorn.

According to Elerie (2004), such Village Surroundings Plans create a framework for cooperation between experts and residents, and between historians and designers. It is crucial that research and design supplement each other, and that a participatory design process involves the residents. Working with landscape biographies and Village Surroundings Plans requires historians and designers to develop new methods. The recent planning methods used in the Netherlands, which have

Fig. 3.8 A pond in an agricultural landscape, reconstructed on the initiative of local residents with the assistance of local farmers, 2007 (Photo: Henk Baas)

a local focus with close attention to interactive planning processes and public-private cooperation, require such an approach. This creates new opportunities for the landscape.

3.8 New Chances for the Local Landscape?

In spite of the variety of instruments that are available for the preservation and further development of the landscape, monitoring suggests that the quality of landscapes is deteriorating (Egmond and Vonk, 2007). Indicators of this process are, for example, the backlog in landscape maintenance and the loss of local identity. This is particularly felt by local residents in the case of newly developed business parks.

The new Spatial Planning Act that was introduced by the Ministry of Housing, Spatial Planning and the Environment in July 2008 gives national and regional governments new opportunities to continue to shape their landscape policy (Ministerie van VROM, 2008). With regard to the assignment of responsibility and authority, the state, provinces, and municipalities are from now on each responsible for their own interests. Provincial or national governments can only interfere with municipal affairs if provincial or national interests make this necessary. On the other hand, under the new Act, the national and provincial governments are authorized to implement their own spatial planning policies, just like the municipalities. National and provincial governments can, moreover, issue general directives and guidelines in order to protect their own interests. This new situation therefore requires a pro-active attitude in which participants take the initiative rather than evaluate afterwards, as was the case prior to July 2008.

Another important element in the Spatial Planning Act and the accompanying Exploitation Act (Ministerie van VROM, 2008) is compensation. New 'red' developments, i.e. detrimental to landscape and/or 'green' values, must be compensated by 'green' investments, with the LDP providing directions for which areas qualify for these green investments. This situation is similar to that in Germany, although the German regulations are more specific and indicate precisely where these green investments should take place. The Dutch LDP is vague on this point, mentioning no specific areas.

A consequence is that the automatic and top-down influence of national policy on regional and local policies no longer exists. If, for instance, a province or other regional authority fails to include detailed national objectives in its own regulations, a local government is no longer obliged to take into account, say, the key qualities of National Landscapes as formulated by the national government. This is a real danger if a provincial government neglects its responsibilities, but it also offers new opportunities. In their regulations, the provinces can stipulate that local governments must formulate new LDPs or include existing ones in their own spatial policy, bringing them into line with provincial policies. This option is strongly promoted by the participants of the Landscape Manifesto. It increases the legal status of LDPs and guarantees their official position in spatial planning.

Provinces can also formulate their own objectives with regard to preserving landscape values in other contexts than National Landscapes. This makes regional regulations a better instrument to strengthen landscape policies than they have been in the past, but whether provincial authorities are willing to use them as such remains to be seen. Landscape Development Plans can also be added to landscape targets adopted earlier, and this makes them a more powerful instrument for local governments.

If both national and regional governments decide not to use these new instruments, the only remaining option is what is termed the 'national stimulation policy' (*nationaal stimuleringsbeleid*). This means that information on 'how to deal with landscape, heritage and preservation' is given, and that local governments are free to use this information as they see fit. Hence, local governments are not obliged to include landscape policy in their spatial planning, nor do they have to meet National Landscape targets in the drafting of their own most important policy instrument, zoning regulation (*Bestemmingsplan*). They are free to do so, but it is doubtful if they will. Without national or regional directives, the preservation of the Dutch landscape becomes a voluntary, local responsibility. Whether this is good or bad remains an open question.

To us, the ideal situation is one in which each municipality has its own Landscape Development Plan, based on the principle of 'local ownership' and supported by such policy instruments as landscape biography, LCA and ECOVAST, and Village Surroundings Plans. Which of these instruments are used is not important, as long as they work (Table 3.1).

Table 3.1 Summary of policy instruments available to municipalities for assessing landscape quality and maintaining and developing landscapes

	LDP	VDP	ECOVAST	BIOGRAPHY
Context	Municipality to Region	Village + around	Village + around	Village to Region
Who is leading	Experts	Locals	Experts	Experts
Local involvement	Yes	Yes	Yes	Yes
Focus upon	Landscape	Livability	Landscape	Landscape history
Legal status	Limited	No	No	No
Fixed set-up	More or less	No	Yes	No
Subsidy	Yes	No	No	No
Method	Mixed	Bottom-up	Mixed	Mixed
Planning tool	Yes	No	Yes	No

LDP Landscape Development Plan
VDP Village Development Plan
ECOVAST European Council for the Village and Small Town
BIOGRAPHY Biography of the Landscape

3.9 Conclusions

The Landscape Development Plans that have been or are being drafted by many Dutch municipalities form a useful framework for policy-making, maintenance, and further development of historical cultural landscapes. The subsidies that are attached to this instrument enable municipalities to formulate their goals at relatively low cost, and to involve other parties in the implementation of the plan. However, it is important that this implementation is approached actively, and that the plans are not left to gather dust at the bottom of a drawer. To ensure an active approach, it has been decided to turn this instrument into one even more focused on implementation.

Another aspect of LDPs is the option to involve the local population in the planning and execution stages. Since this aspect has so far not got off the ground sufficiently, additional policy instruments are being developed, such as the Village Surroundings Plan. This type of plan functions at a more local level and deals to an even greater extent with the local population's wishes with regard to their environment, and with the translation of these wishes into concrete projects. This is local ownership in its purest form.

Furthermore, *Landschapsbeheer Nederland* is currently experimenting with ECOVAST, a variation on the Landscape Character Assessment developed in England. The ECOVAST method can be applied at several levels, such as an LDP or a Village Surroundings Plan.

The concept of landscape biography with its cultural-historical character has been shown to be a welcome addition to the series of landscape policy instruments that are available. Particularly its interdisciplinary approach, its in-depth study of landscape history, and the cooperation of professionals and non-professionals in producing the biography have all added to its value. Professionals and non-professionals differ in their priorities, in the 'histories' they write, and in their assessments of situations. Involving both groups in a landscape planning policy that is intended to have a local basis is essential for creating local support and commitment.

References

Abrahamse JE, Brinkkemper O, Spek T (2008) Integreren met de landschapsbiografie. Rooilijn 41(4):272–279

Beek R (2009) Reliëf in tijd en Ruimte: Interdisciplinair onderzoek naar mens en landschap van Oost-Nederland tussen vroege prehistorie en middeleeuwen. Wageningen University Dissertations, 4733. Wageningen University and Research Centre, Wageningen

Baas HG, Herwaarden GJ (2001) Naar een Landschapsontwikkelingsplan: Een onderzoek naar de criteria ten behoeve van het Landschapsontwikkelingsplan. Landschapsbeheer Nederland, Utrecht

Beek R, Keunen LJ (2006). A cultural biography of the covers and landscapes in the Salland and Achterhoek regions: the aims and methods of the Eastern Netherlands Project. In: van Heeringen RM, Lauwerier RCGM (eds), Berichten van de Rijksdienst voor het Oudheidkundig Bodemonderzoek – Proceedings of the National Service for Archaeological Heritage in the Netherlands, 46. Rijksdienst voor Archeologie, Cultuurlandschap en Monumenten, Amersfoort, pp 355–375

Civilscape (n.d.) www.civilscape.org. Accessed 12 May 2010

Die Dülmener Wilpferde im Merfelder Bruch (n.d.) www.wildpferde.de. Accessed 12 May 2010

Dirkx J (2009). Polderen slecht voor het landschap. Kennis-Online, 6(februari), 5. www.kennisonline.wur.nl. Accessed 12 May 2010

Duineveld M (2006) Van oude dingen, de mensen, die voorbij gaan: Over de voorwaarden meer recht te kunnen doen aan de door burgers gewaardeerde cultuurhistories. Eburon, Delft

ECOVAST (2006). Landscape identification: a guide to good practice. ECOVAST – European Council for the Village and Small Town. http://ecovast.org/papers/good_guid_e.pdf. Accessed 30 Nov 2009

Egmond P, Vonk M (eds) (2007) Natuurbalans. Milieu en Natuurplanbureau, Bilthoven

Elerie H (2004) Instrumenten voor de regionale schaal: Een pleidooi voor toepassing van de landschapsbiografie en het dorpsomgevingsplan. In: Rodermond J (ed) Eén geschiedenis en ontwerp. Stimuleringsfonds voor Architectuur, The Hague, pp 32–40

Elerie H, Spek T (2009) Van Jeruzalem tot Ezelakker: Veldnamen als levend erfgoed in het National Landschap Drentsche Aa. Matrijs, Utrecht

Evaluatie Belvedere (2008) Op na(ar) 2009. Royal Haskoning, Nijmegen

Fairclough G (n.d.) Boundless horizons: Historic landscape characterisation. http://www.englishheritage.org.uk/upload/pdf/boudless_horizons.pdf?1273634891. Accessed 12 May 2010

Groenewoudt BJ (2006) Sporen van oud groen: Bomen en bos in het historische cultuurlandschap van Zutphen-Looërenk. In: Brinkkemper O, Deeben J, van Doesburg J, Hallewas DP, Theunissen EM, Verlinde AD (eds) Vakken in vlakken: Archeologische kennis in lagen. Nederlandse Archeologische Rapporten, 32 Rijksdienst voor Archeologie, Cultuurlandschap en Monumenten, Amersfoort, pp 117–146

Groenewoudt B, Haaster H, Beek R, Brinkkemper O (2007) Towards a reverse image: botanical research into the landscape history of the Eastern Netherlands (BC 1100–AD 1500). Landsc Hist 29:17–33

Groenewoudt BJ, Keunen LJ (2008) Berlewalde: Een verdwenen Achterhoekse wildernis. Vitruvius 2(5):10–17

Hidding M, Kolen J, Spek T (2001) De biografie van het landschap: Ontwerp voor een inter- en multidisciplinaire benadering van de landschapsgeschiedenis en het cultuurhistorisch erfgoed. In: Bloemers JHF, During R, Elerie JNH, Groenendijk HA, Hidding M, Kolen J, Spek T, Wijnen M-H (eds) Bodemarchief in Behoud en Ontwikkeling: De conceptuele grondslagen. Van Gorcum, Assen, pp 7–110

Janssen J, Pieterse N, Broek L (2007) Nationale Landschappen: Beleidsdilemma's in de praktijk. Ruimtelijk Planbureau, Den Haag

Knaap W, Valk A (eds) (2006) Multiple landscape: merging past and present. Selected papers from the fifth international workshop on sustainable land use planning, 7–9 June 2004. Wageningen University, Land Use Planning Group/ISOMUL, Wageningen

Koedoot M (2004) Het Neolithicum van nu: Over de alledaagse betekenis van archeologisch erfgoed in de polder De Gouw en de Groetpolder. Wageningen University, Wageningen

Kolen J (1993). The cultural biography of the landscape: a re-appraisal of history. Leiden (unpublished paper presented to the Theoretical Archaeology Group (TAG), Durham 1993)

Kolen J (2005) De biografie van het landschap: Drie essays over landschap, geschiedenis en erfgoed. Vrije Universiteit, Amsterdam

Kopytoff I (1986) The cultural biography of things: commodization as process. In: Appadurai A (ed) The social life of things. Cambridge University Press, Cambridge, pp 64–91

LanceWad (n.d.) Mapping the landscape and cultural heritage in the Wadden sea region. http://www.waddensea-secretariat.org/lancewad/index.html. Accessed 12 May 2010

Landschapsbeheer Nederland (2006) Landschapsmanifest. www.landschapsmanifest.nl. Accessed 12 May 2010

Landschapsbeheer Utrecht (2000) Evaluatie functioneren gemeentelijke landschapsplannen, gemeentelijke landschapscoördinatoren en de samenwerking met Landschapsbeheer Utrecht. Landschapsbeheer Utrecht, Utrecht

Ministeries van LNV & VROM (2008) Agenda Landschap, landschappelijk verantwoord ondernemen voor iedereen. Den Haag: Ministerie van Landbouw, Natuurbeheer en Visserij and Ministerie van Volkshuisvesting, Ruimtelijke Ordening en Milieu

Ministeries van OCW, LNV, VROM & V&W (1999) Nota Belvedere: Beleidsnota over de relatie cultuurhistorie en ruimtelijke inrichting. Den Haag: Ministeries van Onderwijs, Cultuur en Wetenschappen, Landbouw, Natuurbeheer en Visserij, Volkshuisvesting, Ruimtelijke Ordening en Milieubeheer & Verkeer en Waterstaat. http://www.belvedere.nu/download/nota.pdf. Accessed 12 May 2010

Ministerie van VROM (2008) Wet Ruimtelijke Ordening (WRO). Den Haag: Ministerie van VROM. http://www.vrom.nl/Docs/ruimte/200807_IntegraletekstWro.pdf. Accessed 17 May 2010

Ministeries van VROM, LNV, V&W & EZ (2005) Nota Ruimte: Ruimte voor ontwikkeling. Den Haag: Ministeries van VROM, LNV, VenW & EZ. http://www2.minvrom.nl/notaruimte/. Accessed 12 May 2010

Nationale Landschappen (n.d.) Bekijk Nederland opnieuw. www.nationalelandschappen.nl. Accessed 28 Sept 2009

Overbeek G, Vader J (2008) Genieten van landschap en ervoor zorgen. Landbouw Economisch Instituut, LEI, Wageningen

Renes J (2008) Nationale Landschappen: Bruikbaar concept voor het beheer van cultuurlandschappen?. Geografie 17(3):6–10

Roo G, Schwartz M, Beukema G (2001) Omgevingsplanning, een innovatief proces: Over integratie, participatie, omgevingsplannen en de gebiedsgerichte aanpak. SDU, Den Haag

Rooijakkers G (1999) Mythisch landschap: Varhalen en rituelen als culturele biografie van een regio. In: Kolen JT, Lemaire T (eds) Landschap in meervoud: Perspectieven op het Nederlandse Landschap in de 20ste/21ste eeuw. Van Arkel, Utrecht, pp 2–24

Roymans N (1995) The cultural biography of urnfields and the long-term history of a mythical landscape. Archaeol Dialog 2:2–24

Sager T (1994) Communicative planning theory. Avebury, Aldershot

Swanwick C (2002). Landscape character assessment: guidance for England and Scotland. Countryside Agency & Scottish Natural Heritage, Edinburgh. http://www.landscapecharacter.org.uk/lca/guidance. Accessed 23 Mar 2010

Verhoeven G (2000) Het landschapsbeleidsplan: Van beleid naar uitvoering. Openbaar bestuur 6:33–36

Vos R, Braaksma P, Marrewijk D (eds) (2007) Baas in eigen dorp: Over local ownership als instrument voor ontwikkeling, beheer en behoud van de leefomgeving. http://www.lancewadplan.nl/index.php?id=1377. Accessed 12 May 2010

Woestenburg M (2008) Instrument voor mensen met durf. Landwerk 4:5–9

Ziel PY, Baarslag F (2008) Inventarisatie landschapsontwikkelingsplan gemeenten: Bronckhorst, Lochem, Zutphen, Rapport Royal Haskoning 9S2458.AO. Royal Haskoning, Enschede

Chapter 4
The Participatory Dimension in Nature Conservation Processes: Examples of Ideology and Practice from Norway

Karoline Daugstad

Abstract The dominating legal instrument for area-based protection in Norway has been the Nature Conservation Act. Although protection is directed mainly towards species, ecosystems and landforms, area designations in practice affect landscapes. For this reason, lessons may be learnt from investigating recent policy shifts and managerial changes under the Nature Conservation Act, especially related to participatory aspects relevant to the implementation of the European Landscape Convention. The present chapter provides a schematic view of current approaches to and shifting paradigms in nature conservation on an international

K. Daugstad (✉)
Department of Geography, Norwegian University of Science and Technology, NO-7491 Trondheim, Norway
e-mail: karoline.daugstad@svt.ntnu.no

M. Jones, M. Stenseke (eds.), *The European Landscape Convention*, Landscape Series 13, DOI 10.1007/978-90-481-9932-7_4,

level, with particular focus on participatory approaches. This is followed by the presentation of findings from recent research into the issue of management systems and models related to designated nature conservation areas in Norway. The last part of the paper positions the findings in relation to the European Landscape Convention.

Keywords Nature conservation · Local participation · Management models

4.1 Introduction

Under the European Landscape Convention (ELC), the importance of landscapes emerges as a key concern in planning and management. Participation and engagement with landscape is a crucial point in the convention text, in short communicating that people should have a say regarding the development of 'their' landscapes.

In this chapter, participation in landscape issues will be explored within the context of nature conservation with empirical evidence from recent designation and management processes of national parks in Norway. Nature conservation – in the form of area designations such as national parks – is in practice a landscape matter regulated by national laws. Drawing on research into the participatory dimension of nature conservation processes can give insights into participatory aspects of landscape management and conservation. Even if area-based nature protection according to the Norwegian Nature Conservation Act is more explicit in protecting species, ecosystems, and landforms than landscapes as such (Lov om Naturvern, 1970[1]), area designations obviously include landscapes. While the European Landscape Convention is a political framework for protection, management, and planning, rather than providing practical guidelines, insights from empirical studies in related landscape fields may help to pinpoint principles for management fulfilling the Convention.

The following questions are asked:

- What are the main approaches and paradigms in nature conservation, especially during recent decades?
- What insights can be drawn from recent Norwegian studies of different management models in protected areas?
- How can these insights be relevant to implementing the European Landscape Convention especially related to the participatory aspect?

4.2 Approaches to Nature Conservation

Since the introduction internationally of protected areas in the form of national parks in the nineteenth century, designation processes and resultant management regimes have been largely expert-oriented. Designation has been mostly based on preservation and the idea of pristine nature rather than directed towards approaches integrating use and protection. National parks have also aimed to reflect national values and protect national attributes, meaning that the responsibility for keeping such values has been in the hands of national actors (Daugstad, 1999).

In recent decades however, shifts towards democratization of public management, support for the principle of subsidiarity, and stronger claims to nationally designated areas as an asset for the nation's inhabitants have challenged the hegemonic top-down expert approach to nature conservation. A striking element in these shifting trends is participation, particularly in terms of local participation and/or communities voicing their interests and priorities when nature is designated as being of special value (Daugstad et al., 2006a, b; Vistad et al., 2006).

This trend can be described as a shift between two approaches to nature conservation (Daugstad et al., 2006b).The first is 'the fortress approach' or 'the fences-and-fines approach' and is characterized by the following:

- preservation is enforced by a policy of 'no trespass'
- local inhabitants are granted no rights in the area under protection
- protection of wildlife or ecosystems is the major concern
- conservation is to be handled by specialists, often public bodies or agencies at the national level
- the necessary knowledge base for protection and management is expert oriented and based on ecological science.

The second is 'the community approach' or community-based conservation, with the following characteristics:

- people affected by conservation should take part in decision and management of the resources protected (moral or democratic argument)
- protected areas contain resources necessary for the upkeep of local livelihoods
- protected areas have landscape or nature qualities due to human presence (cultural landscape)
- situated local knowledge is necessary for good management of the areas in question
- local participation may reduce a costly level of conflict in protected areas through increased legitimacy of the protection process.

There are a number of different forms or degrees of community participation, ranging from local actors being informed or consulted, through interactive participation to self-mobilization where local actors take initiative independent of external decisions or actions (Pimbert and Pretty, 1997). A substantial body of literature debating or scrutinizing the fortress approach or the community-based approach – particularly in relation to participatory models – has emerged over the last 20 years in the wake of the concept of sustainable development put forth in the report of the Brundtland Commission (World Commission on Environment and Development) in 1987 and at the Rio Conference (United Nations Conference on Environment and Development) in 1992.

4.3 For and Against Community-Based Conservation

The literature on participatory approaches to nature conservation can be divided into that produced by critics and that produced by proponents.

4.3.1 Critics of Community-Based Conservation

The main line of argument presented is that participatory models for conservation lead to reduced quality of conservation. It is argued that local actors take a utilitarian or anthropocentric approach to resource management, rather than an eco-centric approach. For example, local actors have needs and priorities incompatible to biodiversity conservation, as suggested by the documentation of species loss in protected areas with community participation (Terborgh, 1999; Adams and Hulme, 2001). Some critics are less strict; while they do not rule out models of management in which local actors have a role, they nonetheless argue that community decision-making should be limited so as to not to incur any detrimental effects on the biological assets under protection (see for example Wilshusen et al., 2002).

4.3.2 Proponents of Community-Based Conservation

Arguments in favour of community-based conservation are centred on the moral or democratic right of people to influence their own future. That is, communities should have a say in major decisions regarding resources they depend upon. Further, protected areas are important for economic activity in a local community. Last, making communities in charge of or responsible for 'their own' protected areas activates relevant local knowledge, reduces conflict and results in a cost-effective model of securing nature qualities through activating skilled persons already situated in the areas in question. However, accompanying this general support for participatory approaches is also a critical attitude towards many projects or management models seemingly based on community involvement. Numerous cases in which participatory conservation has resulted in the marginalization of parts of the local community due to ethnicity, class, occupation or gender have been reported (Neumann, 1997; Gurran, 2004; Lane and Corbett, 2005; O'Rourke, 2005; Daugstad et al., 2006b; Svarstad et al., 2006). This has led to calls for fundamental changes in how society views nature conservation, in order for 'real participation' to be successful, and suggests that implementing participatory models alone does not guarantee a real alternative to the fortress approach (see for example Cleaver, 2002; Michaelidou and Decker, 2004; Gerritsen and Wiersum, 2005).

4.4 Recent Insights from Participatory Models in Norway

In line with the international trends of democratization and local or community based models in nature protection, political decisions at the national level in Norway signal a new approach to nature protection. The most concrete of these, so far, has been a national pilot project testing alternative management models of large areas protected by the Nature Conservation Act – mainly national parks and protected landscapes. The pilot project was initiated by Parliament in 2001 (St. meld. 31 (2000–2001)). Through implementing alternative models of management in four newly established protected areas over a period of 5 years, the aim of the project was to gain new insights to serve as a basis for a revised nature protection and management policy. In the four test areas, different models of management have been implemented with varying degree of local participation. All the models are different from the management regime embedded in the Nature Conservation Act, where national parks management is a national responsibility, carried out in practice by the County Governor's Office as the state representative in each county (Lov om Naturvern, 1970). By and large, the areas under protection as national parks in Norway are state-owned and consist of upland areas. To a large extent these areas are commons with resource use rights mainly attached to agriculture and forestry. However, the share of privately owned land in national parks is increasing due to a national protection policy aiming for a more representative selection of nature types under this form of protection (Daugstad and Rønningen, 2004). Despite differences in management models across the four test areas, a key element in all is that the municipal level has a much stronger say than in the general model. The municipality level is the local democratic level in Norway, where the municipal council is elected every 4 years and adherent boards and committees are appointed accordingly.

The pilot project has been evaluated by research and consultancy institutions, and insights presented here are drawn from the evaluation reports. This is complemented with findings from an additional research project comparing one of the pilot areas with another protected area, where a local participation model was implemented but not within the frame of the national pilot study. The evaluation reports are all published and the findings from the research project are also published (Svarstad et al., 2003; Daugstad et al., 2005; 2006a, b; Svarstad et al., 2006; Vistad et al., 2006).

Results from these studies indicate how participatory models work in practice. Three key themes are identified, and are discussed in turn below: (1) the local as a democratic level of decision and management; (2) legitimacy for nature conservation; and (3) representing 'the local'.

4.4.1 The Local as a Democratic Level of Decision and Management

When the municipal level becomes responsible for large protected areas covering several municipalities, this creates challenges both on an inter-municipal level and on an intra-municipal level.

At the inter-municipal level, administrative units not necessarily accustomed to cooperating with neighbouring municipalities must develop and maintain a uniform management policy for shared, large protected areas such as a national park. One issue emerging is the division of 'costs' related to protection. First, for a national park covering several municipalities, the respective municipal shares vary from only a few percent of the total area under protection to a major part of it – thus creating a division between 'small' and 'large' national park municipalities. We find that municipalities with large protected areas argue for more funding (and also more representatives in park councils, etc.) compared to 'small' municipalities, due to the burden of protection being heavier for larger areas under protection. Specifically, it is argued that the number of dispensation applications regarding motorized traffic (which according to national law is very restricted in protected areas), building of cabins and road development is substantially higher in a 'large' municipality. A second line of argument reflects the view that 'we are in this together', suggesting that all municipalities affected by a national park should take joint responsibility for uniform and holistic management, rather than arguing about the number of square kilometres with protected status (Daugstad, 2005).

Another aspect of inter-municipal relations is the potential for clashes between different municipal policies. Even if national law in relation to protected areas and resource management in general applies to all municipalities, there is a certain room for interpretation whereby policies or restriction levels may differ between some municipalities. This becomes evident when municipalities are to be part of a uniform national park management. For example, one of the pilot parks requires that eight municipalities must cooperate on uniform management. Here, one of the municipalities has a substantial number of second homes or cabins in the mountains and is a municipality with long traditions in tourism. This has led to a liberal policy towards various encroachments in the mountain areas and motorized access for cabin owners to their cabins. In another part of the same national park, a different municipality is developing marketing strategies for what can be labelled 'pristine nature tourism' based on landscapes free from encroachments, with the silent walking tour or a 'contemplative hike' in 'non-motorized' mountains. Since both municipalities are part of the national park management council, together with six others, someone needs to adjust. The resultant situation is that the liberal municipality has become less liberal. Community-based conservation has in this case led to more restrictive management for this municipality (Vistad et al., 2006).

The evaluation reports show that with regard to motorized traffic in the protected areas, and especially for snowmobiles during winter, it is very difficult to reach a common policy between municipalities. Dealing with applications or dispensations from the law prohibiting motorized traffic in protected areas takes a substantial share of municipal administration, and established policies are very hard to change – both for bureaucrats and for the general public, who are accustomed to the municipal policy (Falleth, 2004a, b).

New management models also have an effect on intra-municipal relationships, especially between politicians and bureaucrats. Both represent the community in terms of the municipality, but with obviously different roles. The management models in the four pilot case areas have different structures for political vis-à-vis bureaucratic influence. In the most clear-cut political model, one of the parks has a council in which all eight municipalities are represented by their mayor. This is the decisive level. After a few years with this model, there developed some tensions between the political and the administrative levels: the bureaucrats, as the skilled, educated nature managers and planners, did not feel they had a sufficient say as experts in important matters. To address this situation, an inter-municipal administrative co-ordinating committee was established to create a more uniform practice and enable the bureaucrats to have a stronger voice in relation to the politicians in charge (Vistad et al., 2006). In one of the other pilot areas, each municipality has been delegated the authority to manage the municipality's part of the national park according to a pie-slice model. The inter-municipal contact is to be secured by an inter-municipal policy council and an administrative committee. The council has an advisory role only, and it has not functioned. The administrative committee has worked much better, despite the fact that it has very limited administrative resources. The political council mainly follows advice from the administrative committee. According to the evaluator, the pie-slice model, in which inter-municipal bodies have unclear authority, does not promote efficient management. Nor does it enhance legitimacy for nature conservation among the public in general, since the management model and structure seem unclear from the outside (Skjeggedal and Aasetre, 2005).

4.4.2 Legitimacy for Nature Conservation

The almost compulsory traditional reaction to nature conservation from local actors, especially landowners and farmers, is resistance and scepticism (Daugstad et al., 2000). Frequent arguments underpinning their resistance include: fear of losing rights to resources or fear of heavy restrictions on use of resources, especially related to agriculture and forestry; limits to future use or changes in use in general; and a feeling of losing control over land. Scepticism towards nature conservation is also frequently voiced by other local actors, such as municipal politicians, due to an anticipated restriction of municipal government.

Can more emphasis on local participation prevent local scepticism towards nature conservation? Findings from the pilot cases may give some indications, although more studies at a broader scale are needed. In one of the pilot cases, the evaluators surveyed local attitudes (among politicians and leaders of organizations and boards influenced by the conservation status) towards conservation in 2003 and 2005. While there were no clear changes towards less scepticism and hence more support for conservation in the 2 year period, what emerged was a stronger support for local management of the protected areas, especially from the politicians. The evaluators suggested that this increased support might be related to the substantial effort put into making a specific management plan for the area, and especially seeing the plan as paving the way towards an increased role for the local level in the future (Hovik and Falleth, 2005). The same development is documented in one of the other areas, with evaluators suggesting that opponents of protection status tend to give up, because it is very likely that protection will be approved regardless of local protest. Protection is perceived as an inevitable fact that one might just as well make the best of (Aasetre and Skjeggedal 2004).

From the research project in which two areas with differing nature protection status were compared (one of them also a pilot case), there are indications that initial resistance towards protection faded away during the protection process. For the two cases studied in this particular project, the process between making the initial protection applications in Parliament to the final approval of actual designations took respectively 18 and 16 years. During this long time period, local scepticism and, at times, strong resistance from municipalities, primary businesses, and landowners were presented in the local and regional media and in public hearings, but became silent along the way. For both cases, local participation on advisory boards or committees was facilitated by the responsible body in charge (i.e. the County Governor). A combination of different factors can explain this shift: the above-mentioned reaction of giving up; an expected benefit from management in the hands of local actors (municipal politicians); a changed view as negative impacts of conservation proved to be less severe than expected; and selection mechanisms affecting which community members have a voice as local spokespersons in the media, and which may hide their 'real attitudes' – either positive or negative – towards conservation (Svarstad et al., 2003; Daugstad et al., 2005; Vistad et al., 2006).

4.4.3 Representing 'The Local'

A pressing question in participatory processes is 'who participates'? Who gets to represent 'the local' or 'the community' in models where the local level is central? Even if democratization of decisions and management of conservation areas is a defined goal, the mechanisms – explicit or implicit – defining representatives as local are not necessarily democratic. Power relations, roles, history, and position (or even mere chance) influence who really participates.

What insights can be drawn from the pilot case studies in this respect? A striking finding from the research project is the absence of women on boards and committees representing the communities. Across the cases, the representation of women on various boards and committees was between 7.5 and 18% (Svarstad et al., 2006). These groups consisted to a large extent of mayors, farmers' representatives and landowners' organizations. In rural communities, mayors tend to be men and the primary business interests have a male dominated representation. This may explain why local bodies in charge of protected nature are often male dominated. However, there *were* female members of relevant organizations in the municipalities, but they were not approached and asked to be represented on the boards and committees. This lack of female representation can perhaps be explained both in terms of 'supply' (available women) and 'demand' (women asked). Either way, it does not eliminate the problem this type of pattern represents in terms of democracy and local participation.

For the pilot case areas, it was also suggested that a broader representation of local interests is required, beyond that of elected politicians. Specifically, representatives of farmers' and landowners' interests, reindeer management boards and tourism associations argue that they should have a stronger representation in the management bodies (Vistad et al., 2006).

Actors who see themselves as representing local interests may also be located outside the community. As shown by Daugstad et al. (2006) for one of the pilot case areas, the Norwegian Hikers' Association formed an alliance with the mayor in one municipality in a common endeavour to argue for tourism activity (cabin with accommodation) in the national park. This particular activity had been much debated due to the potential disturbance caused by tourist traffic to reindeer, a key species in the national park. In another example from the same area, the Norwegian Hunters' and Fishers' Organization demonstrated divergent views between the national and local levels of the organization. Where the national level opposed developing tourism activity in the national park with a cabin and following infrastructure, the local branch welcomed this development, thus further illustrating how 'localness' influences attitudes. As Daugstad et al. (2006a:14) suggest, 'maybe this is because an ideal organization – at the top level – can act more in accordance with founding principles than can the local level, where decisions tend to be more pragmatic or compromises'. In this case, 'localness' influences attitudes when a more or less 'pro-nature' organization – which may depend on sustainable reindeer management in its own interest – is inclined to act more liberally in a local than in a national context.

4.5 Discussion

Evidence from recent participatory processes in management of nature conservation areas in Norway points in several directions, but does not provide a uniform pool of guidance towards specific landscape policies or measures. However, the findings give some useful indications of variables or issues to keep in mind when implementing participatory models of any kind. These include: the importance of context; exclusion and inclusion mechanisms; an awareness of the extent to which a participatory process can be viewed as 'a clean slate'; and how underlying structures or overarching decisions may or may not override new models.

With specific reference to the pilot cases analysed in this chapter, it is important to keep in mind that areas where the local level has been given a stronger say are at the same time areas protected under the Nature Conservation Act. This implies that the areas have been surveyed and documented within a scientific expert regime, whereby designation has been suggested and decided on the basis of key nature qualities such as species, eco-systems or landforms. As long as the areas are not 'de-conserved', the national legislation sets limits to the decision-making capacity of any locally run management apparatus. It is likely that the national environmental authorities will intervene if a management regime jeopardizes the qualities justifying the protection status. Outside Norway, other countries have tested alternatives to protection via conservation laws – called 'green partnerships'. Such models are only recently emerging in Norway, however, and it will be important to evaluate how these develop in terms of participation, definition of landscape qualities (i.e. defined by whom?), and acceptable levels of landscape change. Another approach to alternative area management is the regional parks well known from for example France, but newly introduced to Norway.

The European Landscape Convention gives a positive message of landscape being 'everybody's concern', and suggests that having a role and having a say leads to responsible and sound decisions for the benefit of all. Likewise, an argument behind the pilot studies in alternative management of protected areas in Norway has been to increase the legitimacy of area designations and thereby contribute to more effective management due to a lower level of conflicts (Daugstad et al., 2006b). Looking at the vast literature on local participation, it might be tempting to characterize some of this as somewhat un-nuanced or idealistic in promoting 'the good local community' (Lane and Corbett, 2005; O'Rourke, 2005). Although it might seem antagonistic to criticize such a position, it is important to acknowledge that participatory processes are not always possible, feasible or within reach. Further, a belief that local resistance can be solved either by education or economic compensation is both unrealistic as well as somewhat patronizing in assuming 'local ignorance'. With reference to opposition towards protected areas in Germany, Stoll-Kleeman (2001: 118, 126–127) summarizes these concerns in the following way:

> The debate on reasons for local opposition ... usually starts from the assumption that opposition is based on incomplete knowledge or concern for the environmental benefits associated with natural area designation. Consequently, solutions tend to rely on strategies like financial compensation or environmental education. [...] Many disputes over nature conservation are rooted in social conditions. [...] There is much talk about deliberative and interactive processes of stakeholder negotiation in resource management. [...] These approaches nevertheless rely on a set of procedures that build trust and assume some basis for agreement. They also assume some sense of civic responsibility, of 'give and take' among negotiating partners to bring about various side-deals in the event of an impasse. [...] ...it is by no means clear that such circumstances will actually exist.

Similar concerns have been voiced by Selman (2004: 372–373), who argues that:

> Genuine participation involves authorities relinquishing a degree of power, and placing trust in lay communities. [...]...there is an implicit assumption that the laity will share a 'polite' view of landscape, and will thus endorse a strategy sympathetic to conventional aesthetics. This cannot be taken for granted.

In the case of area designations in Norway, these are decisions embedded in national law, thereby giving a high status to such areas. For national parks it is hard to imagine the control and management of the parks being handed over to a coalition of business interests, landowners or non-governmental organizations (NGOs) – it is not a coincidence that the pilot studies in practice have defined the democratic level of the municipality as the local level. The situation might be more open with regard to implementing the European Landscape Convention more generally. This is not tied to any specific law, but it is to be applied in every landscape. On the other hand, the lack of a specialized legal instrument dedicated to the Convention might mean that ELC principles tend to be applied in landscapes with fewer conflicts or in areas where the laity share a 'polite' view of landscape, as Selman suggests.

There might still be a lack of real attempts to facilitate local participation. Many approaches are top-down systems disguised under a gloss of community-based rhetoric. Such a view is presented by Cooke and Kothari (2002) and Lane and Corbett (2005), with reference especially to developing countries. The same aspect is pinpointed in Norwegian studies of local participation in the management of protected areas where actors voice a need for 'real' participation, meaning that they actually come to play a decisive role and not just legitimize a seemingly democratic process (Daugstad et al., 2005). In other words, there is not enough 'hard evidence' to dismiss community-based conservation on the basis of cases where community involvement has not always been genuine.

The European Landscape Convention is general and non-specific regarding what measures are needed to take landscape concerns seriously in protection, management, and planning. This opens the possibility of implementing new models of landscape management not only in relation to nature conservation legislation but also in other legal measures involving designations and partnership models.

Note

1. This law has undergone revision and been replaced by the Nature Diversity Act, which came into force on 1 July 2009 (Naturmangfoldloven, 2009). Area-based protection continues in the new Act, but there is a stronger focus on developing management plans with specific measures for upholding the cultural landscape character in areas with protected status.

References

Aasetre J, Skjeggedal T (2004) Lokal forvaltning av verneområdene i Forollhogna: Delrapport 1 i evalueringen av forvaltningsforsøket. Nord-Trøndelagsforskning, Steinkjer

Adams WM, Hulme D (2001) Conservation & communities: changing narratives, policies and practices in African conservation. In: Hulme D, Murphree M (eds) African Wildlife & Livelihoods: The promise and performance of community conservation. James Currey, Oxford, pp 9–23

Cleaver F (2002) Institutions, agency and the limitations of participatory approaches to development. In: Cooke B, Kothari U (eds) Participation: the new tyranny?. Zed Books, London, pp 36–55

Cooke B, Kothari U (eds) (2002) Participation: the new tyranny?. Zed Books, London

Daugstad K (1999) Mellom romantikk og realisme: Om seterlandskapet som ideal og realitet, R-16/99. Centre for Rural Research, Trondheim

Daugstad K (2005) Nasjonalparken møter kommunen. Nationen 25.052005

Daugstad K, Kaltenborn BP, Vistad OI (2000) Vern – planer og prosesser: Identifisering av kunnskapsstatus og -behov, N-3/00. Centre for Rural Research, Trondheim

Daugstad K, Rønningen K (2004) Landskapet som felles gode og privat ressurs. In: Setten G (ed) Det levende landskapet. Tapir, Trondheim, pp 111–129

Daugstad K, Svarstad H, Vistad OI (2005) Geiranger-Herdalen: Lang dags ferd mot vern. Plan 2005 3–4:58–64

Daugstad, K, Svarstad, H, Vistad, OI (2006a) A case of conflicts in conservation: two trenches or a three-dimensional complexity?. Landsc Res 31(1):1–19
Daugstad K, Svarstad H, Vistad OI (2006b) Community-based nature protection. Int J Environ Cult Econ Soc Sustainability 2(3):127–135
Falleth E (2004a) Forvaltningsforsøket i Setesdal Vesthei–Ryfylkeheiane landskapsvernområde og Steinbuskardet–Hisdal biotopvernområde: Kommunenes og sekretariatets arbeid fra 2000 til 2004: Delrapport 4 av NIBRs evaluering av forvaltningsforsøket. NIBR, Oslo
Falleth E (2004b) Hvordan dannes nettverk: Om lokal organisering og kollektiv handling i Setesdal Vesthei–Ryfylkeheiane. Kart og Plan 3:158–164
Gerritsen P, Wiersum F (2005) Farmer and conventional perspectives on conservation in Western Mexico. Mt Res Dev 25(1):30–36
Gurran N (2004) Integrating local people and national parks in rural Australia: community involvement and support for conservation. In: Camarda I, Manfredo MJ, Mulas F, Teel TL (eds) Global challenges of parks and protected area management: proceedings of the 9th ISSRM. Carlo Delfino editore, Sassari-Italy, pp 273–283
Hovik S, Falleth E (2005) Holdninger til landskapsvern og lokal forvaltning i Setesdal Vesthei–Ryfylkeheiane, N-139. NIBR, Oslo
Lane MB, Corbett T (2005) The tyranny of localism: indigenous participation in community-based environmental management. J Environ Policy Plann 7(2):141–159
Lov om Naturvern (1970) Norges Lover 1685–1991. Ad Notam, Oslo
Michaelidou M, Decker DJ (2004) Nature conservation and community viability: implications of European Union policies for mountain communities in the Pafos forest, Cyprus. In: Camarda I, Manfredo MJ, Mulas F, Teel TL (eds) Global challenges of parks and protected area management: proceedings of the 9th ISSRM. Carlo Delfino editore, Sassari-Italy, pp 393–403
Naturmangfoldloven (2009) Lov 2009-06-19 nr 100: Lov om forvaltning av naturens mangfold (naturmangfoldloven). Lovdata. http://www.lovdata.no/all/hl-20090619-100.html. Accessed 26 Mar 2010
Neumann RP (1997) Primitive ideas: protected area buffer zones and the politics of land in Africa. Dev Change 28:559–582
O'Rourke E (2005) Landscape planning and community participation: local lessons from Mullaghmore, the Burren National Park. Landsc Res 30(4):483–500
Pimbert MP, Pretty JN (1997) Parks, people and professionals: putting 'participation' into protected area management. In: Ghimire BK, Pimbert MP (eds) Social change and conservation. Earthscan, London, pp 297–330
Selman P (2004) Community participation in the planning and management of cultural landscapes. J Environ Plann Manage 47(3):365–392
Skjeggedal T, Aasetre J (2005) Lokal forvaltning av verneområdene i Forollhogna: Delrapport 2 i evalueringen av forvaltningsforsøket. Trøndelag forskning og utvikling, Steinkjer
St. meld. 31 (2000–2001) Kommune, Fylke, Stat – en bedre oppgavefordeling. Kommunal- og regionaldepartementet, Oslo
Stoll-Kleeman S (2001) Opposition to the designation of protected areas in Germany. J Environ Plann Manage 44(1):109–128
Svarstad H, Daugstad K, Vistad OI (2003) Hvem vil være enige og tro på Dovrefjell? Et case om framveksten av lokal forvaltning av verneområder. Utmark 4(2), (not paginated). http://www.utmark.org. Accessed 10 Dec 2003
Svarstad H, Daugstad K, Vistad OI, Guldvik I (2006) New protected areas in Norway: local participation without gender equality?. Mt Res Dev 26(1):48–54
Terborgh J (1999) Requiem for nature. Island Press, Washington, DC
Vistad OI, Daugstad K, Svarstad H (2006) Store verneområde med lokal forvaltning: Funn og refleksjonar. Utmark 6(1) (not paginated). http://www.utmark.org. Accessed 5 Mar 2006

Wilshusen PR, Brechin SR, Fortwangler CR, West PC (2002) Reinventing a square wheel: critique of a resurgent 'protection paradigm' in international biodiversity conservation. Soc Nat Resour 15:17–40

World Commission on Environment and Development (1987) Our common future. Oxford University Press, Oxford

Chapter 5
The Implementation of the European Landscape Convention in Poland

Anna Majchrowska

Abstract This chapter aims to examine the implementation of the European Landscape Convention (ELC) in Poland and related landscape issues. Legislative Acts and national policy documents are analysed, and a review of scientific literature is undertaken. It is found that Poland has neither its own legal definition of the term 'landscape' nor regulations specifically oriented towards landscape and the ELC. The term 'landscape' is found in a number of laws pertaining to various sectors: environment and nature protection, culture and national heritage, and spatial planning. No specific national landscape policy document has been drawn

A. Majchrowska (✉)
Department of Physical Geography, University of Lodz, 90-139 Lodz, Poland
e-mail: majchrow@uni.lodz.pl

M. Jones, M. Stenseke (eds.), *The European Landscape Convention*,
Landscape Series 13, DOI 10.1007/978-90-481-9932-7_5,

up, but there is a wide range of sectorial policy documents that address landscape and the ELC directly or indirectly. Although study of landscapes is one of the traditional themes of geographical research in Poland, research on Landscape Quality Objectives is at its initial stage. With respect to public participation, since Poland acceded to the European Union (EU) in 2004, its laws on public participation have been evolving in order to adjust to the EU's legal regulations. However, law enactment alone does not instantly create new good practices of public participation nor does it eradicate bad habits that have formed in earlier years.

Keywords European Landscape Convention · Landscape policy · Landscape research · Participatory landscape decision-making

5.1 Introduction

In recent decades, the challenge of managing landscape change has been considerable in Poland on account of the speed and radical character of current changes in Polish society. Among the main driving forces are economic growth and liberalization of the real-estate market, accompanied by social transformations that often lead to a weakening of people's ties to place. Despite various problems in this period of transformation, Poland is willing to take the opportunity to re-examine national laws, policies, and attitudes towards the land and landscape, and has embarked on implementation of the European Landscape Convention (ELC).

Articles 5 and 6 of the European Landscape Convention (Council of Europe, 2000) place on the signatory states obligations to:

- Recognize landscapes in law as an essential component of people's surroundings
- Establish landscape policy and integrate landscape into all relevant policies
- Establish procedures for public participation in matters related to landscape
- Identify and assess landscapes and define landscape quality objectives
- Educate and raise awareness of landscape issues among the general public
- Encourage international co-operation.

Poland signed the European Landscape Convention on 21 December 2001, and after its ratification in September 2004, the Convention came into force in Poland on 1 January 2005. The text of the Convention was published in the Journal of Laws (Dziennik Ustaw, 2006), and thereby, according to the Constitution of Poland (Dziennik Ustaw, 1997: Article 91), it became a part of the domestic legal system. Initially, responsibility for implementing the ELC was vested in the Department of National Forms of Nature Conservation, under the Ministry of the Environment. On 16 November 2008, the Directorate General of Environmental Protection was created as a central administrative body, supervised by the Ministry of the Environment. The Directorate General has been given charge of the implementation of the Convention, although its main tasks include, inter alia, the designation, protection, and management of protected areas, and control of the environmental implications of decisions by means of Environmental Impact Assessment (EIA) and Strategic Environmental Assessment (SEA) procedures. As the Directorate General has functioned for a relatively short time, the scale of its involvement in the implementation of the ELC has been so far imperceptible.

Besides the Ministry of the Environment, other departments of the government also deal with landscape-related matters, including the Ministry of Regional Development, Ministry of Culture and National Heritage, and Ministry of Infrastructure. Each of these departments focuses its work first and foremost on its own sector. This devolved responsibility for landscape means that no governmental level body has taken the lead in implementing the ELC.

This chapter provides an overview of activities undertaken in Poland to implement the ELC, as well as of a number of other actions, past and present, related to landscape issues although not necessarily directly linked to the Convention. The chapter is based on an analysis of legislative Acts and national policy documents and consultancy reports, and a review of scientific literature. The documents analysed are those that contained the terms 'landscape' and/or 'European Landscape Convention' in their text. The aim of the selection was to gain better insight into the process of implementing the ELC in Poland. The documents were analysed in terms of the intent of the ELC and the key measures set out in Articles 5 and 6. The study does not present a complete survey of performance in all those areas, focusing mainly on: the meaning of the term 'landscape' in Polish law; the existing protective measures dedicated to landscapes; public participation in landscape matters; research projects related to identification and assessment of landscapes; and Landscape Quality Objectives.

5.2 Recognition of Landscapes in Law

5.2.1 The Concept of Landscape in Polish Law

The term 'landscape' does not have a univocal meaning in Polish law. Neither is there specific legislation oriented towards landscape and the European Landscape Convention. The ELC (Council of Europe, 2000: Article 1) defines landscape as 'an area, as perceived by people, whose character is the result of the action and interaction of natural and/or human factors' and calls on the signatory states to acknowledge in law that landscape includes nature, culture, and human perceptions. Such a concept of landscape is inconsistent with current Polish legislation.

The term 'landscape' is found in a number of laws pertaining to various sectors: environment and nature protection, culture and national heritage, spatial planning, and regional development – but there is no clear statement about what landscape is. Thus the meaning of 'landscape' varies in different laws.

The Environmental Protection Act (Dziennik Ustaw, 2001a: Article 3) includes the word 'landscape' in a definition of environment, which is described as 'a totality of natural elements, including those transformed by man, in particular terrain surface, mineral deposits, water, air, landscape, and climate and the other elements of biodiversity, together with their interactions'. Here, landscape is one of several components of nature, equivalent to minerals, water or plants, while its spatial and cultural aspects are not mentioned.

Similarly, the Nature Conservation Act (Dziennik Ustaw, 2004: Article 2) states that:

> . . .nature protection consists in preservation, sustainable use and restoration of resources, formations, and elements of nature, such as: wild plants, animals and fungi; protected species of plants, animals and fungi; migratory species; natural habitats, endangered natural habitats, rare and protected species of plants, animals and fungi; formations of animated and unanimated nature and fossil remains of plants and animals; landscape; green areas in towns and villages; and trees.

The same Act (Article 5) defines landscape protection as 'preservation of characteristics of a landscape', and characterizes the values of landscape as 'ecological, aesthetic and cultural amenities of the area and of the associated terrain surface, natural formations and components created by nature or by activity of man', thereby extending the sense of landscape beyond natural features.

5.2.2 Natural Elements of Landscape

The Nature Conservation Act (Dziennik Ustaw, 2004) specifies the subject, scope, and type of protected areas, thereby regulating the designation and protection of natural elements of exceptional landscapes. According to the provisions of the Act, eight types of protected areas and objects can be designated (Table 5.1) (excluding Natura 2000 areas and the protection of species). Three of them – National Park, Nature Reserve, and Landscape Park – include landscape values in their protection objectives, mainly, though not exclusively, with regard to natural elements of landscape.

A National Park comprises protected areas with particularly outstanding scientific, natural, social, cultural, and educational characteristics, not smaller than

Table 5.1 Status of nature protection in Poland on 31 December 2007

Type of protected area	Number	Area km^2	Area % of country
National park	23	3,172	1.0
Nature reserve	1,423	1,688	0.5
Landscape park	120	25,150	8.0
Landscape protection area	412	69,598	22.3
Area of ecological utility	6,686	461	0.2
Documentation site	153	8	–
Natural and scenic complex	207	935	0.4
Monument of nature	35,074	–	–

Source: Central Statistical Office (Główny Urząd Statystyczny, 2008)

1000 ha, where the environment as a whole is protected, including landscape characteristics.

A Nature Reserve is a protected area with primeval or slightly altered ecosystems, refugia, natural habitats, as well as habitats of flora, fauna and fungi, and abiotic elements of nature, having significant scientific, natural, cultural, or scenic value.

A Landscape Park is an area protected due to environmental, historical, and cultural values.

Two other types of protected area, namely Landscape Protection Area and Natural and Scenic Complex, are designed specifically to protect outstanding landscape features. A Landscape Protection Area is designated to preserve outstanding landscape features, with the aim of protecting diverse ecosystems, valuable especially on account of their potential for satisfying tourist needs, or serving as ecological corridors. Natural and Scenic Complexes are created to protect spectacular fragments of natural and cultural landscapes and to preserve their aesthetic value.

The system of protected areas is well developed and the eight types of nature protection areas cover about 32.5% of the total territory of the country.

5.2.3 Cultural Elements of Landscape

The Protection and Safekeeping of Historical Monuments Act (Dziennik Ustaw, 2003b: Article 3) contains a definition of cultural landscape as 'a space historically formed by human activity, which includes man-made structures and natural objects'. This meaning seems to be somewhat closer to the intent of the Convention than that in the Nature Conservation Act. The definition does not, however, mention the significance of landscape for people and emotional values of the landscape. This Act regulates protection of cultural heritage and provides for the designation of four types of protected objects:

- Listing in the Register of Cultural Heritage (c.270,000 items) (Table 5.2) managed by the National Heritage Board of Poland
- Monument of History (35 monuments)
- Culture Park, aiming at protection of cultural landscape and conservation of areas with outstanding landscape and monuments of traditional local architecture
- Conservation Zone, establishing protection in a local spatial development plan.

Thus, in Polish law landscape is currently treated in a conventional, piecemeal way, where natural and cultural elements are seen as separate from one another, whereas experiential and social values of landscapes are not even mentioned. Landscape regulations focus on designation of exceptional areas and objects and on their conservation, refraining from more proactive statements regarding common landscapes.

Existing legislation needs to be adjusted if it is to reflect more fully the intent and aims of the ELC. Studies on the need and scope of the adjustments were

Table 5.2 Objects listed in the Register of Cultural Heritage in Poland

Objects	Number
Immovable objects (31.12.2008)	**63,368**
Urban patterns	1,021
Places of worship	11,921
Defensive structures	874
Public buildings	4,041
Castles	418
Palaces	2,018
Mansions	2,754
Parks and gardens	6,937
Manors	5,186
Agricultural buildings	2,044
Residential buildings	17,198
Industrial objects	2,026
Cemeteries	3,992
Other	2,938
Movable objects (31.12.2007)	**201,673**
Furnishings of places of worship	139,144
Collections	48,285
Other	14,244

Source: National Heritage Board of Poland (Krajowy Ośrodek Badań i Dokumentacji Zabytków, 2007, 2008)

carried out in 2006, when the Institute of Environmental Protection prepared 'A Report on Legal and Spatial Planning Tools for the Implementation of the European Landscape Convention' (Cichocki and Sienkiewicz, 2006). Having reviewed the chief legislation concerning landscape, the authors concluded that there was no need to formulate a specific law related to landscapes and the European Landscape Convention. They stressed, however, the necessity for strengthening of the protection of landscapes, and suggested revising the current law. They proposed better cohesion of existing protective measures relating to natural and cultural elements of landscapes and to increase the scope of landscape analyses in a preparatory stage of local spatial development plans and other planning documents.

Almost simultaneously, the State Council for Nature Conservation, an advisory body to the Ministry of Environment, produced an overview of necessary modifications of Polish law in order to improve the implementation of the European Landscape Convention (Ochrona krajobrazu... n.d.). Like the previous report, this did not call for a new specific landscape law, but only for amendments to the existing laws, in particular the Environmental Protection Act, Nature Conservation Act, and Spatial Planning and Land Development Act. The Council recommended introducing to the latter two Acts a definition of landscape that would reflect the ELC's intent, and adding a definition of Landscape Quality Objectives to both laws. In the Nature Conservation Act, landscape protection would be separated from nature protection, and developed as a theme in its own right. This would extend the definition of landscape protection to include maintenance of natural, cultural, and aesthetic values, and an economic dimension.

The Council proposed that Landscape Quality Objectives should become protection goals in areas designated for landscape protection, whereas outside designated areas Landscape Quality Objectives would be binding for planning documents and location permits. The Council called on the Ministry of Environment to draw up a national strategy for landscape planning, protection, and management. Its statements would be binding for all sectorial strategies, plans, and programmes related to the use of nature and landscape.

Despite those substantial efforts at analysis, no legislation has thus far been revised towards articulating the spirit of the Convention.

5.3 Establishment of Landscape Policy and Integration of Landscape into All Relevant Policies

There was no response from the Ministry of Environment or other government agencies to the proposal from the State Council for Nature Conservation for the development of an overarching national strategy of landscape planning, protection, and management. As a result, no specific national landscape policy document has been drawn up.

Despite the absence of a national landscape policy, there is a wide range of governmental-level sectorial policy documents that address landscape and the ELC directly or indirectly, mostly within the fields of environmental protection, rural development, and spatial planning. For example, the National Strategy for Sustainable Use and Conservation of Biodiversity 2007–2013 mentions the protection of natural and cultural landscapes and the safeguarding of ecological functions of landscape through a planning system (Krajowa strategia ochrony i umiarkowanego użytkowania różnorodności biologicznej... 2003).

The National Environmental Policy for the period 2003–2006 gave consideration to designated landscapes, rural areas, and aesthetic values of landscapes, as well as to education on the benefits of landscapes (Polityka Ekologiczna Państwa na lata 2003–2006... 2002). Notably, the National Environmental Policy for the period 2009–2012 only recognizes the necessity of transposing the provisions of the ELC to the Polish nature protection law (Polityka Ekologiczna Państwa w latach 2009–2012... 2008).

The Strategy for Rural Development for the period 2007–2013 includes the following themes that are relevant to the ELC: the importance of agriculture for landscapes; the relationship between rural landscapes and traditional agriculture; cultural heritage and local identity; and the negative impact of agriculture on landscapes (Strategia rozwoju obszarów wiejskich i rolnictwa na lata 2007–2013 . . .2005).

The Ministry of Regional Development is currently drawing up the National Scheme of Spatial Development for 2008–2033, which is the key national level document relating to spatial planning. The Expert Draft of this document (Ekspercki Projekt... 2008) includes a chapter dedicated to landscape with an explicit commitment to the ELC objectives, but implicitly gives attention to landscape issues by dealing with: scattered sub-urban and rural settlement; marginalization or new functions of some areas; local development based on local heritage and products; new type of protected areas for support of natural and cultural heritage and local identity; and new functions of historical monuments. The document stresses the need for spatial order and congruence.

The lack of an overarching landscape policy means, however, that a range of policy areas likely to affect landscapes have remained uncoordinated.

5.4 Establishment of Procedures for Public Participation in Matters Related to Landscape

In 1998, Poland signed and, in 2002, ratified the Aarhus Convention on Access to Information, Public Participation in Decision-making and Access to Justice in Environmental Matters (UNECE, 1998). The principles of public participation were defined in the Act on Access to Information on the Environment and Environmental Protection and on Environmental Impact Assessment of 9 November 2000 (Dziennik Ustaw, 2000b). This Act expired on 30 September 2002, when most of its provisions were transferred to the Environmental Protection Act of 2001 (Dziennik Ustaw, 2001a). Some provisions for public participation were included in other regulations, for example: the Water Law (Dziennik Ustaw, 2001b), Spatial Planning and Land Development Act (Dziennik Ustaw, 2003a), Act on Rules of Support of Regional Development (Dziennik Ustaw, 2000a), and the Nature Conservation Act (Dziennik Ustaw, 2004).

After Poland acceded to the European Union (EU) in 2004, it patterned laws and regulations on public participation in decision-making processes after EU legislation, e.g. Directives on the assessment of the effects of certain public and private projects on the environment (EEC, 1985; EC, 2001); on public access to environmental information (EC, 2003a); and providing for public participation in respect of the drawing-up of certain plans and programmes relating to the environment (EC, 2003b). The participation procedures have been evolving in order to adjust to the EU legal regulations. In October 2008, a new law was promulgated on access to information on the environment and environmental protection, participation of society in decision-making in environmental matters, and on procedures for Environmental Impact Assessment (Dziennik Ustaw, 2008). The Act regulates the procedures for assessing the environmental impact of investments and for public participation in such procedures, as well as issues related to providing information on the environment. An important factor that precipitated the introduction of this law, and affected the development of public involvement in decision-making on regional and local level in general, has been the requirement of well-documented public participation in environmental decision-making related to investment projects that would benefit from the European Union funds. The Act replaced the provisions on public participation in the Environmental Protection Act of 27 April 2001, as the latter did not fully accord with certain specified EU regulations and directives (i.e. EC, 2006 and EEC, 1985). Without this amendment, investment projects with inappropriate EIA procedures could be put in jeopardy. As a result of the requirements of the EU regulations, public authorities have become less reluctant to admit citizen participation, which used to be regarded as an obstacle, delaying procedures. Effective opportunities for participation give civic organizations and individuals a sense of co-responsibility and encourage involvement. At present, no mechanisms exist for ongoing involvement of the public in decision-making where landscape is specifically concerned. Public participation is guaranteed in the assessment of the impact of projects on the environment, and in spatial planning that sets out a framework to co-ordinate the interaction of different policies and actions across space, which directly impacts landscapes.

The system of spatial planning in Poland consists of three levels: national, regional, and local, and public participation in relation to planning documents at every level is guaranteed by law. The public has the right to put forward petitions, to be informed and consulted on draft policies and plans, to propose changes to the plans, and to appeal if the proposals are rejected.

Regulations regarding spatial planning at a local level are, paradoxically, the main obstacle to effective public participation. At a local level, the authorities are obliged to draw up a local spatial management policy, termed the 'study on the conditions and directions of development of the commune'. The study is a preparatory document for local spatial development plans. After approval by the local council, the study becomes an internal administrative act, which is not legally binding. It is crucial, however, that local spatial development plans must agree with its provisions. Local spatial development plans, after approval by the local council, form a local law, binding for the commune, other authorities, and all citizens. However, the drawing-up of local spatial development plans is not obligatory, and only about 20% of the area of the country has been included in such plans. In default of local spatial development plans, building permits are issued as mostly discretionary administrative decisions. Thus public participation in the delivery of the spatial policy of a commune can be circumvented. Lack of appropriate coverage by local spatial development plans is a widely acknowledged reason for the spatial chaos and ugliness of Polish landscapes (Bariery i problemy gospodarki przestrzennej w Polsce... 2003).

There are no legal obstacles as such for the general public to become actively involved in matters related to landscape in Poland. However, law enactment alone does not instantly create new social behaviour and good practice of public participation, or eradicate habits that have formed over many years. In practice, public participation often does not conform to domestic and European legal requirements. According to a report of the Polish World Wildlife Fund (WWF Poland) on the quality of public participation in Poland (Kasprzyk et al., 2007), the main reasons for this include:

- Lack of tradition of public participation in decision-making, which stems from a long practice of central planning with no room for public participation and no access to information. On the one hand, old habits include excessive administrative control of decision-making, undermining the role of public participation,

while on the other hand, citizens do not know enough about their right to access information, nor do they wish to get involved. Public participation is limited to minimum activity, or even avoided, and the authorities take advantage of the citizens' ignorance. Additionally, the complicated nature of some of the matters discourages those who may be influenced by the decisions made, while favouring specialists such as experts and administrators.
- Weakness of civil society, which leads to 'superficial participation', i.e. the government or administration complies with minimum requirements only. Weak civil society includes weak non-governmental organizations. These are few, with a small number of members, economically dependent on the authorities, and with little influence.
- Low level of social trust as only 14% of the Poles trust other people. Distrust, manifesting itself as a suspicion that information is hidden or manipulated, and that the authorities are corrupt, diminishes effectiveness of public participation (Kasprzyk et al., 2007).

This lack of proper participation practice may explain the late reactions of citizens, which then take the form of protests against certain decisions. Spectacular examples have been protests against the dam in Czorsztyn in the 1990s, and against the planned international road crossing the Rospuda River valley in the Augustów Primeval Forest Natura 2000 site in 2007.

5.5 Identification and Assessment of Landscapes and Definition of Landscape Quality Objectives

In Poland, the study of landscapes is one of the traditional themes of geographical research. In physical geography, landscape is defined as a spatial and material system consisting of rocks, physical relief, water, vegetation, and atmosphere (Solon and Richling, 1996). Physical geography regularly uses the concept of natural landscape, meaning an area delimited by biophysical features. Natural landscapes can be identified everywhere, also in areas significantly influenced by human activity. They should not be mistaken for primeval landscapes with no trace of human impact. In physical geography, biophysical characteristics of an area are the basis for identification, characterization, and typologies of landscapes.

The earliest typology of natural landscapes of Poland was drawn up by Jerzy Kondracki in the late 1950s (Kondracki, 1960). Kondracki's categorization is based on the relief of the Earth's surface and includes four landscape classes: lowlands, uplands, mountains, and river valleys and terrain depressions. Within lowland landscapes, geomorphological processes forming the physical relief are the main distinguishing feature for landscape genera. Each lowland landscape genus is divided into landscape kinds on the basis of relative elevation and slope inclination of main landforms. Upland landscapes are identified in areas situated 200–600 m above sea level. The main distinguishing feature is geology. Upland landscape genera are divided into landscape kinds on the basis of intensity of dissection (intensity of erosion). Landscapes of highlands and high mountains are identified on the basis of elevation and vertical zonation. Division of landscapes of valley and depressions is based on hydrological conditions and the predominance of accumulation or erosion.

After some modifications, mainly in 1980s, Kondracki's typology was reworked by Andrzej Richling and Andrzej Dąbrowski and presented in the 'Atlas of the Republic of Poland' in 1995 (Richling and Dąbrowski, 1995). The country was divided into 4 classes, 14 genera, and 25 kinds of landscape. Each landscape kind was characterized by three biophysical components: soils, hydrology, and vegetation.

Although the scientific output of Polish landscape researchers amounts to hundreds of publications, geo-ecological landscape studies have had slight influence on practice and governance. Due to a weak relationship between research and practice, the results of research have not been effectively employed by the practitioners and authorities responsible for landscape planning and environmental management, and hence nature is often exposed to the risk of degradation, and humans to deterioration of the quality of their lives.

The reason for the insufficient implementation of research has been lack of a uniform landscape research initiative at the national level. According to Solon (2008), there is an urgent need for standardization of landscape-ecological methods for practical purposes.

In 2005, though without specific reference to the Convention, the Polish Association for Landscape Ecology (a majority of its members consisting of geographers) drew up a nation-wide research project 'Landscapes of Poland' (Drużkowski 2006). The project aimed at a synthesis of biophysical knowledge of landscapes in Poland to serve as a sound basis for practice in nature conservation, environmental management, and land-use optimization. The project aimed to describe the landscape diversity of the country by integrating data on natural and cultural elements of landscapes and architecture, on human impact on landscapes, and on resulting change. Regrettably, this research proposal was rejected in the course of evaluation for funding.

Natural and cultural elements of landscape are typically studied separately in Poland. The delineation of cultural landscapes of Poland was based on a method devised by landscape architect Janusz Bogdanowski (1983 and 1990), who distinguished architectural and landscape units based on relief, land use, and historical information on landscape development and interrelations of its elements.

In 1996, a draft proposal for a National System of Protection of Historical Cultural Landscapes in Poland was prepared under the auspices of the Ministry of Culture and Art. The project was directed by Janusz Bogdanowski and was accompanied by an agreement between the Ministry of Culture and Art and the Ministry of Environment regarding: unified protection of natural and cultural heritage; integration of protection programmes with spatial planning; joint research on values of nature and cultural heritage; and legal and organizational framework for concerted protection and management (Michałowski 1996). Protection of historical cultural landscapes extended protective measures not only over individual and collective complexes of historic property, but also over landscape areas, uniting the diversified conservation issues of the protection of all of its cultural ingredients. The project included a list of historical cultural landscapes to be included in a register of monuments (Łuczyńska-Bruzda and Malinowska 1996).

In 2004 the Ministry of Environment commissioned a pilot study on the most valuable Polish landscapes, termed the Red Book of Landscapes of Poland Project. The project was originally conceived within the State Council for Nature Conservation in 2002. It was probably the first research project undertaken in response to signing the ELC by Poland (Baranowska-Janota et al., 2004). 'The Red Book of Landscapes of Poland' aimed at identification and description of the most valuable landscapes, representing both natural diversity and cultural wealth. For each landscape to be included in the Red Book, a characterization sheet was to be produced with a representative photograph of the important features of that landscape. Landscape characteristics were to be established through interdisciplinary description, including existing threats and risk vulnerability.

The landscapes to be included in the Red Book were assessed against the following criteria:

- Geographical location,
- Degree of naturalness (natural–cultural),
- Physical relief (mountains–plain),
- Main land use or land cover (forest–residential),
- State of preservation (harmonious–degraded),
- Existing threats (in great danger–relatively safe),
- Aesthetic values (exceptionally attractive–quite attractive),
- Degree of uniqueness (unique–frequent).

With respect to geographical location, the landscapes taken into consideration represented the ten main geographical regions of Poland, from the Baltic Sea Shore and Lakeland in the North, through the Central Lowlands and Uplands, to the Sudety and the Carpathians Mountain Ranges in the South.

With regard to naturalness, five landscape types were distinguished:

- Natural landscapes – close to primeval areas, with their character formed and maintained by natural processes.
- Natural changed landscapes – dominated by natural habitats, although the natural processes that formed the landscape are controlled by humans or slightly changed.
- Semi-natural landscapes – containing a balanced proportion of natural habitats and human-made objects, such as settlements, military structures, and infrastructure.
- Cultural landscapes – defined as 'space historically formed by human activity, which includes man-made structures and natural objects' in the Protection and Safekeeping of Historical Monuments Act (Dziennik Ustaw, 2003b: Article 3)

and can be exemplified by the centres of cities, such as Toruń or Gdańsk, or by rural patterns, for example, the village of Chochołów.

- Cultural natural landscapes – defined as space historically formed by human activity, composed or ordered according to certain objectives, consisting of human-made and natural components (Baranowska-Janota et al., 2004), with biotic elements arranged or laid out by humans. Typically, these landscapes include parks and gardens or green areas of former defensive structures.

After being assessed against the above criteria, 198 landscapes (objects and areas) have been chosen to be included in the Red Book. One of these is shown in Fig. 5.1. The Red Book of Landscapes of Poland Project, regrettably not continued, has been the only study linking to any considerable degree natural and cultural elements of landscapes and making a specific reference to the ELC.

Research on Landscape Quality Objectives is at its initial stage. The first and so far the only output of this research was presented by Barbara Sowińska and Tadeusz Chmielewski (2007). The authors used a questionnaire to gather information on various social groups' opinions on natural and cultural features to be protected and preserved in the landscape of the Roztocze–Solska Forest Biosphere Reserve in south-eastern Poland. The research revealed a discrepancy between citizen preferences and the management of parts of the area. Hence, it is crucial to develop ways

Fig. 5.1 Księży Młyn ('Priest's Mill'): nineteenth-century workers' houses in Łódź, 2009 – an example of a Red Book Landscape (Photo: Anna Majchrowska)

of understanding social perceptions of a whole range of landscape features with a view to formulating Landscape Quality Objectives at regional and national levels.

5.6 The Way Forward

It seems that when Poland ratified the European Landscape Convention, it had neither a clear understanding of what it implied nor a strong commitment at ministerial level to make efforts to implement it. Among the many obligations pursuant to the Convention is the task of the government to introduce and clarify the intentions of the ELC in the relevant laws and to modify legislation in order to facilitate the implementation process. Law revisions are necessary as current legislation on landscape is fragmented and outdated, with natural and cultural elements dealt with separately, only outstanding landscapes focused on, and the landscape's significance for everyone insufficiently highlighted. The government should draw up a national landscape policy aimed at landscape protection, management, and planning, and promote the idea of landscape so that it permeates different policy areas and laws, including, for example, sustainable development, which is regulated in the Constitution of Poland. Actions undertaken at national level have been so far limited and piecemeal and thus have not contributed to co-ordination of the implementation of the Convention and its integration into present landscape-related activities.

Participation seems to be a major challenge of the Convention. In themselves, national regulations concerning public participation are compatible with the aims of the European Landscape Convention, and there are no legal obstacles for the general public to become actively involved in the formulation and delivery of projects related to landscape. However, law enactment alone does not instantly create new social behaviour and good practice of public participation, or eradicate habits that have formed over the years. In practice, public participation often does not conform to domestic and European legal requirements.

Among the obstacles to participation is a traditional administrative culture and a relatively weak civic society, which are legacies of the past period of central planning, giving no room for public participation. Further, there is a deficit of concern for landscape protection and management in official and civic circles. Raising awareness of landscape and its values is a prerequisite for responsible public participation in landscape-related decision-making.

It is necessary to overcome indifference towards landscape and the design of public space. There is a need to educate both the general public and the authorities concerning landscapes and of the role of landscape in human lives. The role of landscape research and education is crucial, but links between research and practice are weak. The main reasons include:

- The favouring of basic rather than applied research by the national research funding scheme
- The weak influence of successful applied research on the progression of academic careers

- The lack of formal procedures for information exchange and co-operation between researchers and other groups of society, including public administration and business; current co-operation relies mostly on private contacts.

During the socio-economic transformation process that is still going on in Poland, landscape matters are subordinated to state and individual economic (and political) pursuits. Although communes control land use by virtue of the Spatial Planning and Land Development Act (Dziennik Ustaw, 2003a) and by-laws, ad hoc changes to the landscape can hardly be prevented, as only about 20% of the area of the country has been included in local spatial development plans, and building permits are issued by local administrations. The political emphasis is on housing, infrastructure or agricultural profitability, and landscape is viewed as an obstacle to development plans rather than a positive development output, or it is regarded as something that can be taken care of later, when economic goals have been achieved.

Despite all these difficulties, there are encouraging instances of good practice in landscape matters at the local and regional levels in civic, administrative, and professional circles in Poland. Examples include photography contests, ecomuseums (Ekomuzea n.d.), greenways (Program Greenways w Polsce n.d.), and regional landscape strategies (Kistowski et al., 2005), even although these are often implemented without being associated with the Convention.

Finally, landscape has caught the interest of interdisciplinary research teams, which have begun to investigate its spatial, cultural, environmental, economic, and social aspects in accordance with the intent of the ELC and the holistic meaning of the term.

References

Baranowska-Janota M, Marcinek R, Myczkowski Z (2004) Czerwona Księga Krajobrazu Polski. Projekt pilotażowy tematu. Kraków (manuscript)

Bariery i problemy gospodarki przestrzennej w Polsce (sfera planowania i zagospodarowania przestrzennego): Raport krajowego sekretariatu Habitat. (2003) http://www.spatium.uni.lodz.pl/bariery_i_problemy_gospodarki_przestrzennej_w_polsce.pdf. Accessed 27 Oct 2009

Bogdanowski J (1983) Wprowadzenie do regionalizmu architektoniczno-krajobrazowego. Wiad Ekol 29:183–197

Bogdanowski J (1990) Metoda jednostek i wnętrz architektoniczno – krajobrazowych (JARK – WAK) w studiach i projektowaniu. Politechnika Krakowska, Kraków

Cichocki Z, Sienkiewicz J (2006) Prawno-planistyczne narzędzia ochrony i kształtowania krajobrazu w aspekcie wymogów wynikających z Europejskiej Konwencji Krajobrazowej. Instytut Ochrony Środowiska, Zakład Ekologicznych Podstaw Planowania Prze-strzennego, Wrocław

Council of Europe (2000). European Landscape Convention, Florence, 20.X.2000. ETS No. 176. http://conventions.coe.int/Treaty/en/Treaties/Html/176.htm. Accessed 27 Oct 2009

Drużkowski M (2006) Krajobrazy Polski (koncepcja badań ekologiczno-krajobrazowych). Probl Ekol Krajobrazu 16/1:61–68

Dziennik Ustaw (1997) Konstytucja Rzeczypospolitej Polskiej z dnia 2 kwietnia 1997 r. Dziennik Ustaw, 1997 nr 78 poz. 483. http://www.sejm.gov.pl/prawo/konst/polski/kon1.htm. Accessed 1 Jun 2010

Dziennik Ustaw (2000a) Ustawa z dnia 12 maja 2000 r. o zasadach wspierania rozwoju regio-nalnego. Dziennik Ustaw, 2000 nr 48 poz. 550. http://isap.sejm.gov.pl/Download?id=WDU20000480550+2003%2401%2401&type=3. Accessed 1 Jun 2010

Dziennik Ustaw (2000b) Ustawa z dnia 9 listopada 2000 r. o dostępie do informacji o środowisku i jego ochronie oraz o ocenach oddziaływania na środowisko. Dziennik Ustaw, 2000 nr 109 poz. 1157. http://isap.sejm.gov.pl/Download?id=WDU20001091157&type=3. Accessed 1 Jun 2010

Dziennik Ustaw (2001a) Ustawa z dnia 27 kwietnia 2001 r. Prawo ochrony środowiska. Dziennik Ustaw, 2001 nr 62 poz. 627. http://isap.sejm.gov.pl/Download?id=WDU20010620627+2003%2407%2411&type=3. Accessed 1 Jun 2010

Dziennik Ustaw (2001b) Ustawa z dnia 18 lipca 2001 r. Prawo wodne. Dziennik Ustaw, 2001 nr 115 poz. 1229. http://isap.sejm.gov.pl/Download?id=WDU20011151229+2002%2401%2401&type=3. Accessed 1 Jun 2010

Dziennik Ustaw (2003a) Ustawa z dnia 27 marca 2003 r. o planowaniu i zagospodarowaniu przestrzennym. Dziennik Ustaw, 2003 nr 80 poz. 717. http://isap.sejm.gov.pl/Download?id=WDU20030800717+2003%2407%2411&type=1. Accessed 1 Jun 2010

Dziennik Ustaw (2003b) Ustawa z dnia 23 lipca 2003 r. o ochronie zabytków i opiece nad zabytkami. Dziennik Ustaw, 2003 nr 162 poz. 1568. http://isap.sejm.gov.pl/Download?id=WDU20031621568&type=3. Accessed 1 Jun 2010

Dziennik Ustaw (2004) Ustawa z dnia 16 kwietnia 2004 r. o ochronie przyrody. Dziennik Ustaw, 2004 nr 92 poz. 880. http://isap.sejm.gov.pl/Download?id=WDU20040920880&type=3. Accessed 1 Jun 2010

Dziennik Ustaw (2006) Europejska Konwencja Krajobrazowa, sporządzona we Florencji dnia 20 października 2000 r. Dziennik Ustaw, 2006 nr 14 poz. 98. http://isap.sejm.gov.pl/Download?id=WDU20060140098&type=2. Accessed 1 Jun 2010

Dziennik Ustaw (2008) Ustawa z dnia 3 października 2008 r. o udostępnianiu informacji o środowisku i jego ochronie, udziale społeczeństwa w ochronie środowiska oraz o ocenach oddziaływania na środowisko. Dziennik Ustaw, 2008 nr 199 poz. 1227. http://isap.sejm.gov.pl/Download?id=WDU20081991227&type=3. Accessed 1 Jun 2010

EC (2001) Directive 2001/42/EC of the European Parliament and of the Council of 27 June 2001 on the assessment of the effects of certain plans and programmes on the environment. Official Journal of the European Communities, L 197, 21/07/2001, 30–37. EUR-Lex. http://eur-lex.europa.eu/LexUriServ/LexUriServ.do?uri=OJ:L:2001:197:0030:0037:EN:PDF. Accessed 6 May 2010

EC (2003a) Directive 2003/4/EC of the European Parliament and of the Council of 28 January 2003 on public access to environmental information and repealing Council Directive 90/313/EEC. Official Journal of the European Union, L 041, 14/02/2003, 26–32. EUR-Lex. http://eur-lex.europa.eu/LexUriServ/LexUriServ.do?uri=OJ:L:2003:041:0026:0032:EN:PDF. Accessed 6 May 2010

EC (2003b) Directive 2003/35/EC of the European Parliament and of the Council of 26 May 2003 providing for public participation in respect of the drawing up of certain plans and programmes relating to the environment and amending with regard to public participation and access to justice Council Directives 85/337/EEC and 96/61/EC. Official Journal of the European Union, L 156, 25/06/2003, 17-25. EUR-Lex. http://eur-lex.europa.eu/LexUriServ/LexUriServ.do?uri=OJ:L:2003:156:0017:0024:EN:PDF. Accessed 6 May 2010

EC (2006) Council Regulation (EC) No 1083/2006 of 11 July 2006 laying down general provisions on the European Regional Development Fund, the European Social Fund and the Cohesion Fund and repealing Regulation (EC) 1260/1999. Official Journal of the European Union, L 210, 31/7/2006, 25-78. EUR-Lex. http://eur-lex.europa.eu/LexUriServ/LexUriServ.do?uri=OJ:L:2006:210:0025:0025:EN:PDF. Accessed 7 May 2010

EEC (1985) Council Directive 85/337/EEC of 27 June 1985 on the assessment of the effects of certain public and private projects on the environment. Official Journal, L 175, 05/07/1985, 40–48.

EUR-Lex. http://eur-lex.europa.eu/LexUriServ/LexUriServ.do?uri=CELEX:31985L0337:EN:HTML. Accessed 6 May 2010

Ekomuzea (n.d.) Fundacja Partnerstwo dla środowiska. http://www.ekomuzea.pl/. Accessed 29 Oct 2009

Ekspercki Projekt Koncepcji Przestrzennego Zagospodarowania Kraju do roku 2033 (2008) Ministerstwo Rozwoju Regionalnego. http://www.mrr.gov.pl/rozwoj_regionalny/poziom_krajowy/polska_polityka_przestrzenna/prace_nad_KPZK_2008_2033/Documents/PE_KPZK_STYCZEN_z_ERRATA_NA_STRONIE.pdf. Accessed 1 Jun 2010

Główny Urząd Statystyczny (2008) Ochrona środowiska 2008. Warszawa. http://www.stat.gov.pl/cps/rde/xbcr/gus/PUBL_RLS_ochrona_srodowiska_2008r.pdf. Accessed 25 Oct 2009

Kasprzyk M, Kupper D, Kiejzik-Głowińska M, Machińska H, Pchałek M, Roggenbuck A, Serwecińska D, Strumińska M, Trojanowska K, Wiśniewska M (2007) Jakość konsultacji społecznych w Polsce: Krajowa praktyka a uwarunkowania prawne. WWF Polska. http://wwf.pl/informacje/publikacje/fundusze_ue/jakosc_konsultacji.pdf. Accessed on 28 Oct 2009

Kistowski M, Lipińska B, Korwel-Lejkowska B (2005) Studium ochrony krajobrazu województwa pomorskiego. Urząd Marszałkowski Województwa Pomorskiego. http://www.woj-pomorskie.pl/downloads/studium_ochr_krajobr_Pomorza_22-06-07_A.pdf. Accessed 6 Apr 2009

Kondracki J (1960) Typy krajobrazu naturalnego (środowiska geograficznego) w Polsce. Prz Geogr 30:1–2

Krajowa strategia ochrony i umiarkowanego użytkowania różnorodności biologicznej wraz z Programem działań (2003) http://wwf.pl/informacje/publikacje/natura/strategia_roznorodnosc_biologiczna.pdf. Accessed 9 Jun 2010

Krajowy Ośrodek Badań i Dokumentacji Zabytków (2007) Stan krajowej ewidencji i rejestru zabytków na 31 grudnia 2007. http://www.kobidz.pl/app/site.php5/Show/386.html. Accessed 6 Apr 2009

Krajowy Ośrodek Badań i Dokumentacji Zabytków (2008) Obiekty nieruchome wpisane do rejestru zabytków – charakterystyka. http://www.kobidz.pl/app/site.php5/getFile/13292. Accessed 3 Nov 2009

Łuczyńska-Bruzda M, Malinowska H (1996) Zarys koncepcji krajowego systemu ochrony zabytkowych krajobrazów kulturowych w Polsce: Wstępny wniosek dla Centralnego urzędu planowania do Strategii rozwoju i koncepcji polityki przestrzennego zagospodarowania kraju. Krajobrazy 16(28):21–40

Michałowski A (1996) Ochrona krajobrazu wyzwaniem XXI wieku w Polsce. Krajobrazy 16(28):11–20

Ochrona krajobrazu, implementacja Europejskiej Konwencji Krajobrazowej (n.d.) http://mobi.mos.gov.pl/g2/big/2009_03/712e0af3b3c085208a14dca894f750ed.pdf. Accessed 30 Oct 2009

Polityka Ekologiczna Państwa na lata 2003–2006 z uwzględnieniem perspektywy na lata 2007–2010 (2002) Ministerstwo Środowiska. http://www.mos.gov.pl/g2/big/2009_04/36383d1a880bbc0b65d0a1c501571e73.pdf. Accessed 9 Jun 2010

Polityka Ekologiczna Państwa w latach 2009–2012 z perspektywą do roku 2016 (2008) Ministerstwo Środowiska. http://www.mos.gov.pl/g2/big/2009_11/8183a2c86f4d7e2cdf8c3572bdba0bc6.pdf. Accessed 9 Jun 2010

Program Greenways w Polsce (n.d.) Fundacja Partnerstwo dla środowiska. http://www.greenways.pl/. Accessed 29 Oct 2009

Richling A, Dąbrowski A (1995) Mapa 53.1: Typy krajobrazu naturalnego. Atlas Rzeczpospolitej Polskiej. Główny Geodeta Kraju, Warszawa

Solon J (2008) Kierunki standaryzacji metod badań krajobrazu do celów praktycznych. Prz Geogr 80:39–54

Solon J, Richling A (1996) Ekologia krajobrazu. PWN, Warszawa

Sowińska B, Chmielewski TJ (2007) Polish experience in landscape quality objective identification on the example of the Roztocze–Solska Forest Biosphere Reserve. In: 5th meeting of the workshops of the council of Europe for the implementation of the European Landscape Convention, landscape quality objectives: from theory to practice, Girona, Spain, 28–29 September 2006. European spatial planning and landscape, 84. Council of Europe Publishing, Strasbourg, pp 87–95

Strategia rozwoju obszarów wiejskich i rolnictwa na lata 2007–2013 (z elementami prognozy do roku 2020) (2005) Ministerstwo Rolnictwa i Rozwoju Wsi. http://www.fundusz-estrukturalne.gov.pl/informator/npr2/dokumenty%20strategiczne/rolnictwo.pdf. Accessed 9 Jun 2010

UNECE (1998) Convention on Access to Information, Public Participation in Decision-making and Access to Justice in Environmental Matters done at Aarhus, Denmark on 25 June 1998. United Nations Economic Commission for Europe. http://www.unece.org/env/pp/treatytext.htm. Accessed 11 Mar 2010

Chapter 6
Landscape Regulation in Regional Territorial Planning: A View from Spain

Berezi Elorrieta and Dolores Sánchez-Aguilera

Abstract During the last two decades, Spain's regional governments have developed new regional territorial plans and legal instruments which, in some cases, incorporate the landscape as a new component in accordance with guidelines and documents issued by the Council of Europe and the European Union. The aim of this chapter is to study the role of the landscape in these regional laws and the part

B. Elorrieta (✉)
Faculty of Geography and History, University of Barcelona, 08001 Barcelona, Spain
e-mail: berezielorrieta@gmail.com

M. Jones, M. Stenseke (eds.), *The European Landscape Convention*,
Landscape Series 13, DOI 10.1007/978-90-481-9932-7_6,

it plays in regional territorial planning, first, in terms of the efforts made to improve landscape conservation, management, and planning, and, second, in terms of the extent to which these plans adhere to the new framework of reference provided by the European Landscape Convention. We undertake an analysis of the differences in the legislative provisions made for landscapes in Spain's regions. We also examine how the integration of landscape protection within territorial planning has been established as a new goal by studying the Regional Territorial Plans, which reflect the spatial implementation of these new strategies and policies.

Keywords Landscape policies · Regional planning · Autonomous communities · European Landscape Convention · Public participation

6.1 Introduction

The fundamental changes that are rapidly altering many of Spain's landscapes seem to reflect a lack of awareness as to their real value. Over the last two decades, land-use patterns across the country have changed dramatically. This is so much so that recent studies of the changes affecting land use and land cover reveal Spain to be one of the countries, together with Ireland and Portugal, which has expanded its artificial surface (land covered by structures and the transport network) most. According to the Sustainability Observatory, Spain's artificial surface grew by 29.5% in the period 1987–2000 and now occupies 2.1% of the territory (OSE, 2006). One of the main causes of this expansion has been urban occupation, which has become increasingly more extensive. This trend is most notable in central hinterland regions, such as Madrid, and in coastal regions, primarily on the Mediterranean, but there is evidence of it beginning to affect the Atlantic and Cantabrian coasts too. The economic growth rate (dependent on activities such as construction, transport and tourism, which consume large areas of land), the consolidation and entrenchment of urban sprawl, and the heavy investment in infrastructure during the period 1987–2000 are major causes. Spain also stands out among its European counterparts in terms of the sharp rise in areas of land under permanent irrigation. However, parallel to these changes, the last two decades have witnessed major regulatory progress aimed at protecting landscapes, and there have been significant advances in the legal framework regulating landscapes.

Spain signed the European Landscape Convention (ELC) in October 2000, and ratified it 7 years later, on 26 November 2007. The implementation of the ELC should boost the introduction of new mechanisms and instruments so as to comply with the provisions, both general and specific, contained in the Convention.

The aim of this chapter is to examine whether Spain, at a regional level, has developed the required instruments to guarantee landscape protection, management, and planning, thereby honouring its commitment to the ELC. Following the signing of the ELC, it has been the authorities in Spain's autonomous regions that have adopted the measures provided for by the Convention, and there has been a concerted attempt to include the landscape within the various laws and regional plans that have been passed, with the ELC acting as the main framework of reference. The methodology adopted in this study includes an analysis of existing regional planning instruments, as well as the legislation and other landscape management instruments employed in each of Spain's autonomous communities (regions), in order to examine how they are being used. This information is summarized in tabular form to facilitate a comparative analysis across the regions of the instruments being used to regulate the landscape.

6.2 Landscape in Europe: The Construction of a New Framework

Around Europe, there is an increasing awareness of the value of the landscape, a concern which began to gain widespread attention during the 1980s. The European Union Directive on Environmental Impact Assessment of 1985 was a pioneer measure in this regard as it led to the adoption of a definition of the environment that specifically included the landscape, among other elements (European Commission, 1985).

The European Spatial Development Perspective of 1999 was a further milestone in this process, establishing the landscape as one of the objectives and policy options for the future development of European territory through the principle of 'creative management of cultural landscapes' (European Commission, 1999). The European Landscape Convention a year later became the first (and so far the only) international treaty to focus specifically on landscape protection. Spain signed the ELC on the same day that it was adopted. In 2008, the Council of Europe published more detailed guidelines for implementing the Convention. The text provides examples of instruments that can be used for this purpose. The guidelines stress that 'it is necessary to emphasize that landscape issues should be approached through a systematic landscape planning process adapted to the different administrative levels, from national to local, throughout the whole territory, including urban and extra-urban areas' (Council of Europe, 2008: appendix 1, 1).

The Convention represents a major step forward as far as landscapes are concerned, since it recognizes the need for landscapes to be valued and recognized in their own right, and not solely as complementary assets (Mata Olmo, 2008). All too often landscapes are mentioned in the Articles of national laws and regulations, but in practice the necessary consideration is not paid to them. The Convention seeks to bring about a change in the basic approach to the landscape and to make the landscape pivotal in all land planning (Zoido Naranjo, 2006).

The ELC sets out general measures that oblige the Parties to: (a) to recognize landscapes in law; (b) to establish and implement landscape policies aimed at landscape protection, management, and planning; (c) to establish procedures for the participation of the general public, local and regional authorities, and other parties; and (d) to integrate landscape into its regional and town planning policies, as well as other policies (Council of Europe, 2000: Article 1). This chapter does not aim to undertake a detailed examination of the ELC document itself, but to highlight certain points that might have a bearing on the implementation of the ELC in Spain. The specific measures that we consider in order to determine whether Spain has indeed adopted them are the legal recognition of the landscape, the development of specific landscape policies, the integration of landscapes into land planning processes, and the role of public participation (Caravaca, 2004).

6.3 Landscape Policies in Spain: A Fragmented Picture

The implementation of landscape policies throughout Europe is highly uneven. In some countries good progress has been made, while in others policies are illustrative of a somewhat disorganized approach. Such is the case in Spain, where references to landscapes in legal texts are usually ambiguous and vague.

Nevertheless, Spanish legislation has made some attempt at accommodating the landscape, but rarely has this been put to effective use, being primarily limited to protection-based policies for specific landscapes (Nogué, 2006). At the national level, the most relevant piece of legislation is the Law of 1989 on the Conservation of Natural Areas and Flora and Fauna (Ley, 1989). This Act represents the most significant step taken in the field of nature preservation. One of its 'inspirational principles' is landscape conservation, but its most relevant innovation was the introduction of 'protected landscapes'. The Act perceives the landscape in terms of protection, limited to areas of particular value or presenting a singular quality.

Other public issues have still to be faced up to and solved. The first of these is the assignment of political and administrative powers. The distribution of powers is a complex matter, reflecting the decentralization process that was initiated following Spain's transition to democracy. Upon the passing of the Constitution of 1978, Spain created a unique system of regional autonomy, known as the 'state of the autonomies'. Centralism, nationalism and separatism played an important role in the Spanish transition. For fear that separatism would lead to instability and a dictatorial backlash, a compromise was struck among the moderate political parties taking part in the drafting of the Constitution. The aim was to appease separatist forces and so disarm the extreme right. A highly decentralized state was established, compared both with the previous Francoist regime and with modern territorial arrangements in many Western European nations. Nowadays, political power in Spain is channelled through a central government and 17 autonomous communities. These regional governments are responsible (at least in part) for schools, universities, health, social services, culture, urban and rural development, and territorial planning. All in all, under the system of *autonomías*, Spain is recognized as being an extraordinarily decentralized country. The autonomous communities have wide legislative and executive autonomy, with their own parliaments and regional governments, although the distribution of powers may differ in each community, as laid out in their Statutes of Autonomy (legal texts establishing their rights and powers).

The Spanish Constitution contains not a single reference to the landscape, although it might be claimed that it is included within the general approach to the environment and natural resources. By contrast, the Statutes of some of the autonomous communities do make mention of this concept, which can be interpreted as an assumption of competence in this field. However, this issue remains largely unexplored. Yet, the autonomous communities already have full powers in the spheres of land planning, the environment, and cultural heritage, which (together with town planning) are the main concerns of the ELC.

The Spanish national government, however, retains the right to lay down guidelines for the management of natural resources, including landscapes. Thus begins an interesting debate concerning where the powers of landscape management lie (Cortina Ramos, 2009a). Recently, it would seem that the Government has begun to show some concern regarding this issue, although basic guidelines have still to be drawn up. The only real progress to date has been made at the strictly regional level.

The new generation of Statutes governing the autonomous communities grants greater importance to the landscape. While this does not as yet form part of their environmental powers, the Statutes expressly recognize the right to the landscape and the duty to oversee its responsible use and management. For example, the 2006 Catalan Statute (Estatut d'Autonomia de Catalunya, 2006: Article 27, §1) explicitly provides that:

> All persons have the right to live in a balanced, sustainable and healthy environment (. . .). They also have the right to enjoy the natural resources and the landscape in conditions of equality, and the duty to employ these responsibly and to prevent their deterioration.

Today, the autonomous communities are beginning to develop new normative models for implementing the principles, objectives, and criteria of the ELC. However, Spain cannot yet boast a framework that clearly establishes at national level the basic legislation and regulations for implementing its landscape policies (Cortina Ramos, 2009a).

By taking 7 years to ratify the ELC, Spain delayed confirmation of its initial intentions as stated in 2000. Hence the incorporation of landscapes into regulations and plans has not formed part of a gradual process promoted by the central government encouraging the autonomous communities to adopt measures in this area, but instead a number of communities took the initiative and chose not to wait for the central government to act, and so introduced their own instruments of landscape management. The Convention is an instrument designed to be implemented at both regional and local levels (Zoido Naranjo, 2001), as is apparent in many of its proposed plans of action. We analyse these management instruments at the regional level, i.e. that of Spain's autonomous communities, examining the following:

- Legislation passed by the autonomous communities
- Regional planning regulations
- Other instruments, including landscape observatories.

Table 6.1 seeks to provide a summary of landscape policies, including specific examples, adopted in Spain at the regional level. The autonomous communities included have been selected on the basis of the importance of the work undertaken

Table 6.1 Landscape policies in Spain at the regional level

Community	Landscape Act	Other laws	Instruments of regional planning	Other instruments
Andalusia	–	Act 4/1989 on Preservation of Natural Spaces and Wild Flora and Fauna (Ley, 1989): – Creation of the specific concept of *protected landscape* – The aesthetic value of landscapes is also considered when creating new parks	Andalusia Land Planning Scheme (Plan de Ordenación del Territorio de Andalucía, 2006): – Defines a Land Heritage System (SPT) which includes provisions related to the management and promotion of landscapes – Foresees action plans with effects on landscape management	– Centre for Landscape and Territory Studies – Andalusia landscape maps
		Act 1/1994 on Land Planning (Ley, 1994): – Guidelines for landscape planning in plans at the sub-regional level		
Catalonia	Act 8/2005 on Landscape Protection, Management and Planning (Ley, 2005a): – Use of ELC terminology – Acknowledgment of the 'right' to landscape	Decree 343/2006 (Decreto, 2006b): – Passed in order to develop Act 8/2005 – Regulation of the studies and reports of landscape impact and integration	Catalonia General Land Scheme (Pla Territorial General de Catalunya, 1995)	– Landscape Observatory – Landscape Catalogues
Valencian Community	Act 4/2004 on Land Planning and Landscape Protection (Ley, 2004a): – Design of an action plan for landscape	Decree 120/2006 to approve the Landscape Regulation (Decreto, 2006a): – Passed in order to develop Acts 4/2004 (Ley, 2004a), 10/2004 (on land unqualified for development) (Ley, 2004b) and Act 16/2005 on town planning (Ley, 2005b)	–	

Table 6.1 (continued)

Community	Landscape Act	Other laws	Instruments of regional planning	Other instruments
Basque Country	–	Act 3/1998 on the General Protection of the Environment of the Basque Country (Ley, 1998): – Requirement to draw up the Environmental Framework Programme	Regional Planning Guidelines (Directrices de Ordenación Territorial del País Vasco, 1997): – Landscape is considered an element of the physical context – Introduction of the need for landscape studies and catalogues	– Basque Environmental Strategy for Sustainable Development – Environmental Framework Programme
Galicia	Act 7/2008 on Landscape Protection (Ley, 2008): – Regulation of landscape catalogues and guidelines, studies and action plans in protected areas	Act 9/2001 on Nature Preservation (Ley, 2001)	Guidelines for Land Management (in progress)	– Landscape Catalogues – Landscape Agreements
Canary Islands	–	Act 12/1987 on the Approval of Natural Spaces of the Canary Islands (Ley, 1987) Legislative Decree 1/2000 of the Consolidated Text of the Acts on Land Management in the Canary Islands and on Natural Spaces of the Canary Islands (Decreto Legislativo, 2000)	General Guidelines on Planning and Tourism (Directrices de Ordenación General y del Tursimo de Canarias, 2003): – Landscape as a critical dimension of the identity of the Canary Islands Sub-regional level: Special Plans for Landscape Management	

in this sphere and, in each case, we identify the following aspects: specific legislation regulating the landscape; other laws making reference to the landscape; ways in which the landscape has been incorporated into instruments of regional planning; and any other instruments applicable to landscape management. Having examined these policies, we move on to a study of the Landscape Observatories currently operating, given their highly innovative and functional nature.

6.3.1 Landscape Planning and Legislation in the Autonomous Communities

The territorial plans are planning instruments for the spaces in which people live and act, and as such they are inevitably landscape projects for the future (Esteban Noguera, 2009). The challenge facing planners is to respond to the contradictory demands made by processes of economic and social development, on the one hand, and maintenance of the quality of the visual environment, on the other. Planning, via the implementation of regulations, can ensure that good relations are established between the infrastructure and human activities and the landscape.

Spain's urban and regional planning has been marked by a clear orientation towards the management of physical space, and as such, it has had a direct impact on the landscape. Although there is a long tradition of urban planning in Spain, regional planning has a much more irregular tradition and has produced insufficient results for us to judge its effectiveness. In the 1980s, the autonomous governments often placed an emphasis on regional planning because, it would seem, they saw it as an instrument for asserting themselves within their territories during this new political phase (Esteban Noguera, 2009). However, some plans have still to be implemented, and many autonomous communities are still without a regional territorial plan.

Be that as it may, all the autonomous communities without exception (as well as some of the island governments) have adopted the ELC as a point of reference for their respective regional and landscape policies, although the actual development of these policies is very irregular (OSE, 2009). Table 6.1 shows that virtually all the autonomous communities are currently concerned with landscape planning, albeit in different ways. The first region in Spain to introduce legislation regulating its landscapes was Valencia in an Act passed in 2004 (Ley, 2004a), which incorporated aspects of both landscape and land planning. One year later, in 2005, Catalonia introduced its own Act for the Protection of the Landscape (Ley, 2005a) text which would later serve as a model for many other communities. More recently, in July 2008, Galicia passed its own Landscape Act (Ley, 2008). The Balearic Islands, together with some of the other communities, are also currently drafting their own landscape legislation, with the European Landscape Convention as their point of reference.

In *Andalusia*, the situation is somewhat different: although it has not introduced a specific landscape Act, several regional-level laws include a range of different provisions for its landscapes. Notwithstanding this, a region-wide law is needed so as to regulate all aspects of landscape management and to ensure the integration of all relevant functions. However, the region's recently established Land Planning

Scheme includes provisions regulating landscape management and their promotion, in addition to action plans with a specific focus on landscape management (Plan de Ordenación del Territorio de Andalucía, 2006, 38):

> The landscape is a key element in the protection system and, therefore, the Plan provides for the establishment of a Regional Landscape Programme, whose guidelines need to include such aspects as: general criteria for the protection, preservation and enhancement of the landscape in each of the territories of Andalusia; the management of landscape in urban and regional planning; criteria for the integration within the landscape of infrastructure projects; and the criteria and measures, as regards the landscape, for the adoption of plans, programmes and interventions with territorial impact.

However, Andalusia has gone even further and has developed other interesting tools, including a landscape study centre and a landscape map, included within the second volume of the Atlas of Andalusia (Regional Ministry of Public Works and Transport & Regional Environment Ministry, 2005) as a tool for identifying and assessing the assets constituting the region's landscapes.

Catalonia is one of the communities with its own specific law regulating the landscape. The Catalan Landscape Act, passed in 2005 (Ley, 2005a), takes its inspiration directly from the ELC (Nogué, 2006), a fact that is apparent in the introduction. Indeed, the Catalan Parliament adhered unanimously to the ELC in December 2000, thus showing a pioneering interest in landscapes when at the national level such interest had yet to be shown. This adhesion was recorded in the Landscape Act, adding (Ley, 2005a: Preamble I, §5):

> This law is designed to give positive content to this endorsement. Thus, it gives legal protection to Catalan landscapes and establishes the relevant instruments for their management and improvement.

The Act, remarkable for its markedly innovative nature, gave rise to considerable expectations, since it was the first Catalan law to recognize the 'right' to landscape, thus endowing this asset with true legal value. The Act is short, concise, and relatively straightforward. It assumes the definitions and objectives as set out in the ELC and proposes the integration of the landscape in all public policies that have a territorial impact.

Some of the tools for the management of the landscape that have emerged from the application of this law are the Landscape Observatory and the landscape catalogues, which undertake a diagnosis of the state of Catalan landscapes. The landscape catalogues allow the region's landscapes to be identified and assessed, and for their historical evolution and trends to be understood.

Catalonia has not been so innovative in its land planning policies. Although its land scheme was passed in 1995, it has still to be implemented. Nevertheless, we are given to expect that the landscape will be seen as a basic element in all future territorial plans.

The landscape legislation introduced by the *Valencian Community* is the Act of 2004 on Land Planning and Landscape Protection (Ley, 2004a). This includes a Landscape Action Plan that aims to identify landscapes of regional and local interest. The law states that all territorial action plans and general plans should include a landscape study in order to catalogue, assess, and protect the region's landscapes. It also establishes specific rules for the application of the law in rural and urban environments. However, to date no specific regional planning instrument has been made available. Work is currently being undertaken on this, and the landscape is certain to figure as a key feature in all future plans.

In the *Basque Country*, there is no single landscape Act, although a number of related plans have been implemented, including the Basque Environmental Strategy for Sustainable Development and the Environmental Framework Programme. The regional strategy, dating back to 1997 and hence older than the ELC, recognizes the value of all landscapes including industrial, farming, and urban landscapes. The region's highly industrial past is reflected in the considerable attention paid in the plan to the management of its industrial landscapes. Moreover, the strategic plan (termed guidelines) introduced the need to undertake landscape studies and to map areas of interest (Directrices de Ordenación Territorial del País Vasco, 1997). The plan provides that: 'The Administration must undertake a continuous monitoring of those actions that have most impact on the landscape' (DOT, 2004: §5.2.G.g3).

Galicia has only just recently adopted a Landscape Act (Ley, 2008), and it is perhaps too soon to draw any conclusions. It is, however, deeply rooted in the ELC. The Explanatory Memorandum of the Galician Landscape Act (Ley, 2008: I, §8) states:

> This law, in accordance with the provisions of the European Landscape Convention (. . .) seeks to recognize landscapes in law and promote landscape policies, understanding the landscape as 'an essential element for the individual and social welfare, whose protection, management, and planning entails rights and obligations for all', as defined in the aforementioned convention.

The law stresses the need for cooperation between all levels of public administration in order to ensure the effective implementation of landscape measures. It also seeks to involve local actors in the protection of the landscape through the drawing up of landscape agreements, as instruments of cooperation between public administrations, local organizations, and other socio-economic actors.

As in the Catalan Act, this law provides for the preparation of landscape catalogues for Galicia as an instrument for the management of the landscape. In summer 2008, a new regional plan was also approved, although it has still to be officially passed. According to its promoters, the plan's guidelines are particularly concerned with the preservation of Galicia's cultural heritage preservation and the conservation of its natural environment, including the conservation of landscapes and their intrinsic value.

Finally, the case of the *Canary Islands* is also of interest. Here, government guidelines do not focus specifically on the landscape, but rather the individual islands (i.e. at the sub-regional level) have engaged in the design of Special Landscape Management Plans. The regional Act of 1987 for the Approval of Natural Spaces among the Canary Islands (Ley, 1987) omitted many important areas, thus requiring the introduction of additional legislation to rectify these deficiencies. In 2008, landscape plans were approved for the islands of Gran Canaria and Tenerife, a crucial step in these environments, which are subject to immense pressures from tourism. Tenerife's Special Landscape Plan draws heavily on the ELC, and places great emphasis on the involvement of the local population in defining the island's landscape objectives and policies.

6.3.2 Other Instruments: Landscape Observatories

The legal framework outlined above will undoubtedly change the way in which landscapes will be perceived over the next few years and will facilitate the integration of landscapes into all policies in this area of action. One of the elements that best demonstrates the administration's efforts to implement the principles of the European Landscape Convention is the creation of specialist landscape agencies.

At present, two institutions have been created in Spain, under the auspices of regional governments. These institutions are the Landscape Observatory of Catalonia and the Centre for Landscape and Territory Studies in Andalusia. Their primary goals are to identify and assess landscapes, to integrate landscapes within relevant government policies, to raise public awareness, to promote education and training, and to cooperate at the European level through the exchange of experiences and information. Table 6.2 provides a brief comparison of the characteristics and main functions of both institutions.

The *Landscape Observatory of Catalonia* (*Observatori del Paisatge*) acts as an advisory body to the Government of Catalonia and Catalan society in general on all questions related to the landscape (Nogué, 2007). It was created under the Act for the Protection, Management and Planning of the Landscape (Ley, 2005a) in Catalonia but was legally formed earlier, in 2004, its constitution being published in the Government of Catalonia's Official Gazette (Resolutió, 2004). The main purposes of the Landscape Observatory are to raise the awareness of Catalan society for its landscapes and to support the application of the European Landscape Convention in Catalonia (Nogué, 2007).

Similarly, in 2005, the Andalusian Government created the *Centre for Landscape and Territory Studies (Centro de Estudios Paisaje y Territorio – CEPT*). The centre was established as a framework partnership agreement between the Ministry of Transport of Andalusia and the Universities of Seville, Granada, Malaga, Cordoba, Cadiz, Almeria, Huelva, Jaen, International Andalusia, and Pablo de Olavide. The main goal of the Centre is to strengthen cooperation between the Ministry of Public Works and Transportation and the public universities of Andalusia so as to foster positive interaction between their respective lines of research, scientists and teachers and to promote the development of landscape and spatial planning studies.

Today, the two institutions are the main instruments supporting the implementation of the European Landscape Convention in Spain and have many points in common in terms of how they function. Yet, it is perhaps worth pointing out some differences. In the case of Catalonia, the government opted for the creation of an observatory, while although an observatory is planned in Andalusia the government has yet to establish one. Thus, the CEPT has a marked academic profile, and while

Table 6.2 Comparison between the Landscape Observatory of Catalonia and Centre for Landscape and Territory Studies of Andalusia

		Observatori del Paisatge The Landscape Observatory	Centro de Estudios Paisaje y Territorio Centre for Landscape and Territory Studies
Date of creation		2004	2005
Initiation of activity		2005	2006
Headquarters		Olot (Catalunya)	Sevilla (Andalucía)
Organization chart		Government of Catalonia Local authorities Catalan universities Professional groups Catalan society	Government of Andalusia Andalusian universities
Main functions	Identification	To prepare landscape catalogues in Catalonia in order to identify, classify, and qualify the various landscapes	The scientific study of the Andalusian landscape and territory, both in its natural constituents, and in its use, management, and organization of human activities
		To establish criteria for the adoption of measures for landscape protection, management, and planning	The development of innovative and technological knowledge concerning the landscape and territory, as well as geographical information, especially GIS
		To fix criteria so as to establish the landscape quality targets and the measures and actions necessary in order to achieve these targets	

Table 6.2 (continued)

	Observatori del Paisatge The Landscape Observatory	Centro de Estudios Paisaje y Territorio Centre for Landscape and Territory Studies
Assessment/monitoring	Establishing mechanisms for the observation of evolution and change in the landscape	Systematic territorial knowledge through the creation of a Landscape Observatory in Andalusia, conceived as a tool for innovation and the continuous monitoring of developments and of the dynamic features of the territory and the Andalusian countryside
	Proposals of actions for the improvement, restoration or creation of landscape	To become an instrument to support government activities addressed at managing the landscape
Public awareness and participation	To promote social sensitization campaigns with respect to the landscape, its evolution, functions, and change	To strengthen territorial and geographical information concerning Andalusian culture, through education, appreciation, and enjoyment of the area and its landscapes
		To enhance the image of Andalusia through a better knowledge of its diversity and territorial landscape and its cohesive elements

Table 6.2 (continued)

		Observatori del Paisatge The Landscape Observatory	Centro de Estudios Paisaje y Territorio Centre for Landscape and Territory Studies
	Education and training promotion	Stimulating scientific and academic collaboration in matters of landscape, and the interchange of work and experiences among specialists and experts from universities and other academic and cultural institutions	To cooperate with other scientific institutions in Andalusia, especially with its universities, research centres and institutes, but primarily with the university departments concerned with knowledge of the territory and landscapes
		Preparation of seminars, courses, exhibitions and conferences, as well as publications and specific programs of information and training on landscape policies	To assist in the transmission of knowledge and education regarding the geographical landscape at all levels of education, especially within the university, giving special attention to postgraduate courses
International Cooperation	Following up European initiatives on matters related to landscape	To participate in scientific research networks at national and international levels with similar objectives	

Table 6.2 (continued)

		Observatori del Paisatge The Landscape Observatory	Centro de Estudios Paisaje y Territorio Centre for Landscape and Territory Studies
			In the terms and within the limits set out in the Convention and the rule of internal functioning, this Centre aims also to: (a) Advise and make available to social and economic agents, as well as government agencies, the knowledge acquired about the territory and the Andalusian countryside; (b) Promote and participate in international cooperation initiatives that relate to knowledge, protection and management of landscapes, giving priority to those related to developing countries, mainly from Latin America and the Maghreb
	Other	Creation of a documentation centre to the Catalan public Dissemination of landscape studies and reports, and establishment of working methodologies	
Web		http://www.catpaisatge.net/	http://www.paisajeyterritorio.es/

it shares many of the goals of the *Observatori del Paisatge*, it very much forms part of the Andalusian Research, Development and Innovation Plan. For this reason, its primary goals are those of the scientific study of the landscape, and matters relating to the raising of public awareness and training. Indeed, education and training are the two main pillars on which the Andalusian Centre rests, clearly reflecting the fact that it is jointly managed by the Andalusian universities, which have implemented a Master's Degree in Landscape Protection and Management. In Catalonia, by contrast, the Observatory's contribution to training has focused on the presentation of innovative online educational materials – 'City, Territory and Landscape' – for students enrolled in compulsory secondary education. The project has involved the creation of an educational website on landscape studies and can be accessed via the Observatory's website.

However, the most important activity of the Catalan Observatory has been creating its Landscape Catalogue – a new instrument for the introduction of objectives related to the landscape into town and country planning and into sector policies, and by so doing, adopting the principles and strategies of action established in the ELC.

The Catalan Act on Landscape Protection, Management and Planning defines the catalogues as 'documents of a descriptive and forward-looking character which define the types of Catalan landscapes, identifying their values and state of preservation and proposing the quality objectives which they must meet' (Ley, 2005a: Preamble III, §2). The Observatory is creating seven catalogues to coincide with the seven administrative Catalan regions and has prepared a Landscape Catalogue Prototype, which establishes the basic conceptual, methodological, and procedural framework for the seven catalogues to ensure they are prepared in a coherent and coordinated fashion. The methodology adopted should prove to be an important tool of reference for future work and is set to be followed by other regional governments. In March 2009, the Canary Islands government announced the establishment of its own Observatory, along the lines of those established in Catalonia.

6.3.3 Landscape Protection Foundations: An Example of Private Initiative

Although not specifically provided for in Spanish legislation, there exist various forms of 'green investment' channeled through a number of bodies and foundations involved in the management of natural resources and the landscape. These instruments serve to mobilize collective resources for the financing of the landscape. A good example is provided by the work being undertaken in the field of landscape protection and in the raising of public awareness by the Territory and Landscape Foundation (*Fundació Territori i Paisatge*) in Catalonia. The Foundation forms part of the Social Action sponsored by the *Caixa de Catalunya* savings bank, and its contribution to landscape protection in Spain is truly pioneering. Since 2002, it has dedicated more than 2 million Euros to the delivery of 232 projects in the fields of conservation, awareness-raising, and environmental education. These projects include visits to the countryside, conservation camps, and guided walks. Originally, this foundation operated solely within the territory of Catalonia but it has begun to organize activities in other regions of Spain (primarily Valencia).

The three main lines of action of the *Fundació* are:

- To acquire parts of the territory that represent well-conserved natural environments and subsequently manage them with the co-operation of conservationist institutions, other non-government organizations, and public bodies
- To co-operate in nature conservation projects being carried out by institutions nationwide
- To carry out an ongoing educational task, among children and youths as well as adults, in order to create environmental values in society.

One of the initiatives that is also now well established, thanks to the support of the foundation, is the *custodia del territorio*, a term that has been adapted from US and Canadian models of land stewardship. The initiative involves a voluntary agreement between a landowner and a stewardship body which helps the owner maintain and protect his land using a variety of mechanisms. Termed social property, i.e. lands under private ownership managed as if they were public, its initial points of reference were the work of the National Trust in the UK, the *Natuurmonumenten* in the Netherlands, and the *Conservattoire du Littoral* in France, bodies with which it maintains very close links through the Eurosite network of entities managing sites for nature throughout Europe.

6.3.4 Public Participation in Landscape Policy in Spain

Adopting the concepts that underpin the ELC – the protection, management, and planning of the landscape – is not solely a subject for experts, but rather a process in which all citizens can play a fundamental role. In landscape policy, it is not only the political and institutional bodies that should be taking unilateral decisions, but the citizens should also intervene and offer their perspectives, individually or collectively (organized in citizens' groups).

The concept of public participation implies taking into consideration the social perception of the landscape and the aspirations of the community in decisions regarding the protection, management, and planning of the landscape (Cortina Ramos, 2009b). The defence of one's own cultural traits, manifested as elements or characteristics of the landscape, can boost social participation in common affairs. However, to achieve a model of participative democracy requires that there be an institutional and legal base containing instruments that can give expression to and guarantee this participation through sets of norms or regulations (Romero, 2008).

The number of participative processes involved in the design of Spain's regional polices and landscape management is increasing, as gradually administrative units dedicated to promoting participation are being set up and more human resources and specialists are dedicated to this work. In recent years, in some regions of Spain, there has been a mushrooming of citizen movements defending the landscape heritage. However, to date there have only been timid attempts at incorporating public consultations and participation in landscape management proposals and study tasks.

One manifestation of public participation in landscape management is the development of Landscape Quality Objectives, which are of the utmost importance since they fix the targets that society sets itself in terms of improving its landscapes. Landscape Quality Objectives can become an essential point of reference for regional and sectorial policies, other bodies and society in general (Cortina Ramos, 2009b).

Spain's autonomous communities are beginning to work towards mobilizing public participation, but considerable confusion still reigns regarding the various means of participation. These might include public consultations (questionnaires), specific actions of public participation, or the whole *process* of public participation, which implies citizen involvement throughout a complete cycle from the identification phase through to the eventual planning, action, and impact. Among the autonomous communities, Catalonia once again stands out, since here thanks to its landscape catalogues effective processes of public participation are being instigated. The citizens are present from the identification of the landscapes through to their assessment, and they play an active role through 'Landscape Maps', even to the point of assuming certain commitments.

In other communities the first steps are being taken: the Landscape Protection Act in Valencia (Ley, 2004a) makes it compulsory to draw up public participation plans, which have to be designed in parallel with landscape planning tools and instruments (Muñoz Criado, 2008). In Mallorca in the Balearic Islands, there are plans to inaugurate a Landscape Observatory, which seeks to be a true representative of the citizens. The Community of Madrid and the Region of Murcia have undertaken surveys as a means of involving the public, but in the rest of the autonomous communities no mention has yet been made of such processes (OSE, 2009).

6.4 ELC Developments in Spain: An Evaluation

Recent socio-economic changes have left their indelible mark on the Spanish landscapes. The expansion of the artificial surface area, due above all to urban sprawl, and the enforced, often aggressive, changes to landscapes of great value, mean that the ELC has been introduced in a context of ever-growing concern about the impact of unbridled growth.

Despite this concern, Spanish legislation at the national level has only had recourse to instruments that can regulate changes in land use, with the exception of special protection areas, such as the coastline and areas of natural interest. A significant factor from the legal perspective has been the fragmentation of powers in which the central government is limited in its ability to influence land-use policy. In Spain, it is the regional and local governments that are instrumental in shaping the territory.

The starting-point for the work of each autonomous community is very different, in part due to the transfer of responsibility for landscape planning from the national level to the regional level in 1978. This has given rise, on the one hand, to adaptation of landscape policies and measures to the regional realities, but, on the other hand, in the absence of a common framework of action, it has produced uneven situations that hinder the fulfilment of the objectives set by the Council of Europe (OSE, 2009).

Although the implementation of the measures outlined in the ELC is going ahead, the rhythm of change and the adaptation of legal regulations depend on politically inspired initiatives. The main challenge faced by the ELC in Spain today, therefore, is how to unite these political forces. In a context of marked political fragmentation, together with the sectorializaton of responsibility for the landscape, there is a danger that the landscape is insufficiently protected. Integration necessitates the prior adoption of a common legal concept of the landscape (Cortina Ramos, 2009a).

Although the landscape is mentioned in many legal instruments, its presence so far has been marginal, understood in a much less integrative way than it is in the ELC (Cortina Ramos, 2009a). In Spain's legal system the landscape continues to be an 'indeterminate legal concept'. The adoption of the ELC in state or regional legislation implies a broader, more integrative understanding of the landscape. The regional laws passed to date in Catalonia, Valencia and Galicia and those currently being drafted in other communities corroborate this, since they specifically regulate the landscape and are giving shape to new powers over landscapes.

Spain ratified the European Landscape Convention more than 7 years after it had been officially signed and the central government has little to show so far in terms of concrete results derived from the commitment made in 2000. The autonomous communities, therefore, have found it necessary to take the lead and they are now striving to adapt their actions to the operational framework established by the European Landscape Convention. Today, virtually all the communities are introducing actions that have a landscape focus of one sort or another (Ortega, 2007). The landscape has a growing presence in several planning instruments, although the number of comprehensive plans or laws is still limited. Some communities have included or strengthened explicit mentions to the landscape in their recently reformed statutes of autonomy. Three of them (Catalonia, Valencia and Galicia) have passed specific landscape laws, assuming the definition and the commitments of the ELC, and there exists the will to adopt similar legislation in other autonomous communities. In general, references to the landscape have been strengthened in recently adopted urban and regional planning laws and, gradually, landscape diagnostics and commitments are beginning to be incorporated in the instruments that define spatial planning at a range of scales (OSE, 2009).

The fact of associating the whole of the territory with the landscape means that general landscape policies have to be drawn up, while recognizing specific values. The only community to date that has established a specific government landscape department is that of Valencia, while the community that has taken most initiatives in developing a landscape policy is Catalonia, in this case closely linked to its regional planning policy. The Landscape Observatory of Catalonia is an extremely active body that has become a model for the other communities to follow, and the Catalan

Landscape Act – the first to be passed by a regional government – is seen as the main source of inspiration for the other autonomous communities.

To achieve good landscape planning, the landscape question must be included within regional and urban planning tools. The spatial implementation and the technical experience acquired so far in the use of these planning instruments should help in obtaining the necessary attention for the landscape. Likewise, the compulsory norms contained within the regional planning can provide legal support for the Landscape Quality Objectives (Zoido Naranjo, 2004). It would be useful to incorporate the Landscape Quality Objectives into the instruments of regional planning, since, from the point of view of sustainable development, there are obvious points of overlap between the landscape and the region. If regional planning can take landscapes into consideration, it might well find in them a point of reference for establishing new environmental, economic, and social goals.

Acknowledgments This paper has been prepared as part of the research project entitled *Nuevo turismo y desarrollo territorial sostenible*, funded by the *Dirección General de Investigación y gestión del Plan de I+D+i* (Research Project: CSO2008-03315), and by an *Ajut de Suport a la Recerca dels Grups Consolidats de la Generalitat de Catalunya* (2009SGR-253). Berezi Elorrieta has worked with the support of a scholarship provided by the Basque Government.

References

Caravaca C (2004) El paisaje europeo: Un patrimonio común. Ambienta: Revista del Ministerio de Medio Ambiente 35:50–56

Cortina Ramos A (2009a) La regulación jurídica del paisaje. In: Busquets J, Cortina A (eds) Gestión del paisaje: manual de protección, gestión y ordenación del paisaje. Ariel, Barcelona, pp 317–346

Cortina Ramos A (2009b) La participación, mediación y concertación en paisaje. In: Busquets J, Cortina A (eds) Gestión del paisaje: manual de protección, gestión y ordenación del paisaje. Ariel, Barcelona, pp 347–378

Council of Europe (2000) European Landscape Convention, Florence, 20.X.2000. ETS No. 176. Council of Europe, Strasbourg. http://www.coe.int/t/e/Cultural_Co-operation/Environment/Landscape/. Accessed 1 Nov 2009

Council of Europe (2008) Recommendation CM/Rec (2008)3 of the Committee of Ministers on the guidelines for the implementation of the European Landscape Convention. Council of Europe, Strasbourg. https://wcd.coe.int/ViewDoc.jsp?id=1246005. Accessed 1 Nov 2009

Decreto (2006a) Decreto 120/2006, de 11 de agosto, por el que se aprueba el Reglamento de Paisaje de la Comunitat Valenciana. http://www.cma.gva.es/comunes_asp/documentos/legislacion/cas/651-3849-Decreto%20120-2006.pdf. Accessed 28 May 2010

Decreto (2006b) Decreto 343/2006, de 19 de septiembre, por el que se desarrolla la Ley 8/2005 de protección, gestión y ordenación del paisaje de Cataluña. http://www.gencat.cat/eadop/imatges/4723/06255087.pdf. Accessed 28 May 2010

Decreto Legislativo (2000) Decreto Legislativo 1/2000, de 8 de mayo, por el que se aprueba el Texto Refundido de las Leyes de Ordenación del Territorio de Canarias y de Espacios Naturales de Canarias. http://www.gobiernodecanarias.org/juriscan/ficha.jsp?id=19368. Accessed 28 May 2010

Directrices de Ordenación General y del Turismo de Canarias (2003) http://www.gobiernodecanarias.org/boc/2003/073/boc-2003-073-002.pdf. Accessed 9 Jun 2010

Directrices de Ordenación Territorial del País Vasco (1997) Gobierno Vasco – Departmento de Ordenación del Territario, Vivienda y Medio Ambiente, Vitoria – Gasteiz

DOT (2004) El Modelo Territorial de las DOT: La Ordenación del Medio Físico. Departamento de Medio Ambiente, Planificación Territorial, Agricultura y Pesca, Gobierno Vasco. http://www.ingurumena.ejgv.euskadi.net/r49-565/es/contenidos/informacion/dots/es_1165/punto8_c.html. Accessed 9 Jun 2010

Estatut d'Autonomia de Catalunya (2006) http://www.gencat.cat/generalitat/cat/estatut/. Accessed 28 May 2010

Esteban Noguera J (2009) El paisaje en la planificación territorial y urbanística. In: Busquets J, Cortina A (eds) Gestión del paisaje: manual de protección, gestión y ordenación del paisaje. Ariel, Barcelona, pp 379–396

European Commission (1985) Environmental assessment: Council Directive of 27 June 1985 on the assessment of the effects of certain public and private projects on the environment, 85/337/EEC. Off J L 175:40–48, 05/07/1985. http://ec.europa.eu/environment/eia/full-legal-text/85337.htm. Accessed 3 May 2010

European Commission (1999) European spatial development perspective. Office for Official Publications of the European Communities, Luxembourg

Ley (1987) Ley 12/1987, de 19 de junio, de declaración de Espacios Naturales de Canarias. http://www.gobiernodecanarias.org/juriscan/ficha.jsp?id=2055. Accessed 28 May 2010

Ley (1989) Ley 4/1989, de 27 de marzo, de Conservación de los Espacios Naturales y de la Flora y Fauna Silvestre de Andalucía. http://www.juntadeandalucia.es/medioambiente/web/aplicaciones/Normativa/ficheros/l4_89.rtf. Accessed 28 May 2010

Ley (1994) Ley 1/1994, de 11 de enero, de ordenación del territorio de la Comunidad Autónoma de Andalucía. http://www.juntadeandalucia.es/viviendayordenaciondelterritorio/www/servlet/descargacopt?up=7109. Accessed 28 May 2010

Ley (1998) Ley 3/1998, de 27 de febrero, general de protección del medio ambiente del País Vasco. http://www.euskadi.net/bopv2/datos/1998/03/9801344a.pdf. Accessed 28 May 2010

Ley (2001) Ley 9/2001, de 21 de agosto, de Conservación de la Naturaleza de Galicia. http://www.boe.es/boe/dias/2001/09/25/pdfs/A35529-35542.pdf. Accessed 28 May 2010

Ley (2004a) Ley 4/2004, de 30 de junio, de Ordenación del Territorio y Protección del Paisaje de la Comunidad Valenciana. http://www.boe.es/boe/dias/2004/07/20/pdfs/A26294-26316.pdf. Accessed 28 May 2010

Ley (2004b) Ley 10/2004, de 9 de diciembre, del Suelo No Urbanizable de la Comunidad Valenciana. http://www.boe.es/boe/dias/2005/01/19/pdfs/A02036-02049.pdf. Accessed 28 May 2010

Ley (2005a) Ley 8/2005, de 8 de junio, de protección, gestión y ordenación del paisaje de Cataluña. http://www.catpaisatge.net/fitxers/ley_paisaje.pdf. Accessed 28 May 2010

Ley (2005b) Ley 16/2005, de 30 de diciembre, Urbanística Valenciana. http://www.boe.es/boe/dias/2006/02/21/pdfs/A06957-07027.pdf. Accessed 28 May 2010

Ley (2008) Ley 7/2008, de 7 de julio, de protección del paisaje de Galicia. http://www.boe.es/boe/dias/2008/08/20/pdfs/A34968-34973.pdf. Accessed 28 May 2010

Mata Olmo R (2008) El paisaje, patrimonio y recurso para el desarrollo territorial sostenible: conocimiento y acción pública. Árbor: Ciencia, pensamiento y cultura 729:155–172

Muñoz Criado A (2008) La política de paisaje de la Comunitat Valenciana. Cuadernos Geográficos de la Universidad de Granada 43:99–121

Nogué J (2006) Per una consciència de paisatge: L'aportació de la Llei de protecció, gestió i ordenación del paisatge. Activitat Parlamentària 8–9:32–49

Nogué J (2007) Observatorio del paisaje de Cataluña: La emergencia de territorios sin discurso y de paisajes sin imaginario. Ambienta: Revista del Ministerio de Medio Ambiente 63:27–34

OSE (2006) Cambios de ocupación del suelo en España: Implicaciones para la sostenibilidad. Observatorio de la Sostenibilidad en España, Madrid

OSE (2009) Patrimonio natural, cultural y paisajístico: Claves para la sostenibilidad territorial. Observatorio de la Sostenibilidad en España, Madrid

Ortega M (2007) El Convenio Europeo del Paisaje: Claves para un compromiso. Ambienta: Revista del Ministerio de Medio Ambiente 63:18–26

Pla Territorial General de Catalunya (1995) http://www10.gencat.cat/ptop/AppJava/cat/plans/general/ptgeneral.jsp. Accessed 28 May 2010

Plan de Ordenación del Territorio de Andalucía (2006) http://www.andalucia.ccoo.es/comunes/temp/recursos/2/64939.pdf. Accessed 28 May 2010

Regional Ministry of Public Works and Transport & Regional Environment Ministry (2005) Atlas of Andalusia volume II: ecological and territorial mapping. Government of Andalusia, Sevilla

Resolució (2004) Resolució PTO/3386/2004, del 7 de desembre, per la qual es dóna publicitat a l'Acord del Govern del 30 de novembre de 2004, pel qual es constitueix el Consorci de l'Observatori del Paisatge i se n'aproven els Estatuts. http://www.catpaisatge.net/fitxers/resolucio.pdf. Accessed 28 May 2010

Romero EM (2008) El papel de la sociedad civil organizada en los procesos de desarrollo sostenible. In: Simancas M, de Souza A (eds) Sociedad civil organizada y desarrollo sostenible. Consejería de Presidencia, Justicia y Seguridad del Gobierno de Canarias, Las Palmas de Gran Canaria, pp 57–88

Zoido Naranjo F (2001) La Convención Europea del Paisaje y su aplicación en España. Ciudad y territorio: Estudios territoriales 128:275–282

Zoido Naranjo F (2004) El paisaje: patrimonio público y recurso para la mejora de la democracia. PH: Boletín del Instituto Andaluz del Patrimonio Histórico 50:66–73

Zoido Naranjo F (2006) Principales retos de adaptación de la Convención Europea del Paisaje a las políticas de ordenación del territorio en Europa. In: Mata R, Tarroja A (eds) El paisaje y la gestión del territorio, vol 1. Diputación de Barcelona, Barcelona, pp 359–374

Chapter 7
In Search of the Greek Landscape: A Cultural Geography

Theano S. Terkenli

Abstract Although Greece was among the first European countries that signed the European Landscape Convention, it has only recently ratified it, while the landscape, generally speaking, has been absent from most expressions of everyday private and public life in Modern Greece. Moreover, irreparable

T.S. Terkenli (✉)
Department of Geography, University Hill, University of the Aegean, Lesvos 81100, Greece
e-mail: t.terkenli@aegean.gr

M. Jones, M. Stenseke (eds.), *The European Landscape Convention*,
Landscape Series 13, DOI 10.1007/978-90-481-9932-7_7,

destruction of the Greek landscape, dating back to prehistoric times, has recently been exacerbated through widespread neglect, misuse, or even outright destruction, much accelerated since Greece's era of rapid urbanization in the 1950s and 1960s. This chapter begins with a brief illustration of trends and facts that point to the problematic relationship of Greek society with its landscapes. It traces the roots of this relationship in the cultural make-up of the Modern Greek nation-state and in a series of historical particularities and social-institutional deficiencies, much amplified during the post-war period. The objective of this chapter is to attempt to understand and explain this shortcoming by exploring the lack of a well-developed landscape conscience in Modern Greece.

Keywords Landscape · Culture · Landscape conscience · Landscape history · Landscape destruction · Greece

7.1 Introduction: The Greek Landscape in Light of the European Landscape Convention and the Greek Institutional Context

According to the European Landscape Convention (ELC), landscape must become a mainstream political concern, since it plays an important role in the well-being of Europeans who are no longer prepared to tolerate the alteration of their surroundings by technical and economic developments in which they have had no say. Indeed, although national landscape initiatives in many European countries are relatively new and not yet fully implemented (Wascher, 2001), Greece is in the unfortunate position of being far behind most other European countries in landscape protection, as in all landscape matters. If the European landscape is in crisis, the Greek landscape is – to put it mildly – in an even deeper, perhaps irreversible, crisis, in stark contrast to most other southern European countries of the Mediterranean (Grenon and Batisse, 1989; Pettifer, 1993; King et al., 1997; Höchtl et al., 2007; Vogiatzakis et al., 2008).

Greece signed the European Landscape Convention in 2000, but only recently ratified it (16 February 2010). If the institutional context may be outlined in a few words, the country does not to date have a Landscape Department or Directorate at the ministry level, nor landscape institutions at the regional and local levels. The landscape is absent from most expressions of everyday private and public life in Modern Greece, whereas, in the European context, it has repeatedly been attributed the properties of 'an essential component of a community's well-being, and of visitors' enjoyment' (Pedroli et al., 2007: 11).

Institutionally, the Greek landscape's existence is legally acknowledged properly only in the context of environmental legislation, where it is defined in the Act for the Protection of the Environment as 'any dynamic entity of biotic and abiotic environmental factors and elements that either separately or interactively compose a visual experience in a given space' (FEK, 1986: Article 2). The extent of its legal existence lies in its appearance in various environmental laws, master plans, and regulatory statutes concerning the protection of archeological spaces, and in legislation on traditional settlements, aesthetic forests, and national parks. It is implicitly or explicitly dealt with in environmental legislation as 'areas of high biological, ecological, aesthetic or geomorphologic value' (FEK, 1986: Article 1). Two categories of protected natural landscapes have so far been established in Greece: 'aesthetic forests' and 'landscapes of natural beauty' – but their existence plays a minimal role in for instance forestry planning. Besides international organizations active in Greece, such as the European Union (EU), International Council on Monuments and Sites (ICOMOS), World Wildlife Fund (WWF), etc., the most significant stakeholders in landscape policy-making and management have been the Archeological Service of Greece, the Ministry of Culture, the Ministry of Environment, Planning and Public Works, non-governmental organizations (NGOs), local government, private actors, and some civic societies. A major handicap is the precedence given to priorities put forth by the State Archaeological Service in legal matters and decision-making concerning landscape planning, policy, management, and land-use determination for industrial, agricultural, or other general development purposes.

Generally speaking, the current state of affairs as regards the Greek landscape may be described as follows: systematic physical planning interventions have been restricted to metropolitan and urbanized areas and have predominantly been a longstanding tradition of the design sciences. Mobilization in matters pertaining to the agricultural landscape in Greece has only been very recently instigated through European Union legislation and subsidized interventions (through the Common Agricultural Policy or CAP) that enforce rural landscape protection and preservation (Louloudis et al., 2005). The lack of institutional support, in terms of landscape planning, policy, and management, is evident in the absence until very recently (October 2009) of a separate Ministry for the Environment. Instead, all environmental matters have so far been dealt with by the Ministry of the Environment, Regional Planning and Public Works, with a long history of prioritizing the built environment, urban growth, residential development, and public works. With pressure mounting from the EU and the Organisation for Economic Co-operation and Development (OECD) for initiatives for landscape research, planning, and policy, Greece is currently finding itself in a position of having to struggle to meet its own landscape problems and challenges and to develop its own landscape agenda for the future.

With regard to Greek landscape education and science (Terkenli, 2004), landscape education is still lacking at all levels of the educational system despite serious but rather sporadic and fragmented efforts in tertiary education institutions around the country. Only very recently have there been signs of consolidation in technical and graduate studies programmes, such as in the Graduate Programme in Landscape Architecture offered by the Aristoteleian University of Athens. Processes of establishing landscape science, research, and practice have been only slowly gaining ground in Greece in very recent years. There is still for instance a total lack of Greek landscape maps, with the exception of a recently completed atlas of cultural landscapes of Greece, by the Department of Geography at the Harokopeion University of Athens (Greekscapes, 2009).

In the 1990s, Greek landscape science underwent a qualitative shift. Previously the engagement of the design sciences (architecture, landscape architecture, and urban and regional planning) with practical landscape issues, as they developed out of related design and planning initiatives and spatial interventions, was fragmentary, peripheral, and haphazard. More recently, there has developed a more concerted, focused, and systematic landscape approach by several disciplines and practitioners (including landscape historians, rural sociologists, geographers, environmentalists, etc). However, this shift has been characterized by its very limited extent and impact on actual landscape problems and issues in Greece. It is also suffering from disciplinary limits and from the lack of communication and cooperation between academics, practitioners, and administrators, as well as from lack of effective application in landscape policy.

In most circumstances of local or public life, as regards cultural trends, economic activities, political initiatives, social issues, urban and regional development, and planning and management, the Greek landscape seems almost to be a nonentity – and its appearance is correspondingly nondescript (Figs. 7.1 and 7.2). It normally does not constitute an issue, concern, or matter of interest for most laypeople. As a rule, local interests, input, and decision-making concerning the landscape are normally ill-informed, marginalized, or – more commonly – non-existent (Terkenli, 2004; Manolidis, 2008). Under these conditions, landscape matters tend to remain overwhelmingly dependent on public or private economic or political interests. Such facts and tendencies unfailingly characterize a people's and a state's priorities vis-à-vis its landscape and these priorities are, in turn, engraved in its landscapes. Simply put, they mirror the society that created them; they become its representation. This points to a very problematic relationship of Greeks with their landscape, a relationship that this chapter explores.

Fig. 7.1 The main street of a mid-sized contemporary Greek city, Mitiline, Lesvos, 2007 (Photo: Theano S. Terkenli)

Fig. 7.2 The hinterland of Hersonissos, Crete, 1999 (Photo: Theano S. Terkenli)

The chapter proceeds from a brief illustration of indicative trends and facts that point to this problematic relationship of Greek society with its landscapes to their analysis and interpretation, with a bearing on Greece's position with regard to the European Landscape Convention. It traces the roots of this relationship in the cultural make-up of the Modern Greek nation-state and in a series of historical particularities and social-institutional deficiencies, much amplified in the post-war period. The objective is to attempt to understand and explain this shortcoming by exploring the lack of a well-developed landscape conscience in Modern Greece.

7.2 Looking for the Causes

Greece's problematic relationship with its landscape can be traced to lack of a defined and well-developed landscape conscience in the country as compared to other modern (European or not) nation-states. If 'conscience' is defined as the mixture of perceptions, thoughts and emotions, it presupposes the existence of an external world (Sutherland, 1989). Landscape conscience refers then to the distinctive bonds (conscious or subconscious) that characterize a person's or a people's relationships with their landscapes. Undoubtedly, the causes of lacking landscape conscience for Greece are many; some that seem to have played a crucial role will be examined in the following.

The legal, historical, aesthetic, and socio-cultural trajectory of the relationship of Greece with its landscape will be traced through the past 150 years in search of the *urban* origins of a landscape conscience. In the process, some elements will be constructed of an unfulfilled cultural geography of the Greek landscape that has its origins in a multitude of factors, such as: the late industrialization of the country;

the prevalence – according to the historian William McNeill (1978) – of a 'marketplace principle' among its populace; the role of Greek Orthodox ecclesiastical art in landscape representations; and the lack of a sense of commons concerning environmental resources. These are only a few, but critical, pieces of the puzzle. Finally, a brief account will be given of the changes now unfolding in the reconfiguration of the country's urban and agrarian identities, as new notions of urbanity and rurality emerge through the growth of foreign and domestic tourism.

7.2.1 Historical Roots of the Current Situation

Greece, under Ottoman occupation and cultural stagnation from the mid-15th to the early or mid-19th century, never went through any of the stages of landscape formation and landscape conscience formation that modern European cultural landscapes had gone through by the 17th century – notions that accompanied the development of Western European landscapes up to our times (Cosgrove, 1998; Olwig, 2001). Rather, it adopted from the West aspects of modernity in certain realms of life a posteriori, by implanting and overlaying them on to pre-existing cultural particularities and local ways of life. Moreover, upon becoming 'urban', Greeks lost the old connection with the land, nature, and the landscape, which had traditionally been handed down from one generation to another. The few already existing urbanites and the children of the first and subsequent generations of rural migrants into the big cities never developed a sense of landscape in the first place.

Since antiquity, compared to other European people, Greeks have tended to be predominantly urbanites. The ancient Greek world constituted a web of city-states, where citizens were considered only those free individuals in possession of landed property. Cosgrove (2001: 25) writes:

> In the Greek polis citizenship derived initially from ownership of cultivated land, and ownership of immobile property—'real' estate—remained for millennia the foundation of political franchise. ... A hierarchical order that mapped space, society, the idealized body, and its faculties to a scale of humanity and opposed human 'culture' to nature has been continuously reworked in Western thought and practice. The city was regarded as the spatial expression and locus of a fully developed humanity.

Greek thought and culture continued to thrive throughout the Byzantine era, and on through the Ottoman occupation, mainly in the urban centres of southeastern Europe and the Balkans, where most of the Greek population tended to cluster. Perhaps one final indication of this trend is the overwhelming primacy of present-day Athens in the context of the Greek urban system, representing the outcome of enormous centripetal forces on the post-war Greek rural population.

Although any sort of spatial conscience generally attributed to a cultural system tends to find its roots in the history of a modern nation-state, caution must be exercised in generalizing and totalizing as regards whole cultures or social systems. Before the post-war era of rapid Greek urbanization, Greek people of rural, mountain, or island pre-industrialized communities tended to live under conditions tightly interwoven with their particular environments and landscapes. They used to relate to their landscapes through much more organic, multilayered and reciprocally intertwined cultural, environmental, economic, institutional, political, ideological, and legal bonds. These bonds tended to tie people symbiotically to the land, which ensured their livelihood; they also protected and paid homage to the landscape and carved their cultural systems into it and through it. The local housewife used to sweep the street in front of her yard, while her husband would regularly whitewash the village square. The community would assume the clearing or planting of the forest land around the village. Generations of subsequent communities would name the hills, the ravines, the springs, and the mountaintops, and attribute sacred or divine properties to parts of the landscape in the name of protector saints, nymphs and elves, and old legends. As was the case throughout pre-industrial Mediterranean Europe, they would build 'traditionally' in harmony with the landscape and its natural inhabitants, the trees, the beaches, and the watersheds (Manolidis, 2008).

What happened since then? In the following an attempt is made to trace and elucidate this evolution of the relationship of Greeks with their landscapes, starting with images and representations of the landscape in Greek culture and beyond.

7.2.2 Greek Landscape Depiction and Representation

An especially eloquent and revealing view of the Greek landscape in its symbolic and representational perspectives emerges through landscape painting at the time of the formation of the new modern nation-state of Greece, after its war of independence in the 19th century. If landscape is a 'way of seeing' closely connected to the development in Europe of modern urban, commercial life, then landscape representations in art are renditions into some form of image of the ideological construction of the newly-emergent European nation-states (Cosgrove, 1998). Analysis of 19th-century Greek landscape painting exposes the construction of the Greek landscape as a context of human life and experience in accordance with romantic ideals. At the basis of the emergent Greek cultural identity were two ideals: (a) connections to classical antiquity; and (b) Orientalism (Terkenli et al., 2001). Such ideals, originally introduced in Greek landscape painting by Western painters and Western views of the modern Greek state and identity, sought to reconcile the 'Other' with the 'Self' of Western culture in representations of the Greek land for the eyes and the psyche of the Western observer (Terkenli et al., 2001). They were deeply embedded in Western conceptions of the local landscape and were only gradually replaced by indigenous depictions and local landscape ideals – in both formal and naïve renditions of the Greek landscape – in the course of the 20th century.

The first theme, the connection of the Greek landscape with classical antiquity, demonstrates the alleged direct descent of modern Greeks from the Hellenes of the classical period, considered the progenitors of modern European civilization. Incontrovertible witnesses in the Greek landscape to this newly-formed national identity were, among others, the archaeological monuments scattered over and under Greek soil. It is with these that the cultural landscape has until recently been almost exclusively equated by the Greek state and intelligentsia (Doukellis, 1998). For the ordinary Greek subject, however, the Greek landscape was quite different (Stathatos, 1996: 20). In contrast to such glorious depictions, the actual landscape tended to be plainer, even drab, poor, ravaged by war and pillage, and of a less monumental scale. It was the ordinary landscape of a Mediterranean country coming out of four centuries of foreign occupation.

The second landscape theme is the pervasive theme of Orientalism, which has, since the inception of the Modern Greek state, infused its cultural identity as perceived from the West. The Orient (Near East), according to Edward Said (1978), is an idea that has a history and tradition of thought, imagery, and vocabulary that have presence in and for the West. Upon this is constructed the hegemonic relationship of the West with the Orient. More significantly, the 'Other' in the post-Reformation West, argues Vassilis Lambropoulos (1993), is always defined as an integral part of the dominant ideology. This was the case with the idea of 'Greekness', inherent in the definition of Western civilization, yet with the Greek remaining as the 'Other' for Westerners. The 'Other', however, always remains at a safe distance of 'difference' from the hegemonic culture, and this is where the theme of Orientalism comes especially handy in Western constructions of 'Greekness'. Difference can benefit from its intrinsic relation to sameness, especially powerful here in the articulation of the cultural hegemony of the West, which has existed in relation to and to the detriment of various 'Others' (Terkenli et al., 2001).

In the 20th century, Greek landscape painting gained only a partial and gradual emancipation from the influences of foreign schools through the development of various indigenous forms of expression (Kambouridis, 2009). The landscape ideal and form of representation most influential upon the Greek psyche and most characteristic of the Greek cultural realm remained the two-dimensional, apparently flat, but actually inverted, perspective of Greek Orthodox art (Fig. 7.3). The human figure tends to dominate in ecclesiastical iconography, rather than the landscape per se. This inverted perspective pulls the viewer into the painting, rendering the viewer the centre of the world in the work of art and thus exerting a great power of suggestion over him or her. Much inspired by Greek Orthodox ecclesiastical art, El Greco's manner of landscape depiction was perhaps the closest Greek art came to Western landscape depiction and articulation until the creation of the Modern Greek state and the importing of foreign painting and painters to the Greek landscape and all manner of spatial intervention. Nonetheless, Greek Orthodox art, surviving and flourishing amidst the deeply religious populace under Islamic domination, seems to have imprinted its highly influential worldview on the Greek mind and psyche, providing an ideal and way of relating to the world still pervasive in Greek life and art.

7.2.3 The Destruction of the Modern Greek Landscape

European landscapes, products of human-environment interrelations over time, have long faced a variety of forces of transformation, which were accelerated

Fig. 7.3 The baptism of Christ, Nea Moni of Chios, Greece, 2009 (Photo: Courtesy of Gareth Roberts)

through the Industrial Revolution. Currently, however, they are acquiring a series of new attributes under conditions of rapid unprecedented change on a global scale (Terkenli and d'Hauteserre, 2006; Vogiatzakis et al., 2008). Simply put, 'this diverse landscape is in a deep crisis', according to Pedroli et al. (2007: 11). In this context, concerted and integrative intervention in landscape protection, management, and planning becomes essential, in cooperation with the public and all involved stakeholders.

The irreparable destruction of the Greek landscape dates back to prehistoric times (about 1000 BC). Since then, the landscape has been plagued by much neglect, misuse, or even outright destruction, much accelerated since Greece's era of rapid urbanization in the 1950s and 1960s. The distinguished Greek architect Dimitris Pikionis, in his inaugural speech in 1963 for the founding of the First Exposition of the Committee for the Hellenic Landscape in Zappeion, Athens, cried out against the irreparable destruction of the Greek landscape, which was gradually disappearing as a reality while beginning to be imprinted as an image through the photographic lens. Semaioforidis, another accomplished Greek architect, talks about the immaterialization of the landscape of Attica between 1953 and 1963, in the name of the magic words of the times, namely 'urban planning' and, a little later, 'regional planning' (Semaioforidis, 2005: 121–122). These admonitions foretold trends that were to overtake most of continental and insular Greece in the latter part of the 20th and first part of the 21st century, and are still dominating the Greek landscape (for a more detailed history of the evolution of Greek landscapes and the ways these historical impacts have been 'inherited' by contemporary Greek landscapes, see Terkenli, 2002, 2004). Table 7.1 diagrammatically presents the main phases of the evolution

Table 7.1 The historical phases of Greek landscape evolution (Terkenli, 2004)

	Main landscape types		
Chronology	Island	Coastal	Inland
3rd millenium BC to 5th century BC	Urban Revolution, establishment of first cities, emergence of great cultural centres of antiquity, beginning of natural landscape degradation		
5th century BC to 4th century AD	Rise and fall of ancient Greek city-states, accelerated destruction of natural landscape, idealized Arcadian landscape of antiquity		
4th century AD to 14th/15th century AD	Christianity and middle ages, consolidation of previous landscape impacts and characteristics and slow, gradual introduction of a few new ones		
14th/15th century to 1820s–1830s	Occupation of Greece by Ottoman Empire		
	Defensive hilltop settlements, life of hardship	Highest impact of Ottoman occupation, stagnant societies	Idealized landscape of the mountains, freedom, 'good life'
1820s–1830s to 1950s–1960s	Foundations of the modern Greek state		
	Poverty, isolation, abandonment, desertion	Slow, steady rates of growth, new cultural identity and landscape	Gradual liberation of all Greece, cycles of growth and decline
1950s–1960s to today	Industrial revolution and urban development and growth		
	Tourism, growth, development, island landscape idealization	Unparalleled urban growth, landscape destruction	Desertion, isolation, poverty, loss of social dimension of rural landscapes

of Greek landscapes, distinguished in three categories – island, coastal, and inland landscapes – while highlighting characteristics and levels of landscape deterioration during each chronological period from prehistoric times to the present (Terkenli, 2004).

The wide variety of dangers and problems facing the contemporary Greek landscape is well documented (Terkenli, 2004; Louloudis et al., 2005; Hadjimichalis, 2008; Manolidis, 2008; Stathatos, 2008; Trova, 2008; Vlachos and Louloudis, 2008). These reports have not been effective in influencing public or state opinion and not led to a reaction to the issues at hand, due to long-entrenched cultures of corruption, nepotism, and unfavourable administrative structures. The dangers and problems include the following processes, inducing variable degrees of landscape impact: illegal construction; rampant land-use change and subdivision; lack of comprehensive and rational planning and law implementation; insufficient documentation and deficient restoration of historic landscapes; unchecked urban development; intensification of agriculture; landscape homogenization; loss or degradation of natural, aesthetic, and cultural landscape character (e.g. through soil erosion, collapse of old structures, interventions incongruous to local landscape identity, etc); desertion of mountain and remote rural landscapes through abandonment of traditional rural activities; unequal development or geographical exclusion and discrimination; lack of protection measures from illegal interventions; fires; and floods.

If we attempted to distinguish the two or three most detrimental forces affecting the contemporary Greek landscape, we would perhaps point to: unplanned recreational use of the countryside; widespread rural-urban migration and consequent abandonment of agriculture and livestock raising; and the proliferation of second-home construction throughout the country (Terkenli, 2004; Hadjimichalis, 2008; Manolidis, 2008; Stathatos, 2008; Vlachos and Louloudis, 2008). The exodus of rural populations from the Greek countryside robbed it of its guardians and stewards, with strongly negative impacts on its physical regimes and its cultural wealth. The burgeoning growth of secondary residences (legal and illegal) by an increasingly affluent middle class has been:

> . . .exacerbated in recent years by the country's adhesion to the European Union and the influx of North Europeans in search of holiday homes; the inevitable effect has been a rapid and continuing change in the population of certain areas, particularly but not exclusively along the coast. . .. In conjunction however with the Greek government's inflexible policy on real estate taxation, whereby a single transfer of land from a farmer to an incomer can, irrespective of use, bring about a hundred fold increase in the value of the entire area, the new colonizers sooner or later replace the original inhabitants of these supposedly privileged regions (Stathatos, 1996: 18).

Other apparently negative impacts on the landscape only tend to be recognized as such when they become catastrophic or lead to calamities, after which some form of balance is eventually restored in the physical landscape, as, for example, after the fire disasters and human deaths in the Peloponnese during the summer of 2007, and on the outskirts of Athens in August 2009.

7.2.4 Urban-Industrial Deficits and Socio-Cultural Constraints

Varying modern landscape spatialities and varying manifestations of landscape conscience have been identified in Europe at different times and places (Cosgrove, 1998; Bunce, 1994). One common factor that appears to play a significant role in the development of a landscape conscience in the modern European realm is the advent of the Industrial Revolution. 'It was precisely this urbanization, and the increasing distancing from nature to which were subjected the population of societies in the process of industrialization, which almost simultaneously created the need for contact with some substitute, however false' (Stathatos, 1996: 16). The resulting loss of place and landscape particularity was an inevitable outcome of social-structural adjustments instilled by industrial capitalism. Instead, the bourgeoisie reinvented the landscape concept, initially closely tied to the English picturesque landscape school (Stathatos, 1996). A series of new landscape spatialities ensued through the newly emergent contradistinction between the rural and the urban, and through the nostalgia of urbanites for the 'lost' countryside. Thus, the countryside ideal and the rediscovery of the rural landscape was a social construction of the times, best exemplified in the case of the UK, the first nation to experience these trends and the development of a deep landscape conscience six generations ago (Bunce, 1994).

In contrast, Greece never went through a fully fledged industrial revolution. Pettifer links this to the weakly developed environmental movements and environmental conscience: 'There has been no real industrial revolution in Greece and consequently no Romantic movement in literature to see nature threatened by man's activities' (Pettifer, 1993: 172). In lieu of an urban-industrialized socio-cultural system, the country retained its rural character until the post-war mass rural migratory movements into the large urban centres. Many vestiges of the rural ways of life imported into the Greek cities in the 1950s and 1960s still remain strong. Such, for example, is the persistence up to the present of a 'market-place principle' in Greek social life (McNeill, 1978). The historical centrality in Greek society of exchange and of market-place skills emerges as a crucial feature in the lives of all modern Greeks (McNeill, 1978); the market-place principle and material wealth seem to hold a place of uncontested primacy in Greek life. McNeill (1978: 12) develops this idea further:

> Skill in bargaining for the best possible prices, skill in deciding the exact moment at which to make a deal... these were the ways to wealth and success. These were also the skills that won the respect of others in the village, even if such respect might be a little grudging. For if one man was able to do even a little bit better than others in such negotiations, it meant that he had somehow outsmarted everyone else. Privacy and deception play a large role in the successful conduct of such negotiations. Deception must be practiced against one's fellows, who, if they crowd round at the critical moment, might spoil the advantageous deal by trying to get in on it too... Deception must also extend to the person with whom one is dealing... The effort to deceive is of course reciprocal... Suspicion of one's contractual partner therefore remains near the surface... The idea that a deal might be mutually beneficial is hard for a Greek to accept.

From ancient times to the present, Greek life has revolved around market negotiations, a tendency which not only reflects on all aspects of current everyday life and thinking, but has also greatly accommodated Greece's general eagerness to fit, so to speak, into the global economy and Western development models. Accordingly, consumption figures for modern amenities (i.e. household expenditure) have tended to exceed mean European figures (Eurostat Cultural Statistics, 1995, 2006). The uncontested dominance of the market-place principle in contemporary social life gives precedence to economic rather than cultural, environmental or aesthetic concerns regarding spatial construction, planning, and management of any sort – with grave repercussions on Greek nature and landscape.

Hence, in post-war Greece, modernization and development have been defined mostly in economic terms, often to the detriment of environmental, socio-cultural or civic values, as seen in the grave lack of green areas in Greek cities. Economic development, of a quasi-capitalist character, unfolded on the basis of a mainly agrarian society. Alongside many facets and factors of Greece's idiosynchratic economic development, major long-term cultural particularities, such as clientelism and patronage, have been responsible for an atrophic civil society (Demertzis, 1997: 110), with serious repercussions on community life. Legg and Roberts (1997: 72) expand on how, 'despite decades of social and economic change, the state still overwhelms civil society, and personal and family ties remain significant in most areas of life. The domination of civil society by the state is an overwhelming fact of Greek economic, social, and political life'. Consequently, in contemporary urban Greek society, environmental and landscape matters were relegated to the jurisdiction of the state and absolved of individual, personal responsibility. There rightly reigns a cynical and sceptical attitude regarding the role of the various governments in these matters, however. Concerning 'the public good', Greeks tend to think that if the government does anything it will be done badly – or that possibly it should not be done at all – but, 'applied to environmental matters it is very unhelpful as many of the environmental threats need countering by long-term policies that often demand the sacrifice of short-term private interests' (Pettifer, 1993: 173).

One outcome of this trait has been a long-standing lack of a sense of the landscape as a common good in the Modern Greek society. A common good is defined as the integrated set of material and non-material dimensions and features of the landscape at the disposal of a particular social group, where its use by one user diminishes the amount available to all others, but where the exclusion of additional users is difficult or impossible (Bromley, 1991). According to recent social-scientific thought, rationally optimal behaviour favours a cooperative, ethically active and vigilant strategy of generous mutuality (community) (Tuan, 1986; Ostrom, 1990). Generally speaking, one of the most resistant ramparts of the old ways of life has been the nuclear family (Eurostat Cultural Statistics, 1995, 2006). 'Unlike in many other European villages, in rural Greece, most of the time the work unit as well as the marketing and consumption units coincide with the boundaries of the nuclear family' (McNeill, 1978: 15). Since modern urban life also tends to make the nuclear family the primary unit of consumption and mutuality, these rural cultural patterns were readily transferred to Greek cities and implanted into urban ways of life. As elsewhere in the post-war world, however, the nuclear family has been challenged by newer forms of capitalism, ecumenical culture, and a surge in individualism (Karapostolis, 1983).

No matter how materialist the conditions of contemporary socio-cultural life are, however, the 'good life' would be impossible without reference to non-material (cultural) conditions (Tuan, 1986), such as provided by landscape, as a mirror of society and as a stage set for everyday life. In Greece, the disintegration of the traditional environmental conscience of formerly rural populations with regard to outdoor resources, including the landscape, has been replaced by rampant laisser-faire capitalism, land speculation, illegal construction, and short-term profit in most entrepreneurial activity domains. According to Stathatos (1996: 16), 'as far as the Greek perception of natural space is concerned, the problem is exacerbated by a peculiarly Greek form of parochialism, whereby allegiance is pledged to extremely small territorial subdivisions, down to the level of neighborhood or village'. Thus, landscape never constituted a collective good for most Greeks, and especially in the case of urban Greeks. There is no sense of the landscape as part of a common home – the sense of home tending to be narrower in larger cities than in small towns or villages (Terkenli, 1995).

7.3 Tourism: The Changing Scene

All of the above obstacles to the development of a landscape conscience among Greeks are slowly coming under scrutiny or transformation. Some of the most significant reasons for this recent trend may be traced in the following three broad contexts:

1. International pressure on Greece's environmental policy, e.g. from the EU, in order to conform to its agricultural policy (CAP) and to address climatic change and environmental impacts of this by adopting environment-friendly measures of resource use and protection
2. Recent catastrophes and irrevocable loss of large extents of priceless landscape by forest fires (summers 2007 and 2009), and a growing public awareness of landscape loss through uncontrolled growth, illegal construction, and 'development'
3. The need for 'nature', and nostalgia for 'Greece as it used to be' culturally, physically, aesthetically, ethically, historically, and symbolically, as opposed to life in the city; this need has mainly materialized through internal tourism, often in search of personal and collective identity, and ancestral roots.

The Greek landscape, generally speaking, had been taken for granted by the state and Greek society at large until the end of the 1970s. It started to be acknowledged at that time through growing awareness of interconnections emerging between agricultural modernization and change in the rural landscape, and through tourism. Before that, there existed a fairly well-articulated relationship of Greeks with their landscapes in organically developed and long-standing 'traditional' pre-industrial settlements throughout rural and small-town Greece. This close and well-structured coexistence between humans and their landscapes was disrupted with the post-war advent of rapid urbanization, with unchecked growth and development. Since then, perhaps the most significant factor in a slowly emerging dynamic of return to the Greek landscape has been internal domestic tourism. Tourism effectuated in the case of Greece, as it had previously done so in other parts of the world, a rediscovery of the Greek landscape.

On the basis of its visual and relational or experiential character, the landscape constitutes a crucial medium in the nexus of relationships that develop between tourist and visited location. These relationships are obviously highly complex, as well as place-, time- and culture-contingent; in Greece, they represent the most effective ways in which the public at large has been rediscovering the country's landscapes. All landscape aspects and elements – human and natural – are involved in tourism development (Williams, 1997; Lickorish and Jenkins, 2004; Vogiatzakis et al., 2008). At the basis of any ensuing discussion vis-à-vis the landscape, however, stands its environmental nature. The Aegean landscape, for instance, has been much romanticized in recent decades as an idyllic island paradise, isolated and free of the demands of modern life, blessed with perfect climate and characterized by its small-scale, intimate settings ideal for romantic adventures in the land of the 'Greek gods'. The 'four S's' (Sun, Sea, Sand, and Sex) collectively constituted a powerful pole of tourism attraction for the Aegean from its onset in the 1960s. Landscape elements, both natural (the sea, the beach, and sunshine) and human-made (such as the whitewashed cubic houses in real or imitation stone-paved streets), exemplify and reinforce such images of the Aegean and are preserved and highlighted in popular culture (e.g. motion pictures such as *Shirley Valentine* and *Summer Lovers*).

Among its various impacts on place and landscape, the tourism industry has been greatly responsible for the worldwide diffusion of specific landscape forms, functions, and symbolism (Towner, 1996). In place of a fully fledged industrial revolution, tourism has been the main source of the development of awareness of the countryside and the generation of a landscape conscience among Greeks. This was accomplished through the intentional seeking out of a contraposition to the urban industrial contemporary way of life. It was achieved through the escape from congested, suffocating Greek cities and the return to 'nature', to cultural and historical references, and to family or national 'roots'. This is manifested in the retreat to the village family home, the revisiting of ancestral lands, or the construction of a second, 'holiday' home. The country has been selling images of itself (Figs. 7.4 and 7.5) in which 'the sun always shines brightly, where the sea is always blue and placid, the houses – of a uniformly Cycladic style – are invariably freshly whitewashed, and all of whose inhabitants are permanently cheerful, welcoming and colourful' (Stathatos, 1996: 38). Foreign and external tourism activated among contemporary Greeks an increased awareness of the aesthetic, natural, and cultural richness, and the diversity and uniqueness of Greek landscapes. It also instilled the lifestyle of ease, leisure, and generalized consumption, from Coca-Cola to landscape. As a result of thriving Aegean tourism, post-war economic decline and

Fig. 7.4 Sounion: an imaginary landscape depiction from the campaign of the National Tourism Organization of Greece, 2006 (Photo: Courtesy of Yiannes Patellis)

population depletion are now in the process of being reversed in most parts of this region.

As a case in point, the landscape of the Aegean islands has been widely conceived as a cultural image of tourist consumption for its visitors, besides being viewed as a national symbol and as a cultural and family hearth – a historical construct in collective Greek imagination. It has been perceived as an essentially uninhabited landscape during most of the year, while, during holidays and especially in summer, it becomes 'vacationland', the playground of both Greek and international tourism (Tsartas, 1989; Terkenli, 2001). These perceived qualities of the Aegean landscape are mainly derived from its visual characteristics. For example, mainly for purposes of attracting tourism or preserving 'traditional' landscape identity, the facade of urban landscapes has largely been preserved, whereas all else considered 'superfluous' in modern life and tourism has been dispensed with. Visual Aegean landscape characteristics have also been expropriated and exploited for various 'development' purposes, often with negative impacts on their appearance, and undermining the very essence of the landscape that attracted development there in the first place.

Initially, it was islands with cultural heritage of archaeological, religious, or general historical interest that attracted most visitors – both foreign and local. During the 1970s and 1980s, however, these islands were transformed into the quintessential tourist havens of Greece through their establishment as conventional summer tourism destinations. The Greek islands in general constitute the stereotype of an

Fig. 7.5 View of Port Hesonissos, Crete, summer 1999 (Photo: Theano S. Terkenli)

island tourist paradise, with their 'perfect' physical environment (warm, sunny, and beautiful beaches), ancient history interwoven into long-standing 'traditional' ways of life, and hospitable, friendly locals inviting visitors to enjoy an easy way of life. Tourism has boosted the economy of the Aegean islands, changing their main income bases from agriculture to service activities, stemming population out-migration, and creating conditions for new construction and development in the form of tourism infrastructures – catering to the boom in organized charter air transportation systems (Williams, 1997; Minca, 1998; Lickorish and Jenkins, 2004) – or in the form of second-home development. Kizos et al. (2007: 341–342) describe the unequal development of contemporary tourism and its impacts on the Aegean Islands:

> Seasonally, approximately 3.5 million tourists visit, almost exclusively in summer; mostly by charter flights (67% in 2001); and this fact causes intense seasonal changes in transport frequency and environmental pressures. Spatially, most of the beds (250,000 in total) are found on a small number of islands.... In addition to tourists, holidaymakers in general are very important economically and in terms of land use, since the amount of new housing is one of the most intense problems confronting the landscape and the environment. The local economy has benefited greatly from building works, and the consequent rise in land prises, but this development is temporary, whereas the environmental and social impact is permanent.

7.4 Conclusions

Contemporary Greece seems to suffer from a lack of a sense of the significance of one's surroundings for the quality of life, exemplified in the case of landscape as the stage set of everyday life (Terkenli, 2004; Manolidis, 2008). As shown in this chapter, the historical roots of this deficiency lie in Greece's inability to develop its own landscape spatialities alongside Western European models of spatial organization and governance, and to experience a fully fledged industrial revolution – mainly due to the prolonged Ottoman occupation. This era bequeathed the country with a series of problematic socio-cultural trends that compounded the lack of its sense of landscape as a common good: clientelism, state patronage, atrophic civil society, individualism, parochialism, and mistrust of governmental institutions. Moreover, the lack of development of a lay landscape conscience may be seen to have been exacerbated by the fact that Greeks historically have always tended to be urbanites, operating on the basis of a very strong market-place principle and strong social competition ethic, compounded by the influence of the unique way of constructing and depicting the world by Greek Orthodoxy, as illustrated in ecclesiastical iconography. The basis for public participation in matters pertaining to the landscape, in accordance with the intentions of the ELC, is far from existent in Greece, where short-term private interests are routinely prioritized over long-term collective goals of sustainability and a sense of the landscape as a public, common good. Instead, the attitude of the vast majority of the country's citizens to the landscape is one of ignorance, neglect, apathy, disinterest, and distrust in those institutions and media that play a role in its planning, management, and general sustenance.

A country blessed with a high degree of landscape variability and diversity, a source of cultural inspiration since antiquity, presently finds itself under grave threat of loss, with social, cultural, economic, environmental, ethical, aesthetic, and spiritual consequences. Questions that need to be urgently addressed are: whether valued features of the Greek landscape can be protected and saved, before it is too late; whether and how a landscape conscience may be instilled and developed among laypeople and authorities likewise; whether concerted efforts towards landscape planning, development, and management may be instigated and implemented in the country at large; whether knowledge from the experiences of other countries may prove useful here; and whether Greece can transcend its own distorted tourist image exported to the world. Stathatos (1996: 38) argues that there is reason to believe that this image may be becoming innate, perhaps because 'that is what the ultimate loss of innocence consists of: the curse whereby, when one has lived a long time with falsehood, it becomes increasingly difficult to distinguish truth'.

This chapter has suggested that the absence of a well-developed landscape conscience among contemporary Greeks lies at the basis of the country's landscape problem. The task of redefining and developing lay landscape conscience is long and arduous, but for Greece it is imperative. It needs to rest on knowledge and education, active participation in decision-making and, most of all, immediate action in reconfiguring our landscape geographies – a task long overdue. Nonetheless, through reaction to and mobilization against environmental and human disaster, including climatic change, as well as falling rates of growth in tourism, some first signs of such a development are presently emerging. Fully acknowledging the need for serious efforts in this direction, however, still has a long way to go.

References

Bromley D (1991) Environment and economy: property rights and public policy. Blackwell, Cambridge, MA

Bunce M (1994) The countryside ideal: Anglo-American images of landscape. Routledge, London

Cosgrove D (1998) Cultural landscapes. In: Unwin T (ed) A European geography. Longman, London, pp 65–81

Cosgrove D (2001) Apollo's eye: a cartographic genealogy of the earth in the western imagination. The Johns Hopkins University Press, Baltimore, MD

Demertzis N (1997) Greece. In: Eatwell R (ed) European political cultures: conflict of convergence? Routledge, London, pp 107–121

Doukellis P (1998) Mentalities, ideologies and modern Greek landscape management. In: Unwin T (ed) A European geography. Longman, New York, NY, pp 78–79

Eurostat Cultural Statistics (1995) Household consumption expenditure. Eurostat, European Commission. http://epp.eurostat.eu.europa.eu/statistics_explained/index.php/Household_consumption_expenditure. Accessed 7 Nov 2009

Eurostat Cultural Statistics (2006) Household consumption expenditure. Eurostat, European Commission. http://epp.eurostat.eu.europa.eu/statistics_explained/index.php/Household_consumption_expenditure. Accessed 7 Nov 2009

FEK (1986) Gia tin prostassia tou perivallontos 1650/16-10-1986. Fyllo Efimeridas tis Kybernisis. Etniko Typografeio, Athina, pp 3257–3258

Greekscapes (2009) Greekscapes: a "Bird's Eye View" atlas of Greek landscapes. Department of Geography, Harokopio University & John S. Latsis Public Benefits Foundation, Athens, http://www.greekscapes.gr/index.php. Accessed 20 May 2010

Grenon M, Batisse M (eds) (1989) Futures for the Mediterranean Basin: the blue plan. Oxford University Press, Oxford

Hadjimichalis K (2008) Geographikes anaparastaseis/fantasies topion tis ypaithrou. In: Manolidis K (ed) Proceedings of the scientific meeting on "I diekdikissi tis Ypaithrou". Panepistimio Thessalias, Volos, p 27

Höchtl F, Terkenli TS, Plieninger T (2007) Editorial: the European Mediterranean region in the focus of landscape research. Die Erde 138(1):1–6

Kambouridis H (2009) H Neoelliniki Topiographia apo ton 18o eos ton 21o Aiona: Orama, Embeiria kai Anaplassi tou Chorou. Idryma Eikastikon Technon kai Moussikis V. kai M. Theocharaki, Athina

Karapostolis V (1983) I Katanalotiki Symberifora stin Elliniki Koinonia 1960–1975. Ethniko Kentro Koinonikon Erevnon, Athina

King R, Proudfoot L, Smith B (eds) (1997) The Mediterranean: environment and society. Arnold, London

Kizos T, Spilanis I, Koulouri M (2007) The Aegean islands: a paradise lost? Tourism as a driver for changing landscapes. In: Pedroli B, van Doorn A, de Blust G, Paracchini ML, Wascher D, Bunce F (eds) Europe's living landscapes. Wageningen: KNNV Publishing, Zeist, pp 333–348

Lambropoulos V (1993) The rise of eurocentrism: anatomy of interpretation. Princeton University Press, Princeton, NJ

Legg KR, Roberts JM (1997) Modern Greece: a civilization on the periphery. Westview Press, Oxford

Lickorish LJ, Jenkins CL (2004) An introduction to tourism. Kritiki, Athens

Louloudis L, Beopoulos N, Troumbis A (eds) (2005) To Agrotiko Topio: Tto Palimpsisto Aionon Agrotikou Mochthou. Ktima Merkouri, Korakohori Hleias, Athens

Manolidis K (2008) I Ypaithros kai ta Traumata tis Metapolemikis Elladas. In: Manolidis K (ed) Proceedings of the scientific meeting on "I Diekdisisi tis Ypaithroue". Panepistimio Thessalias, Volos, p 54

McNeill WH (1978) The metamorphosis of Greece since world war II. The University of Chicago Press, Chicago, IL

Minca C (1998) Mediterranean metaphors and tourist space: a theoretical approach. In: Conti S, Segre A (eds) Mediterranean geographies. Societa Geographica Italiana and CNR, Roma, pp 257–274

Olwig KR (2001) Landscape as a contested topos of place, community, and self. In: Adams PC, Hoelscher S, Till KE (eds) Textures of place: exploring humanist geographies. University of Minnesota Press, Minneapolis, MN, pp 93–117

Ostrom E (1990) Governing the commons: the evolution of institutions of collective action. Cambridge University Press, Cambridge

Pedroli B, Wascher D, de Blust G, Paracchini ML, van Doorn A (2007) Introduction: Europe's living landscapes at a turning point. In: Pedroli B, van Doorn A, de Blust G, Paracchini ML, Wascher D, Bunce F (eds) Europe's living landscapes.Wageningen: KNNV Publishing, Zeist, pp 11–18

Pettifer J (1993) The Greeks: the land and people since the war. Penguin, London

Said E (1978) Orientalism. Vintage books, New York, NY

Semaioforidis G (2005) Dieleuseis: Keimena gia tin Architektoniki kai tin Metapoli. Metapolis Press, Athina

Stathatos J (1996) The invention of landscape: Greek landscape and Greek photography 1870–1995. Camera Obscura, Thessaloniki

Stathatos J (2008) Anthropos kai Topio sti Sygchroni Elliniki Photografia. In: Manolidis K (ed) Proceedings of the scientific meeting on "I Diekdissi tis Ypaithrou". Panepistimio Thessalias, Volos, p 30

Sutherland NS (ed) (1989) The international dictionary of psychology. Continuum, New York, NY

Terkenli TS (1995) Home as a region. Geogr Rev 85(3):324–334

Terkenli TS (2001) Towards a theory of the landscape: the Aegean landscape as a cultural image. Landsc Urban Plan 57:197–208

Terkenli TS (2002) Landscapes of tourism: towards a global cultural economy of space? Tourism Geographies 4(3):227–254

Terkenli TS (2004) Landscape research in Greece: an overview. BELGEO 2004(2–3):277–288

Terkenli TS, d'Hauteserre A-M (eds) (2006) Landscapes of a new cultural economy of space. Springer, Dordrecht

Terkenli TS, Tsalikidis I, Tsigakou F-M (2001) The physical landscape of Greece in nineteenth-century painting: an exploration of cultural images. In: Anagnostopoulos GL (ed) Art and landscape: International Federation of Landscape Architects, Panhellenic Association of Landscape Architects – IFLA symposium Athens – 1998. Panayotis and Effie Michelis Foundation, Athens, pp 618–632

Towner J (1996) An historical geography of recreation and tourism in the western world 1540–1940. Wiley, New York, NY
Trova V (2008) Apo tin agonia tis ellinikotitas stin eutychia ton tropikon. In: Manolidis K (ed) Proceedings of the scientific meeting on "I Diekdissi tis Ypaithrou". Panepistimio Thessalias, Volos, p 39
Tsartas P (1989) Koinoniko-oikonomikes Epiptoseis tis Touristikis Anaptyksis sto Nomo Kykladon kai kyrios sta nissia Io kai Serifo, kata tin periodo 1950–1980. EKKE, Athens
Tuan Y-F (1986) The good life. The University of Wisconsin Press, Madison, WI
Vlachos G, Louloudis L (2008) Apo to chorafi sto choro: I anadissi tou agrotikou topiou sti meta-paragogiki ypaithro. In: Manolidis K (ed) Proceedings of the scientific meeting on "I Diekdissi tis Ypaithrou". Panepistimio Thessalias, Volos, p 24
Vogiatzakis IN, Pungetti G, Mannion AM (eds) (2008) Mediterranean island landscapes: natural and cultural approaches. Springer, Dordrecht
Wascher D (2001) European landscapes in transition: levels of intervention. In: Green B, Vos W (eds) Threatened landscapes: conserving cultural environments. Spon Press, London, pp 129–138
Williams A (1997) Tourism and uneven development. In: King R, Proudfoot L, Smith B (eds) The Mediterranean: environment and society. Arnold, London, pp 208–226

Part II
Participatory Methods in Practice

Chapter 8
Landscape in Participatory Processes: Tools for Stimulating Debate on Landscape Issues?

A Conceptual and Methodological Reflection from Research-Action Projects in France

Yves Michelin, Thierry Joliveau, and Claire Planchat-Héry

Abstract This chapter discusses the advantages and limits of different tools in participatory processes regarding landscape projects. It presents a brief theoretical review of the relations between landscape policies and landscape concepts by focusing on the French situation. Four study projects concerning local participatory planning conducted by the authors have provided observations on the ways several techniques of landscape representations can be used in order to facilitate the expression of landscape preferences and aspirations. By revisiting the main methodologies and results obtained in these projects, the chapter builds a functional typology of representation techniques for supporting landscape mediation. This typology is based on a series of distinctions: visual and literary media; descriptive, analytic and synthetic techniques; and views 'from the top' and 'from the inside'. The final part of the chapter is devoted to a general discussion of two main topics: the opposition between geo-referenced and ego-referenced space, and the question of the visualization of cultural landscapes.

Y. Michelin (✉)
Clermont University & UMR Metafort, VetAgro Sup, F-63370 Lempdes, France
e-mail: y.michelin@vetagro-sup.fr

M. Jones, M. Stenseke (eds.), *The European Landscape Convention*, Landscape Series 13, DOI 10.1007/978-90-481-9932-7_8,

Keywords Landscape representations · Landscape mediation · Participation · Massif Central

8.1 Introduction

The fifth Article of the European Landscape Convention (Council of Europe, 2000) stipulates that each country has to 'establish procedures for the participation of the general public, local and regional authorities, and other parties with an interest in the definition and implementation of the landscape policies'. However, it is not easy to apply this principle in a concrete way. Experts often dislike discussing their production, and local people are often afraid to express their disagreement in front of these specialists. The aim of this chapter is to discuss the advantages and limits of different tools in participatory processes regarding landscape projects. This discussion is based on a methodological review of four experimental projects conducted in different areas of the French Massif Central over a 20-year period (Michelin et al., 2002). We believe these experiments can help open new perspectives for the organization of local participatory discussion groups concerning landscape topics. By revisiting the observations made in these projects, we attempt to present a synthesis of the strengths and weaknesses of the numerous tools we have used to improve communication with actors regarding landscape. There are close relations between landscape policies and landscape concepts and theories. Our projects are grounded in the French landscape policy framework and in the French landscape theoretical approach. A brief discussion of the conceptual background of our experimentations is provided before the presentation of the actions.

The goals of our projects were in accordance with the intentions of the European Landscape Convention. We aimed to include the local population in the decision process as completely as possible. Hence, we used a large diversity of tools in a large variety of conditions in order to elaborate and communicate useful, clear and explicit landscape representations that local people and external experts could discuss in order to help and document decisions. The main part of this contribution presents an original classification of landscape representation tools and techniques that can be used in a participation context for supporting landscape mediation. Although this preliminary framework has not been established by following rigorous scientific evaluation protocols according to predefined criteria, it is a first proposal coming from systematic, close and careful analysis of our experience. We are still in a pragmatic phase where we try to list the tools that seem to work well in precisely defined situations as well as those that seem to be less useful. It is important to note that it is not easy to carry out a purely experimental study on this topic without disturbing the observation process. In the last part of the chapter, we revisit some classical questions of landscape representation by introducing new ideas. We first consider the way visualization techniques can help to articulate geo-referenced (objective) and ego-referenced (subjective) views of space. Finally, we address the question of the visualization of cultural landscape by submitting a new classification of visual representations. This classification can be used for discussing landscape preferences on three axes: concrete/abstract, physical/virtual, and scientific/artistic.

8.2 Landscape Policies and Landscape Concepts

To be efficient, a landscape policy needs to define and institute its object of action (Guttinger, 2007). However, the landscape is not always clearly defined, even if those who proposed the policy have to some extent been informed by theory. For instance, Daniel and Joanne (1983) define five models of landscapes: ecological, esthetical, psychophysical, psychological and phenomenological. In France, Luginbühl (1991) proposes a socio-cultural model in which people search for well-known references in a landscape they have seen previously in postcards. Sautter (1991) adopts a geo-anthropological perspective and defines four kinds of landscape interpretation related to personal experience, land-use practices, feelings and symbols. Brossard and Wieber (1984) combine in a global model three components: material processes, social construction, and, between these, the visible landscape, i.e. the part of space that is more or less visible. We propose integrating all these propositions into three types of concepts that seem to have guided French landscape policies.

8.2.1 Landscape as the Visible Spatial Extension of Bio-physical or Bio-technical Processes

Many authors in different domains, such as landscape ecology (Forman and Godron, 1986), agronomy (Deffontaines, 1998), and pedology (Girard, 1993), as well as physical geography, consider the landscape as existing in itself. It is considered to be concrete living matter, shaped by biophysical processes that can be modified by human activities or practices. The policies put forward aim to produce an impact on specific visible elements of the landscape related to these processes, for instance hedge networks, ecological corridors, mountain streams and so on. Unfortunately, the options of the experts are sometimes rejected by local people. In France, the Mountain Soils Restoration Act (Loi RTM, 1882), enacted by the government at the end of the nineteenth century, has generated violent conflicts with local farmers in the mountains, including murders, army interventions, and individuals sentenced to hard labour. The landscape was not the target of the policy, but engineers used landscape photographs to promote their work and justify their policy (Brugnot and Casseyre, 2003). This example illustrates well the gap that often exists between the top-down and compulsory conception that an expert has of a public action and the opinion of local people about what is good for their own territory.

8.2.2 Landscape as Cultural Construction

This conception regards landscape as a human construction that cannot exist without people and out of space and time. Some philosophers, such as Roger (1978), and geographers, such as Berque (1990), consider that the idea of landscape itself is specific to particular cultures and is related to the art of painting. In France, the Natural Site Protection Law of 1930 (Loi relative à la protection des sites, 1930) was influenced by the arguments promoted a few decades before by famous writers such as Victor Hugo and painters such as Théodore Rousseau. More recently, some economists have suggested considering landscape as a medium for functions (ecological, quality of life, economic) that correspond to human needs and expectations (de Groot, 2006). This way of thinking has inspired some French regional parks, which aim to protect and to keep alive parts of territories considered as representative of local heritage and thought to be cherished by the inhabitants. However, these objectives are sometimes in conflict with the expectations of local inhabitants who consider that such a policy transforms them into 'Indians living in a reserve'. Such a perspective appears to be backward-looking and in contradiction with the aim of the European Landscape Convention.

8.2.3 Landscape as Personal Experience

This proposition is supported by anthropologists (e.g. Lenclud, 1995) and human geographers (Leighly, 1963; Meinig, 1979). It maintains that each person cultivates their own mental landscape through their daily personal experience, in accordance with their own history, social practices, and cultural background (Sautter, 1979). This approach differentiates 'surface landscapes', related to immediate perceptions, from 'embedded values', related to people, space and time experience. Ingold (2000) considers the landscape in a dwelling perspective. More clearly than the 1993 French Landscape Law (Loi paysage, 1993), the European Landscape Convention is a political translation of this general idea. Stefenson (2008) has conceived a general model of cultural landscapes that can be used as a basis to define how landscape policies can be categorized. She considers that cultural landscapes have three highly interacting components: forms, processes, and experiences. Forms are the visible aspect of the landscape. When they are perceived, forms produce an immediate response and represent the surface of the landscape. These forms are shaped by processes. Some of those processes are biophysical, while others are related to human activity. In the landscape forms, it is possible to find the influence of past processes. The third part of the landscape is the relation between each person and the landscape forms (that person's experience) and the relation between each person and the processes that create a connection with time. Crang and Tavlou (2001) speak of 'subterranean' landscape, the non-visible part of the landscape hidden in each mind and related to personal experiences. As a consequence, the discussion in a participatory process related to landscape planning must not only focus on concrete landscape elements such as spectacular trees, hedge networks or noticeable buildings, but also has to enlarge its purpose to include the feelings, meanings and values that people attach to what they perceive. An old oak in the landscape is not only imposing and venerable, but can also be a landmark for travellers, a playground for children in summer, and the symbol of the permanence of the local society. If, for instance, road planners forget this 'subterranean' tree landscape and do not take the time to negotiate when they propose to fell this old oak to make the road safer, it may be more difficult than they expected, even if there is a popular demand for the road to be made safer.

8.2.4 Different Ways of Integrating Landscape into Policies

Four ways of considering landscape in planning can be considered: (1) landscape gardening, (2) landscape architecture, (3) landscape management, and (4) territorial landscape. In landscape gardening, the main goal is to create new concrete forms and objects, mainly in parks or gardens. It is more a question of aesthetics and design than of management. The approach to landscape architecture is more global. Even if it includes a strong aesthetic and design background, the way a piece of land looks has to be considered, including roads, buildings, and planted areas. We can speak of landscape management when a visual or aesthetic goal is incorporated as one of the objectives of a planning procedure, whether it is a city plan, protection zone, area development plan, etc. The landscape is the end to be achieved through the plan. We use the notion of territorial landscape when the landscape is no longer the first goal of the action but a means for analysing, understanding, planning, and acting upon a territory. Each of these modes is characterized by a specific scale and a specific level of perception. They differ in the area of the zone taken into consideration, the size of the project, the number of people involved, the governance mode, the amount of the budget, the techniques used to collect and communicate data, and so on. In fact, these different contexts may be more or less intertwined and there is a logical continuity between the different approaches. Landscape professionals often work on projects of different kinds without differentiation.

8.2.5 How to Take into Account Landscape Aspirations

In the European Landscape Convention, 'Landscape Quality Objective' means the formulation of the aspirations of the public for a specific landscape by the competent public authorities. The word 'aspiration' is also used in the French version. Applying this notion to landscape planning drives the discussion toward general ideas of landscapes, in terms often disconnected from local places, and often ending in ideological battles (good vs. evil; beautiful vs. ugly, etc.).

Hence planners or lawmakers tend to favour the notion of 'landscape preferences'. However, this 'slippage of concept' changes the manner in which the landscape is considered. On the one hand, an approach by preferences focuses on criteria that can be more clearly assessed. Therefore, it can help in connecting local opinions to specific and located places. On the other hand, these criteria contribute to 'coagulation' of the landscape into something concrete and independent of the different possible perceptions. Many recent studies have developed techniques to approach landscape preferences by applying methods from various fields, such as marketing with the use of focus groups, economics with the use of willingness to pay (Rambonilaza and Dachary-Bernard, 2007) or hedonic prices (Sayadi et al., 2009), or sociology with the use of opinion surveys (Höppner et al., 2007). All these approaches postulate that the landscape is an object like any other, without taking into account the 'subterranean ideas' hidden behind its material appearance. Therefore these methods can gather outsider opinions but they often fail to reach insider ones.

More generally, it is difficult to manage a discussion about what people expect from the landscape of their place. The landscape is a general idea related to feelings, heritage and knowledge. It is constructed in the mind through time, spatial personal experience, and concrete perceptions of territories. We consider that the landscape is the result of a dialectical interaction between signals received from the outside and ideas produced inside the mind. In order to develop participatory approaches, it is necessary to consider the landscape both as a visible part of a territory and as an interpretation of perception signals (including smells, sounds, and other sensory items) through a memory process. To avoid ideological debates about 'the' landscape in general, we prefer speaking about the 'landscape of this place', more related to current situations, where the discussion aims to confront local opinions in order to prepare a specific landscape plan or policy for a specific place. Moreover, we consider that the discourse on landscape expectations must be produced not only through words but also by using specific media such as maps, drawings and photographs that make the opinions locatable, understandable and debatable by everybody involved in a discussion process.

8.3 How to Speak About Landscape Aspirations

In accordance with the European Landscape Convention, our work aims to facilitate public participation in order to define Landscape Quality Objectives. During a consultation or participatory process, it is considered possible to improve the quality of the exchanges by combining face-to-face interviews with other approaches. We identify three directions: indoor group discussion; discussion in the field; and discussion by social immersion.

8.3.1 Indoor Group Discussion

One way to avoid the risk of having only general opinions disconnected from the local context is to combine face-to-face interviews with discussion groups.

Gathering several people representing various opinions reduces the risk of forgetting some opinions that are less shared in the local society. The social regulation process of mixing people together can avoid giving too much importance to minority ideas defended by activists, and, on the other hand, it can contribute to bringing out some original points of view. However, if the participants have no other medium than discourse, the risk is either of drifting towards rough ideological debate or of overshadowing the opinions of those who are not comfortable with speaking in public. For this reason, we consider this approach more efficient if the main goal is to clarify the social relationships within the local social networks. If the objective is to find out different opinions regarding the landscape of a specific area, an approach through individual face-to-face interviews is better.

8.3.2 Discussion in the Field

The great advantage of this method is to be able to stand before the scenery, enabling a link to be made between what is said and what is seen. However, two factors limit its applicability. One is practical – the availability of participants for meetings during the day, weather conditions, etc. The other is that the debate may be limited to discussion of the local places concerned without broader generalization. While this method can be easily adapted to the restoration of specific places (such as viewing points or famous places), it is not well-adapted to regional landscape planning. Hence, some authors ask participants to take photographs, or draw maps or sketches, and continue the discussion with these documents in meetings after the trip (Kalibo and Medley, 2007). In a participatory process, this way of working can be the first step of a mediation process.

8.3.3 Discussion by Social Immersion

This method, inspired by anthropologists, consists in staying in one place for a long time and participating in local activities. Sharing the life of people makes it easier to ask them their opinions and feelings concerning the landscape. In theory this is the most accurate way to obtain precise and faithful points of view, but it cannot be applied in all situations. It requires a substantial amount of time and considerable anthropological knowledge. However, in some conditions, for instance in very traditional and closed-minded societies, this method can be used as preliminary work to establish a climate of confidence before starting more classical interviews. It has been tested in Sweden in order to give a better understanding of the place of landscape management in farmers' practices in a local community (Orth and Coquillou, 2005).

8.4 Combining Discourses and Visual Representations

Various techniques can be used to describe or depict the landscape: oral or written discourses, field trips, and different kinds of images and geospatial tools (photographs, drawings, sketches, 3D diagrams, maps, aerial photography, and geographical information systems (GIS)). More sophisticated means are emerging, such as virtual reality technology. As landscape is visual or, more generally, sensory, it seems clear that visual representation techniques express more directly and easily a feeling or a judgment about the landscape. However, drawing or taking pictures demands specific skills and people are often reluctant to use a pen or a camera. On the other hand, it is not easy to speak or write about landscapes. This necessitates a rich and precise vocabulary and some literary talent.

Because of these difficulties, some researchers and practitioners use such visual representations as media to help people express their opinion. Some use documents produced by artists or professionals skilled in both landscape description and depiction. Those experts can be useful in helping participants to write down, illustrate, or clarify their thoughts. However this method can be misleading if the images do not represent local opinions, especially if they look attractive. Thus professionals have to take care not to impose, even unconsciously, their own view of the landscape. Other practitioners prefer to ask people to create their own representations. This unusual way of working can frighten some participants and sometimes entails a training period. An example is given by Freire (1974), who used photographs taken by locals to help poor people living in shantytowns to describe social conflicts and to imagine solutions to their problems. He taught them the basic photographic techniques, and thus even the poorest and the least educated could express what they had endured. More recently, Soini (2001) discussed the use of mental maps and concept mapping to explore the human dimension of landscape representations. Whatever solution is chosen, many human geographers insist that landscape realities and perceptions should be named with local words, located in the area, related to daily uses and practices, and associated with personal or common values, whether aesthetic, functional, or cultural (e.g. Cosgrove, 1980).

Finally, it is important to remember that landscape perception is not limited to vision. Sounds also play an important role in the perception of landscape, and not only from a poetic point of view. The concept of 'soundscape' is used by several landscape architects to analyse local perceptions (e.g. Irvine et al., 2009). Hence, the choice of the kind of representation can be tricky. During a number of experimental projects in the French Massif Central, we chose to limit our investigations to visual representations, with the aim of giving a common basis for a shared understanding of the landscapes. We used different combinations of visual media in face-to-face situations and in group discussions, after having more precisely defined with our partners which visual tools were suitable for landscape planning. However, this has not limited the discussion to the visual aspects of the landscapes. The tranquility of the place, the 'music' of the landscape, as in the chirping of birds, river songs, or road noises, entered the debate without difficulty.

8.5 The Four Case Studies

The places where we undertook our investigations had common features:

- The landscape was not clearly an issue and did not represent great economic value in itself, even if it was sometimes well-known
- They were rural areas at some distance from towns, with a continuous decline of agricultural predominance in the local economy
- There was a lack of actions or projects relating to landscape, despite collective projects or co-operation policies between communities
- Elected representatives were interested in the experimentation because they were looking for ways to make the inhabitants partners in a local development process, or the development advisers were looking for a second wind to boost current projects.

Despite local differences, the four places (Fig. 8.1) all belong to peripheral areas considered marginal by planning managers and/or local populations. In three of them – although the altitude is not very high – steep slopes, poor soils, and a severe climate limit the possibilities of development. Until the 1980s, the population had been decreasing for a long time. However, new inhabitants are now starting to repopulate these areas, especially in places not too far from the main towns. For Billom municipality, the situation is more favorable from an agronomic point

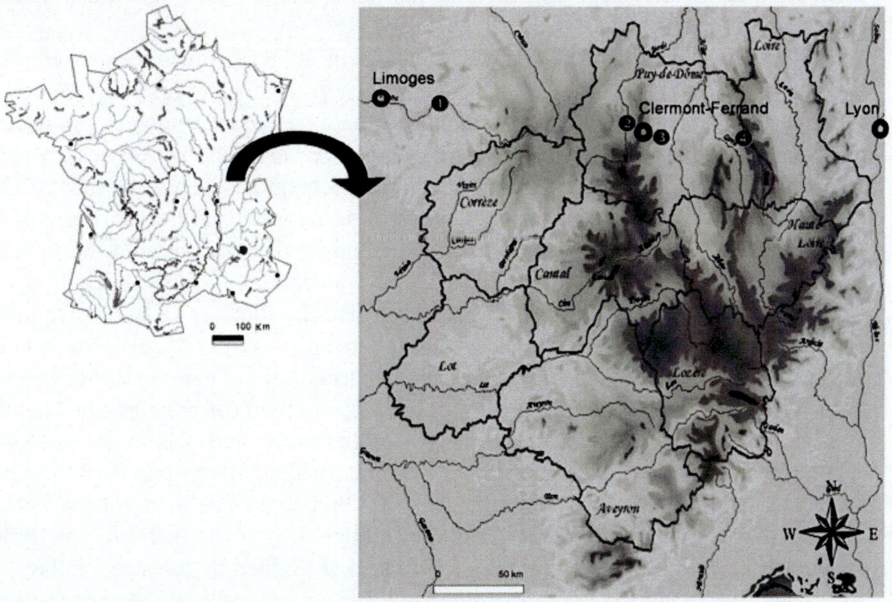

Fig. 8.1 Location of the four case studies in France and in the Massif Central. (1) Monts et Barrages, (2) Chaîne des Puys, (3) Billom, (4) Montagne Thiernoise

of view. The main problem here is related to the peri-urban context and the demand for housing, which competes with agriculture. The main features of the four cases are summarized in Table 8.1.

In the four places, we carried out several research projects with different partners over a long time period. During the work in the Chaîne des Puys region in 1986, we used very little in the way of visual media, except a photographic exhibition that was open to the public in each municipality before a public meeting and a video report, presenting interviews with different stakeholders in the field. This report was used to start the debate. In this project, the landscape was considered only as a background. However, people often used landscape arguments in expressing their expectations.

A more diversified media set was tested in the Montagne Thiernoise region. It combined various visual representations, some of which were realistic while others were more schematic. We also used digital 3D views. At that time, in the mid-1990s, this technology required sophisticated equipment and software and was not as widely disseminated as it is now.

At Billom, the main original feature was to start the debate by inviting people to react to the 'worst case scenario'. Participants were then invited to explain what they were expecting for the area.

In the Pays Monts et Barrages, we utilized maps, photographs, and 3D diagrams to provide a common basis for managing the discussions in a focus group with a bottom-up participatory approach. The main elements concerning the application of our methods in the four cases are found in Table 8.2.

We experimented with several combinations of landscape representations, which were adapted to the local context and partner's goals. These are illustrated in Figs. 8.2, 8.3, 8.4, 8.5, and 8.6.

Figure 8.2 shows landscape sketches, made in 1996, to illustrate two alternative scenarios for an area in Montagne Thiernoise in 2010. The first, 'the tendency scenario', depicts continuation of established trends. Here every owner can decide to clear-cut the plantations whenever he wants. The landscape appears as a patchwork of dark spruce plantations combined with rectangular open areas that will be replanted a few years later. From a number of scenarios, local people were asked to consider two scenarios: the most probable and the least acceptable. In the second scenario, 'the expected landscape scenario', a voluntary policy has enabled the elimination of some plantations. This has reshaped the forest in order to give more openness to the landscape. The clear-cuttings are managed collectively to reduce costs. If the elected officials appreciate this scenario, they are nonetheless conscious of the difficulty of applying this type of policy, especially if the price of wood is low.

Figure 8.3 shows three different ways of representing landscape types in Monts et Barrages region for different purposes. The landscape units map is used to locate four landscape types, which are defined by 3D diagrams. The landscape structure map is a sketch used to explain that the mountain area is divided into small disconnected landscape units while the plateau area is divided by the steep valleys. A schematic representation is used to show the distant influence of the expansion of Limoges on the plateau area, while the mountain area is becoming more and more reforested due to the decreasing rural population.

Table 8.1 Main geographical features of the four case studies

Name of region	Chaîne des Puys	Billom	Pays Monts et Barrages	Montagne Thiernoise
Location	Middle of Massif Central, Puy de Dôme	Middle of Massif Central, Puy de Dôme	North-west Massif Central, Limousin Region	North-east Massif Central, Puy de Dôme
No. of municipalities	18	1	32	9
Area	624 km^2	17 km^2	663 km^2	200 km^2
Population	59,035 (1982)	4,575 (2006)	21,160 (1999)	8000 (1999)
Altitude min./max.	600–1464 m	350–503 m	270–777 m	320–1285 m
Nearest main city and distance	Clermont-Ferrand (10 km)	Clermont-Ferrand (25 km)	Limoges (50 km)	Clermont-Ferrand (55 km)
Landscape policies in project	Local development charter	Local urban plan	Territorial project	Architecture and landscape charter

Table 8.2 Main questions and operational results in each case study

Name of region	Chaîne des Puys (Regional Park)	Billom	Pays Monts et Barrages	Montagne Thiernoise
Working period	1984–1986	2006–2008	1998–2001	1996–2001
Methods	Public exhibition	Face-to-face interviews	Face-to-face interviews	Face-to-face interviews
	Special issue of Regional Park magazine	Focus groups	Public exhibition	Surveys of inhabitants
	16 Public meetings	Public meeting, worst case scenarios	Tourist enquiries	Focus groups
			Focus groups	
Visual documents for discussions	Maps, statistics, video report	Maps, 3D diagrams, photos, 3D model	Maps, surveys using disposable-camera photography, 3D diagrams, sketches, scenarios with virtual maps and 3D landscape simulations	Surveys using disposable cameras, 3D diagrams, sketches, maps of change, simulations on aerial photos, 3D visualizations of scenarios, virtual flights
Main partner	Regional Park	Municipality (Commune)	Agricultural Council (*Chambre d'agriculture*)	Community of Municipalities (*Communauté de communes*)
Actors associated with the process	Elected officials	Agricultural Council (*Chambre d'agriculture*)	Local public administration (*Pays Monts et Barrages*)	Elected officials
		Regional park	Professional associations (foresters, farmers)	Regional Park
			Fishers, hunters, nature conservation associations	
Positive results	Large public participation in a period when it was unusual	A better understanding of farmers' constraints	Agri-environmental measures signed with farmers (1999)	Good involvement of elected officials
			Local development charter (2004)	
Failures	No charter signed (change in Regional Park political strategy after elections)	Final document provides many opportunities for housing, which goes against agricultural protection	Difficulties in continuing local participation process after end of research program	Difficulty to implement agreed landscape priorities in concrete policies

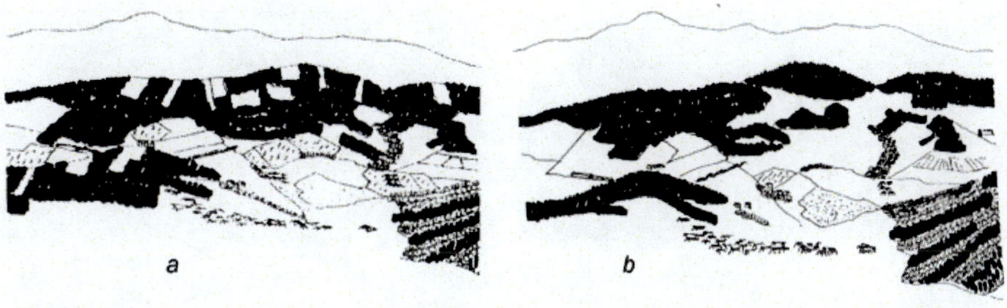

Fig. 8.2 Using landscape sketches to visualize scenarios, applied in the Montagne Thiernoise region. **a** 2010 tendency scenario, **b** 2010 expected landscape scenario (Grawings by landscape architect students from ENSP, Versailles, 1996)

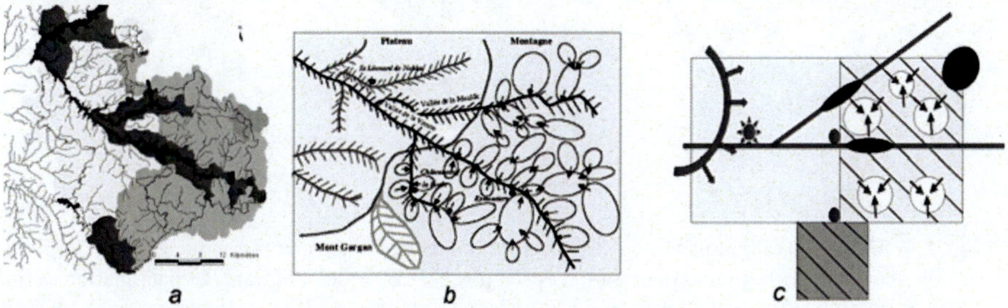

Fig. 8.3 Three types of map representation, applied in Monts et Barrages region. **a** landscape units map, **b** landscape structure map, **c** schematic representation (Drawings by Y. Michelin, 1999)

In Fig. 8.4, a photograph of a landscape in the Chaîne des Puys region is shown alongside a 3D block diagram, which in turn is used to visualize three types of opinions, respectively those of dairy farmers, tourists, and environmentalists.

Figures 8.5 and 8.6 present some combinations of different landscape representations. In the first case, this combination has been applied to discussion of the future in focus groups. In the second case, the aim was to incorporate the farmers' expectations into a local planning map (*Plan Local d'Urbanisme*, PLU).

8.6 Toward a Functional Typology of Representation Techniques for Supporting Landscape Mediation

In the following, we evaluate the results of more than 20 years of practical work fostering participation in landscape planning. As we were associated with concrete programmes, mainly conceived and managed by local authorities, we have not been able systematically to integrate strict scientific evaluation protocols in these experimentations. Therefore, our present purpose is preliminary analysis and pragmatic thinking based upon our experience and dedicated to future investigations.

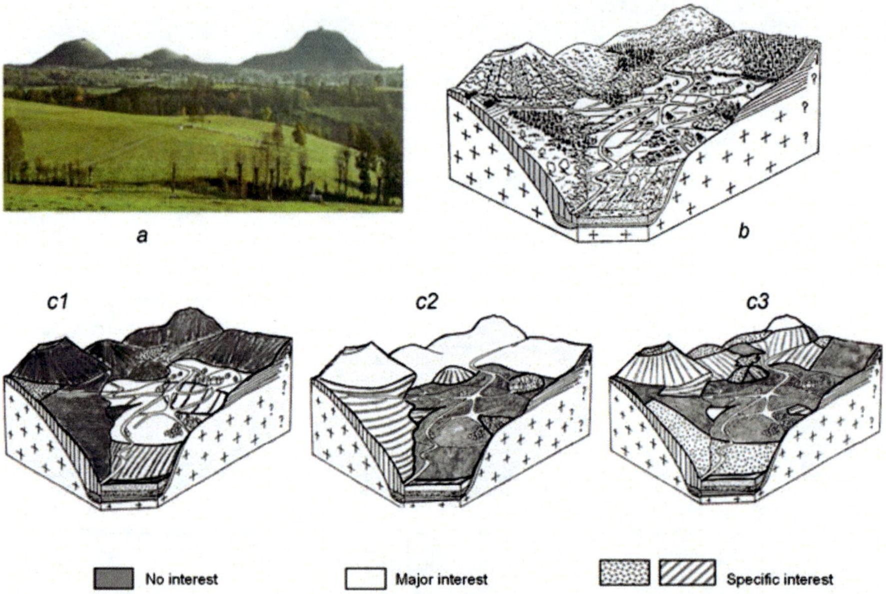

Fig. 8.4 Using 3D diagrams to express different points of view about the same territory, applied in the Chaîne des Puys region. **a** photograph of the place, **b** 3D block diagram, **c** visualization of three types of opinions – c1 dairy farmers, c2 tourists, c3 environmentalists (Drawings by Y. Michelin, 1995)

In our four case studies, we considered that it was impossible to use just one tool to stimulate landscape discussion. Each tool had a role and a functional effect depending on the nature of the project, the audience, the step in the process, etc. It is difficult to propose a general classification of the tools we used. Three factors are interwoven: the type and aspect of the tool, the task it is dedicated to, and the way people react during the participation process and some time after. The same tool, a photograph for instance, can be used (1) at the beginning of the process, during the territorial analysis, to allow someone to express his or her own feelings about the landscape, (2) in a following step to illustrate a landscape type, and (3) at the end as a basis for an illustrated scenario. We use the terms 'medium' for the physical type of landscape representation (map, sketch, discourse, or photograph) and 'technique' for the combination of medium and task during the process (e.g. characterizing landscape with images, or illustrating an opinion or a scenario with a 3D diagram). We provide a summary of the use and reception of these techniques according to the context.

8.6.1 Visual vs. Literary Landscape Representation Techniques

It should be pointed out that the place of visual media in our projects has evolved over time. In the Chaîne des Puys region, visual representations were only used as a background to illustrate the objectives of the project. After the presentation

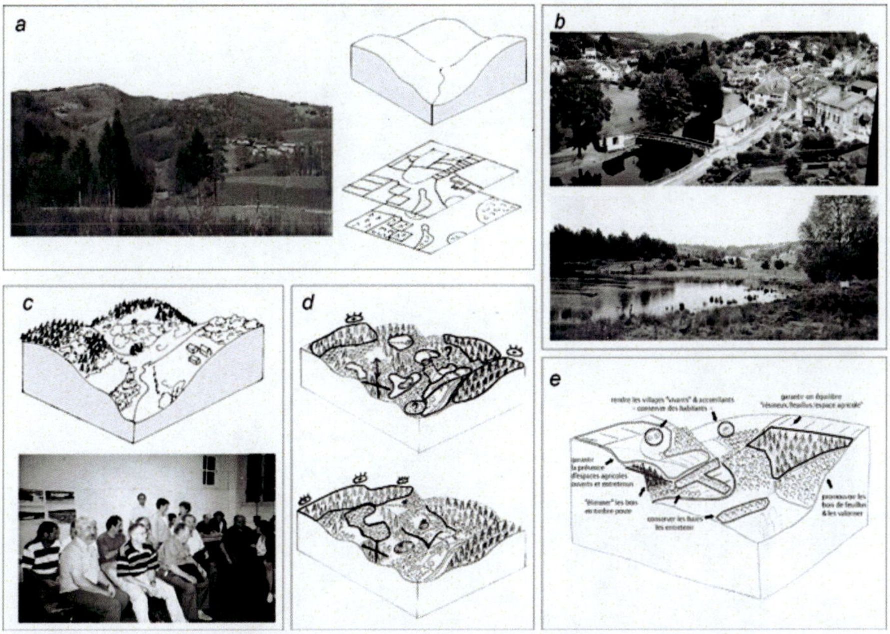

Fig. 8.5 Combining several types of landscape representations to discuss the future, applied in the Monts et Barrages region. **a** landscape analysed by landscape architects using fieldwork, photographs, maps, sketches and drawings, **b** pictures taken by the inhabitants with disposable cameras, combined with interviews and comments, **c** landscape practices analysed by agronomists using interviews and statistical analysis, **d** landscape scenarios produced by a focus group and drawn on a 3D diagram, **e** elaborating a local policy that combines landscape wishes and economic stakes, illustrated in a 3D diagram

of the report, although the debate went well when speaking about local development issues, the discussion about landscape expectations remained too general and ideological, except concerning the top of the Puy-de-Dôme volcano, which all participants considered to be a high-value place.

In the three other regions, visual representations had descriptive and analytical goals, and these documents helped people express better their feelings and exchange opinions about the landscape. In the Monts et Barrages and Montagne Thiernoise regions, we conducted the work in two steps with photographs. First, we asked a small group of people representing a large variety of opinions to take photographs with disposable cameras (Michelin et al., 2005). We had a discussion with them and ask them to explain the reasons why they took these pictures. Each photograph was associated with a personal comment. Second, we created an exhibition with a selection of pictures and comments and asked visitors to vote for the 'image + comment' with which they were most in agreement. If they did not find one that was acceptable, they could indicate a more appropriate location and add a comment. After these two steps, a local landscape culture, both visual and verbal, started to grow among the stakeholders.

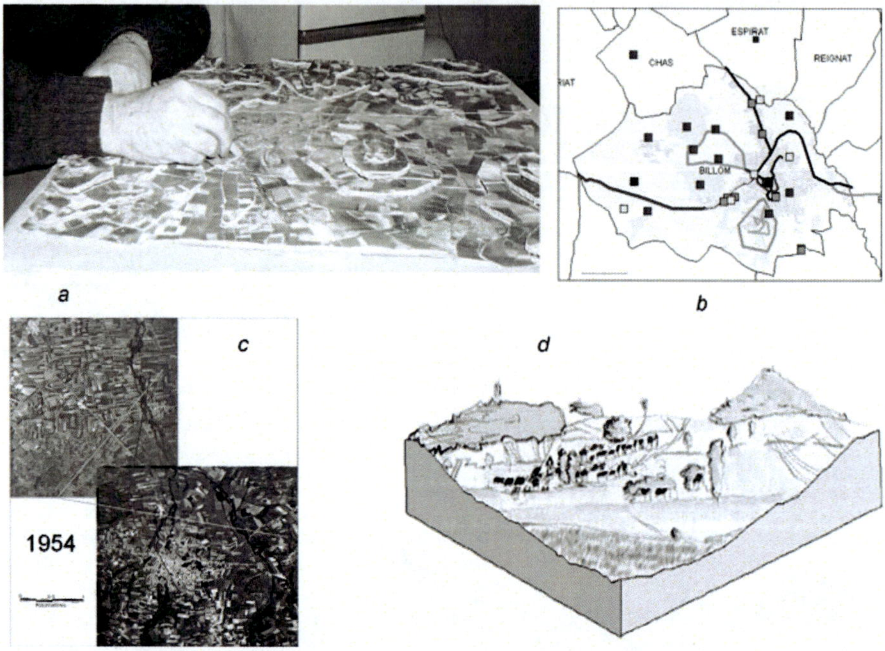

Fig. 8.6 Various landscape representations used as a basis for local participation, applied in Billom municipality. **a** a participant locating his usual itinerary on a 3D model, **b** the result translated on a map, **c** visualization of landscape change – air photographs from 1954 permit people to gain an idea of both what the landscape looked liked 50 years ago and where the different land uses were, **d** block diagram of the area (Drawings by C. Planchat, 2008)

The relevance of 3D diagrams and 3D models seem to be greatest for discussions of planning because such representations are more general and less linked to specific places where private interests or competition can exist. In the Billom case, we found that these representations clearly helped farmers to explain better their constraints to elected leaders who did not know anything about farming systems and practices. However, when the moment came to include these expectations in the planning documents, the visual representations were less adequate than oral or written expressions for affirming, discussing, and disputing the written version of the document (Planchat et al., 2008).

8.6.2 A Framework for Organizing Representation Techniques in a Landscape Project

After careful examination of our experimentations, we consider that it is possible to propose a comprehensive list of ways of using different representation techniques in a territorial landscape project. On one axis, we distinguish three ways of representing landscape issues in relation to the main goal they are supposed to achieve: *descriptive* techniques, *analytic* techniques and *synthetic* techniques.

Descriptive techniques aim to render as objectively and comprehensively as possible the state of the landscape at a specified date. Analytic techniques are useful to interpret a situation, to point out relations between factors, or to explain the problems observed. Synthetic techniques are used to summarize ideas, judgments, or actions about the landscape, which can then be part of an integrated report or document. The distinction between these three techniques is, however, highly theoretical. It is impossible to divide a project into clear and separate phases as description, analysis and synthesis (Sfez, 1992). All techniques can be mobilized at the same moment but, as the project evolves, the relative importance of each task changes, with a larger place dedicated to description at the beginning and to synthesis at the end. The actor-network theory approach (Latour, 1989), and more specifically the boundary-object concept, are highly pertinent in this matter, since actors coming from different social worlds have to address scientific and non-scientific problems and establish 'a mutual modus operandi' (Star and Griesemer, 1989).

On a second axis, we consider the relation between the observer and the landscape. Is he or she inside or outside the landscape? Does he or she look at it vertically or tangentially? When using a landscape representation medium, it is important to take account of this relation. Brossard and Wieber (1984) differentiate between 'views from the inside' and 'views from the top'. The first is from within the landscape; the viewing point is situated in a place that can be occupied by an inhabitant or a visitor. The second approach refers to viewing points located outside the landscape, such as the zenithal signal acquired by a satellite. Rougerie (1985) summarizes this distinction with reference to the opposing views of the visitor and the aviator. It is not so simple, of course. Even if the first approach aims to be more concrete, physical, and subjective, while the second is more abstract, geometric, and objective, a view from the top such as an aerial photo can present a powerful aesthetic, while a personal photograph can represent in a relatively objective manner a specific part of a landscape. According to Rougerie, they have to be combined in order to analyse a landscape as a whole. A large part of the French landscape approach is based on a combination of these two views (Brossard and Wieber, 2008). Our objective here is to facilitate the choice of particular techniques of landscape representation according to the stage in the participatory process and the expected results.

By combining these two axes, we derive six different representation techniques depending on the type of goals and the type of media. It is possible to characterize every technique by its global function, its interest and its limits.

Descriptive techniques simplify and exemplify the complex and inexhaustible world when presenting the landscape. 'Views from the top' (vertical views) refer to a tradition of mapping and aerial or satellite imagery. It is not possible to enumerate the entire list of maps that could be useful for rendering various sorts of phenomena at different scales. Table 8.3 contains some examples.

Views from the inside and tangential views aim to render a specific point of view. Even if sometimes the best way to show a landscape is to stay out of it, viewing it from the top of a mountain for instance, the target of the representation is to illustrate

Table 8.3 Interests and limits of top-down descriptive representations

Examples of medium	Tasks assigned to the medium	Interest	Limits
Small-scale topographic and thematic maps (1:50,000–1: 100,000)	Visualizing the project area in its general environment. Delineation of large landscape units based on morphological factors	Very common. Information simple enough in most cases	Information sometimes too simple (road maps) or not easily interpreted (large-scale vegetation maps). Systematic but documented bias (small plots are missing)
General information, vegetation, roads, land cover, etc.			
Satellite images representing main land covers			
Middle-scale topographic and thematic maps. (1:25,000–1:50,000). Geological or soil maps, land policy plans	Precise location of problems, stakes, and actions decided or planned. Delineation of precise landscape units	Rich and detailed information	Complex documents. Possibly outdated because not continuously updated
Large-scale documents, Land Registry, aerial photographs and ortophotographs	Very precise location. Adapted to a forestry and farming approach or to landscape architecture	Very rich and detailed information. Ortho-photographs contain all visible information. Useful in listing and locating precise actions	Richness and detail of information needs long and tedious information processing. Information on ownership can be highly controversial

Table 8.4 Interests and limits of landscape descriptive representations 'from the inside'

Example of medium	Task assigned to the medium	Interest	Limits
Photographs	Recording a landscape view at one time in one place	Fast. Precise. Easy to take.	Distortion of relief and relative distances (depth of field). Framing effect. False objectivity
Drawings	Emphasizing the elements that are important for the designer	Understandable. Supposes and provokes a careful observation of the landscape	Subjectivity not always explicit. Sometimes voluntary omission or distortion
Sketches	Describing a specific element or presenting a general organization of the landscape	Complementary to general observations	Not always easy to understand. Needs a caption
3D Visualizations	Rendering a landscape view according to a computing model	Explicit and general. Possibility of fine-tuning the level of detail	Requires computing resources. Expensive. Possible confusion with a real view

how the landscape is perceived from inside. We used four types of media in our experimentations, presented in Table 8.4.

Analytic techniques help to propose a general interpretation of important phenomena in relation to landscape problems. GIS is nowadays the main tool used to process spatial and temporal data for the landscape considered 'from the top'. An obvious limitation of these techniques is the necessity to have computing hardware and GIS software. However, the resources and skills are becoming more and more widely available. Rather than making a comprehensive list of analytical function of GIS that can be used in this field, some common examples of analytical tools are provided in Table 8.5.

Analysis of specific points of view from the inside is not as common as from the top. In Table 8.6 are listed some techniques that we have not used often in our experimentations, but which can be useful on some occasions.

Synthetic techniques to summarize ideas and prepare actions include various types of map providing 'views from the top' (Table 8.7) and 3D diagrams and modified photographs providing 'views from the inside' (Table 8.8). We maintain an open definition of 'from the inside', widening it to include representations that are not necessarily exactly from the ground but that aim to express this viewing point. For instance, in order to get a general view of a specific landscape, it is often necessary to produce a 3D block diagram that locates the virtual observer with a tangential view from a low altitude above the ground.

Table 8.5 Interests and limits of top-down analytic landscape representations

Example of medium	Task assigned to the medium	Interest	Limits
Classical distance and overlay tools	Current GIS tools can work out the accessibility of some areas and to combine different information layers to find places corresponding to different criteria	Easy and accessible selection and combination requests	Methods often too simple to provide interesting results in landscape analysis
Topographical analysis	Working out slope, aspect, topographic indices	Getting a better qualification of relief factor	The accuracy of the production depends on the resolution
Analysis of spatial structure of landscape (Landscape Ecology Analysis)	Extraction of indicators, landscape metrics	Quantifying the description of landscape	Rather abstract computation. Not easy to communicate or to use in mediation
Visibility analysis	By automatically delineating 'viewsheds', it is possible to map the level of visible impact of any actions	Produces objective maps of visibility from any place to all other places. Useful for elaborating priorities	Maps not easy to understand because of difficulty to differentiate between passive and active sights.
Change analysis	Simple ways for locating and quantifying change in land cover, population density, or other dynamic phenomena	Checking out and locating the place where the landscape is changing and the speed of this change	Necessitates a multi-temporal data set, usually difficult and time-consuming to collect
Maps from simulations as a result of scenarios of landscape changes	To give an idea of where and how the criteria taken into account will change (i.e. openness, forested rate)	Scenarios allow participants to anticipate what the future of the landscape might be in a more concrete way than with only a discourse	Establishing scenarios needs resources and a scientific team to validate the methodology. In order for scenarios to be a useful mediation tool, their criteria and factors have to be chosen in a participative way. Understanding the maps needs a learning process

Table 8.6 Interests and limits of landscape analytic representations 'from the inside'

Example of medium	Tasks assigned to the medium	Interest	Limits
Analysis of photographs	Interpreting landscape objects from a collection of photographs. These documents might have been taken by scholars (sample), photographers (postcards, artistic work) or by locals (e.g. with disposable cameras)	Very useful for creating a common list of objects that are clearly and precisely identified and for building a common language	Difficult to find a balance between too much detail and oversimplification
Digital 3D views of analysis or scenario results	See from within the landscape some analytical results	Help to visualize an abstract phenomenon	Hybrid visualization (abstract result vs. concrete point of view) is not always efficient

Table 8.7 Interests and limits of top-down synthetic landscape representations

Example of medium	Tasks assigned to the medium	Interest	Limits
Landscape unit map	Delineation of landscape units according to an explicit typology. Defining different functional landscape areas in order to achieve a descriptive or a prescriptive goal	Facilitate spatial thinking and planning in a zoning approach	Risk of oversimplified perspective of the landscape's complex system of relations. Problem of transition zones
Landscape organization scheme	Schematic map showing main forms of spatial organization of the landscape and the structure of their relations	Emphasizing the functional relations between the units	More abstract documents. Needs explanation for most participants
'Chorematic' map	Schematic map based on assembly of elementary structure of geographic space (Brunet and Ferras, 1992)	Revealing the processes that are at the origins of territory and landscape structure	Even more abstract than organization scheme. Quite far from landscape in some aspects

Table 8.8 Interests and limits of synthetic landscape representations 'from the inside'

Example of medium	Tasks assigned to the medium	Interest	Limits
3D diagrams	Built as a representation of clearly defined types of landscape units that can be observed in the area	Useful for thinking collectively, locating problems and ideas for solutions, discussing interventions in a concrete way	Necessitates a serious preliminary typology study to ensure a good sampling of situations to discuss. Drawing skills required on the team
'Modified' photographs	Visualization of the consequences of a scenario on a landscape	More concrete than a drawing or a 3D diagram (well-adapted for well-known places)	The realism can limit the possibilities of discussion with non-experts

8.6.3 What Works and What Does Not

Generally people express at the same time both what the landscape is and what it should be. Hence, it is not easy to place a clear boundary between a purely descriptive and a purely analytical document. The task assigned to the medium is the basis of our typology. Further, it is important to differentiate clearly between the descriptive and the prescriptive phases in the project. We found that inside representations were more suitable for starting the discussion about landscape preferences and values, but also, at the end, they could give materiality to the expected results of a policy. On the other hand, representations 'from the top' gave more objectivity in the analytical phase of the work. Even though we have tried to establish a typology of representations, it is very difficult to give an absolute statement about which representation techniques should be chosen. It depends on the kind of project (gardening, architecture, landscape planning, or territorial landscape), and who the participants are.

The same representation can have a different status from one case to another. For instance, we have used 3D diagrams at the beginning of a project to describe a type of landscape and at the end to visualize the consequences of a policy. As a consequence, we observe the relevance of combining different techniques in order to get as many diverse opinions as possible and the highest sensitive richness in the project. The search for diversity is essential for the success of a participatory process.

Finally, none of the visual representations 'speaks' by itself. All visual media have to be discussed and explained. The articulation of the verbal expression (who speaks, when, for what purpose, how the discourse is recorded, etc.) has to be clarified.

8.7 Final Discussion and Conclusion

Whatever kind of project we consider, a debate on landscape has to answer three needs simultaneously:

1. The different people sitting around the table have to elaborate a personal view and position about landscape matters. Landscape is not an obvious and natural topic to discuss. Local stakeholders are not used to speaking easily about landscape. It is necessary to help them to elaborate their own opinions and perspectives.
2. These viewpoints have to be constructed in a process of exchange and dialogue between the different actors. This dialogue necessitates bringing together people who are able to express disagreements as well as agreements about the topics they are discussing.
3. The mediation process does not have to target a general consensus about all the points considered in the discussion, but one should try to reach an agreement about some specific questions and possibly propose some collective propositions or even statements about the landscape of the area.

This process supposes an exchange and a dialogue between people involved in the landscape planning process. It is not a simple communication problem. One has to set up a mediation process including four fundamental aspects:

1. What are people speaking about? Are the words and concepts clear and fully understood?
2. What relationships are built between the discussion and the planning process?
3. Which values, meanings, symbols, etc. do participants attach to the topics they are speaking about?
4. How to facilitate the discussion in a practical way (tools, methods, organization)?

Beside this general framework, our work contributes to clarify two main points that are developed below: geo-referenced vs. ego-referenced space; and visualization of cultural landscapes.

8.7.1 Geo-referenced vs. Ego-referenced Space

In participatory planning, landscape is not only the visible part of a territory, considered as a set of resources. It is also a relation between each person and a place, as well as with the other people living, visiting, and enjoying this place. Couderchet and Ormaux (2003) build upon Straus's opposition between what he calls landscape space and geographic space (Straus, 2000). Landscape space is the space of daily life and perception, essentially sequential and organized tangentially to the Earth's surface, associated with sensations and always with reference to a horizon. Geographic space is the space of networks and maps, projected from the ground and fitted with coordinates. Couderchet and Ormaux (2003) extend this conception to an opposition between two spaces. They use the term *geo-referenced space* for 'objective' space, projected from a fixed origin, which is the space of planning. The term *ego-referenced* space is subjective and based on proximity; it is the space of individual projects. Every type of space produces its own chain of representations, which are logically incompatible with the other. The main object of mediation is to find a way to articulate these two chains and to integrate personal views into a collective project in order to build up what we could call *socio-referenced space*. More than articulating an opposition between these conceptions of the landscape, our experience invites us to imagine a continuum between geographical and anthropological polarities.

In the perspective of the European Landscape Convention, we face a double challenge. On the one hand, planners and other experts must leave the comfortable attitude of managing geo-referenced space, where everything exists in itself and has a specific and located place. Even if scientifically correct, they have to make an effort to change their attitude by accepting the fact that local people may have other opinions that must be taken into account. Of course, we do not argue that experts should deny their scientific knowledge – but they have to accept the need to explain it and to give people a demonstration of what they propose. In participatory processes, we have observed that if the scientific knowledge was presented in a clear way, with all the explanations asked for by participants, it could play an interesting regulation role for two reasons. First, the documents produced by the project (e.g. maps, GIS simulations, or landscape descriptions) and the explanations that are provided during meetings create a common body of knowledge about the local landscape among the participants. Second, as the visual representations are discussed within the local group, they provide a basis for further discussions about the future of the landscape. This facilitates negotiation between the actors. An example from the Pays Monts et Barrages region illustrates this. At the beginning of the programme, strong opposition existed between foresters and farmers. The latter were complaining in a general way about the reforestation of their territory, while the foresters said that not a single spruce had been planted since the 1980s. Both were partially right and partially wrong. Faced with the map of the evolution of the forested area from 1850 to 1999, they discovered the importance of the phenomenon. Thanks to the 3D diagrams they were able to identify the places where the enclosure of the landscape was problematic for daily life, and they started to think about solutions acceptable to both factions. Farmers and foresters agreed to maintain some open landscapes close to the villages with the help of agri-environmental measures. The foresters asked for regional funds to open some forestry roads in the new plantations to facilitate the clear-cutting of the young trees. They proposed to allow farmers to use them for access to the meadows remaining inside the forests.

It is not because the subjective approach is related to personal feelings and opinions that it is difficult to find a common landscape space. We worked with local communities that had the same heritage and the same cultural references, even if some people were natives and others new incomers or outsiders. However, in our four cases, the actors were not able to work together at the beginning. We observed that the main difficulty was not the individualization of landscape opinions but the misunderstandings between people when they started to debate without any physical medium, in a too general and ideological way and totally disconnected from the local context. Hence, we believe that the opposition between objective and subjective room is not relevant in the context of the European Landscape Convention. We argue for the necessity of developing techniques that permit each actor to reconnect with the territory in a conscious way. The priority is to help each stakeholder to locate, explain and discuss their own point of view. We think that it is preferable to let the participants choose the media they want by proposing a set of diversified representations (maps, drawings, sketches, 3D diagrams, and photographs), accompanied by personal or technical comments. Some people refer mainly to maps or other similar geographical representations. Others prefer pictures or sketches. During the participatory process, the balance between these two types evolved, depending on the topics and the aims of the discussion. Behind these differences, the essential thing is that all the media have been conceived, discussed and validated collectively. During discussion each participant should have the opportunity to pick from the common visual media toolbox the item needed to express an opinion or make a proposition. This is a method of producing common discursive knowledge through the participation process.

8.7.2 Visualization of Cultural Landscapes

At the beginning, our work had a pragmatic goal. We wanted to give neglected populations the opportunity to express their opinion about their landscapes. Even if many points have to be verified in a more systematic way, these first results enable us to propose a classification framework for the visual representation of landscapes that can be used for discussing landscape preferences. We organize this classification along three polarities: concrete vs. abstract; physical vs. virtual; and scientific vs. artistic. In Table 8.9, each dimension is a continuum, and we propose intermediate states between the initial polarities. From left to right, columns represent a gradient of abstraction, while from top to bottom there is a gradient of virtuality. The continuum between scientific and artistic representation is indicated by identifying those representations that can be used in both cases.

In a practical way, representations of located places that aim to imitate reality are fully adapted to concrete negotiations about specific projects or for well-known areas. When they are used for more general purposes, however, they often drive the debate towards conflicts of interest and make more difficult the emergence of landscape expectations. In the Chaîne des Puys region, the photographic exhibition combined with a video report was very well appreciated. During the 18 meetings, several hundred people attended and expressed their interest. However, it has proved difficult to transform this interest into concrete proposals, except for the restoration of the top of the famous Puy-de-Dôme volcano.

Abstract or symbolized representations are difficult to use because they are disconnected from the local situations. However, they can be useful in a long-term

Table 8.9 Proposition for a classification of visual landscape representations

	Existing and located places, imitation of reality	Simplification of reality	Abstraction, symbolization of reality
Physical 'objects'	LANDSCAPE *LAND ART*	3D MODEL *GARDEN*	? *Sculpture*
Images	PHOTOGRAPHS *DRAWINGS*	3D DIAGRAM *CLASSICAL LANDSCAPE PAINTING*	CHOREMATIC MAPS *Surrealist paintings*
Virtual representations	*VIRTUAL REALITY*	*DIGITAL 3D VIEWS*	? *Digital Art*

Text in CAPS = examples of scientific representation, text in *italics* = artistic representation, text in CAPS and *italics* = representation used in both situations

perspective to help people to approach their own landscape culture in a conscious way. Some experiments of land art in three regional parks open new perspectives in this direction (Trakas et al., 2008). In the Monts et Barrages area, we wanted to combine our investigations with an artistic approach. Eymoutiers, one of the municipalities involved in the program, was the birthplace of the famous contemporary painter Paul Rebeyrolles, who gave a large part of his personal collection to this commune. Unfortunately, he refused our proposition, explaining us that his painting was too violent to be able to facilitate the participation process. The combination of 'naïve', expert, and artistic approaches can be done only with people who are ready to accept the challenge.

We have found in our case studies that simplified representations of the landscape provide a compromise between local specificities and generic opinions. Such representations seem to be very suitable when the discussions concern a local charter or at the beginning of local planning procedures. Three positive results illustrate this. In the Montagne Thiernoise region, the use of disposable cameras permitted local people to explain that it was not possible to think of future landscapes without addressing the question of the demise of the local knife industry. In order to demonstrate this fact, one participant took a photograph of shrubs, and explained to us that these shrubs had grown at the place of a small knife workshop that had been abandoned. When the mayor saw this photograph and read the comment, he understood why he met such great difficulties in motivating his own council to start a landscape charter while the local economy was dying. In Billom municipality, the representatives were not very conscious that the farming system was facing serious problems due to the housing development of the place. By thinking about the non-expected landscape, represented by a 3D diagram, they better understood which type of difficulties farmers had to deal with. Part of this new knowledge was incorporated in the final planning document (Planchat et al., 2008). In the Monts et Barrages region, we were surprised to observe how local people had incorporated the 3D landscape representations in their discourses. A few years later, when they wrote a local charter, they clearly mentioned the landscape units with the geographical term *alvéole* (landscape with an alveolar or honeycombed shape), which we had defined with them. However, this acculturation process invites us to become aware of the risk of influencing neglected populations while believing that they are expressing their own opinion. As knowledge is always being constructed, it is important to note that local people can adapt their knowledge during the mediation process. Nevertheless, it is necessary to make sure the weakest categories of population are not influenced unduly. Over time, we have become more and more suspicious of our own scientific knowledge that we had to consolidate by clear understandable arguments. Finally, we had also to undertake our own personal analysis in order to distinguish better our scientific knowledge from our personal feelings, preferences, and opinions. It seems evident, but it is not so easy to respect this in the domain of the landscape. For instance, an agronomist will often prefer a 'beautiful' field crop without any bad seeds, in reference to its potential of production, while an ecologist will appreciate a diversified meadow as an expression of a biodiversity reservoir. If he does not know his own background, a specialist can easily project his personal opinion behind an appearance of objective scientific knowledge.

Further verification of our ideas needs to be undertaken by measuring their impact in the documents and policies that are a result of this participation and by comparing the effectiveness of different combinations of representations in participatory processes. Nonetheless, even at this stage, we hope this contribution will give other practitioners ideas for developing local participation processes for applying the European Landscape Convention in the countryside.

References

Berque A (1990) Le paysage comme forme de sensibilité à l'environnement. In: Atlas des paysages ruraux. De Monza, Paris, pp 108–121

Brossard T, Wieber JC (1984) Le Paysage, trois définitions: Un mode d'analyse et de cartographie. L'Espace géogr 13(1):5–12

Brossard T, Wieber JC (2008) Paysage et information géographique. Hermes-Lavoisier, Paris

Brugnot G, Casseyre Y (2003) De la politique française de restauration des terrains en montagne à la prévention des risques naturels, XIIe Congrès Forestier Mondial, Québec city, Canada. http://www.fao.org/DOCREP/ARTICLE/WFC/XII/0034-C3.htm. Accessed 21 Apr 2010

Brunet R, Ferras R (1992) Les mots de la géographie, dictionnaire critique. GIP RECLUS et La Documentation française, Paris

Cosgrove D (1980) Images of landscape. In: Gold J (ed) An introduction to behavioural geography. Oxford University Press, Oxford, pp 115–125

Couderchet L, Ormaux S (2003) L'évaluation environnementale: Entre la standardisation des protocoles et la spécificité des milieux. In: L'évaluation environnementale: Vers une harmonisation internationale? Actes du 7ème Colloque des spécialistes francophones en évaluation d'impact Liège, 10 au 14 juin 2002, coll. "Evaluations environnementales", vol 3. Bibliothèque Nationale du Québec, Québec, pp 113–120

Council of Europe (2000) European Landscape Convention, Florence, 20.X.2000. ETS No. 176. http://conventions.coe.int/Treaty/en/Treaties/Html/176.htm. Accessed 6 Mar 2007

Crang M, Tavlou P (2001) The city and topologies of memory. Environ Plan D, Soc Space 19: 161–177
Daniel TC, Joanne V (1983) Methodological issues in the assessment of landscape quality. In: Altman I, Wohvill JF (eds) Behavior and the natural environment. Plenum Press, New York, NY, pp 39–83
de Groot R (2006) Function-analysis and valuation as a tool to assess land use conflicts in planning for sustainable, multi-functional landscapes. Landsc Urban Plan 75(3–4):175–186
Deffontaines J-P (1998) Les sentiers d'un géoagronome. Éditions Arguments, Paris
Forman RTT, Godron M (1986) Landscape ecology. Wiley, New York, NY
Freire P (1974) Théâtre de l'opprimé. François Maspero, Paris
Girard CM (1993) Le pédo-paysage: concept permettant de segmenter spatialement le milieu naturel à des fins agronomiques. DERF, Ministère de l'agriculture, Paris
Guttinger P (2007) Approche du paysage en droit français. Cahiers d'économie et sociologie rurale 84–85:11–60
Höppner C, Frick J, Buchecker M (2007) Assessing psycho-social effects of participatory landscape planning. Landsc Urban Plan 83(2–3):196–207
Ingold T (2000) The perception of environment: essays on livelihood, dwelling and skills. Routledge, London
Irvine K, Devine-Wright P, Payne S, Fuller R, Painter B, Gaston K (2009) Green space, soundscape and urban sustainability: an interdisciplinary, empirical study. Local Environ 14(2): 155–172
Kalibo HW, Medley KE (2007) Participatory resources mapping for adaptive collaborative management at Mt. Kasigau, Kenya. Landsc Urban Plan 82(3):145–158
Latour B (1989) La science en action: introduction à la sociologie des sciences. La Découverte, Edition Folio Gallimard, Paris
Leighly J (1963) Man and life: a selection from the writings of Carl O. Sauer. University of California Press, Berkeley, CL
Lenclud G (1995) L'ethnologie et le paysage: questions sans réponse. In: Voisenet C (ed) Paysages au pluriel: Pour une approche ethnologique des paysages. Collection Ethnologie de la France, Cahier 9. Edition de la Maison sciences de l'homme, Paris, pp 3–17
Loi RTM (1882) Loi du 4 /04/ 1882 sur la restauration et la conservation des terrains en montagne (RTM) J off 22/04/1882, décret d'application 11/07/1882
Loi paysage (1993) Loi du 08/01/1993 sur la protection et la mise en valeur des paysages, qu'ils soient naturels, urbains, ruraux, banals ou exceptionnels. J Off 09/01/1993
Loi relative à la protection des sites (1930) Loi du 02/05/1930 relative à la protection des monuments naturels et des sites à caractère artistique, historique, scientifique, légendaire ou pittoresque. J Off 04/05/1930
Luginbühl Y (1991) Le paysage rural: La couleur de l'agricole, la saveur de l'agricole, mais que reste-t-il de l'agricole?. Etud Rurales 121–124:27–44
Meinig D (1979) Symbolic landscapes. In: Meinig D (ed) The interpretation of ordinary landscapes: geographical essays. Oxford University Press, New York, NY, pp 164–192
Michelin Y (1995) Les jardins de Vulcain: paysages d'hier, d'aujourd'hui et de demain dans la Chaîne des Puys du Massif Central français. Editons de la MSH, Paris, 155p
Michelin Y (1999) Des appareils photo jetables au service d'un projet de développement: représentations paysagères et stratégies des acteurs locaux de la Montagne Thiernoise. Cybergeo 65
Michelin Y, Joliveau T, Breuil J, Vigouroux L (2002) Le paysage dans un projet de territoire, démarche et méthode expérimentées en Limousin. Chambre d'agriculture de Haute-Vienne-ENITA-CRENAM, Limoges
Michelin Y, Paradis S, Lelli L (2005) When inhabitants photograph their landscapes to prepare a local sustainable development project: new perspectives for the organization of local participative discussion groups. J Mediterr Ecol 6:19–32
Orth M, Coquillou J (2005) Bråbygden project report plan. ENITAC, Clermont-Ferrand

Planchat C, Lelli L, Michelin Y (2008) Quels modèles de paysage pour quels types de simulation? Premières pistes de réflexion méthodologique à travers l'exemple du bloc-diagramme paysager. In: Colloque du Réseau thématique pluridisciplinaire: "Paysage et environnement". Modélisation paysagère spatialisée: Des approches dynamiques aux évaluations fonctionnelles. Toulouse. 18 pp (unpublished)

Rambonilaza M, Dachary-Bernard J (2007) Land-use planning and public preferences: what can we learn from choice experiment method? Landsc Urban Plan 83(4):318–326

Roger A (1978) Nus et paysages: Essai sur la fonction de l'art. Aubier, Paris

Rougerie G (1985) Le dessus, ou le dedans. Rev géogr Est 25(4):355–363

Sautter G (1979) Le paysage comme connivence. Hérodote 16:40–67

Sautter G (1991) Paysagismes. Etud Rurales 121–124:15–20

Sayadi S, Gonzalez-Roa C, Calatrava-Requena J (2009) Public preferences for landscape features: the case of agricultural landscape in mountainous Mediterranean areas. Land Use Policy 26:334–344

Sfez L (1992) Critique de la décision. Presses de la fondation nationale de sciences politiques, Paris

Soini K (2001) Exploring human dimensions of multifunctional landscapes through mapping and map making. Landsc Urban Plan 57(3–4):225–239

Star SL, Griesemer JR (1989) Institutional ecology, "translations" and boundary objects: amateurs and professionals in Berkeley's museum of vertebrate zoology, 1907–39. Soc Stud Sci 19(3):387–420

Stefenson J (2008) The cultural values model: an integrated approach to values in landscapes. Landsc Urban Plan 84(2):127–139

Straus E (2000) Du sens des sens: contribution à l'étude des fondements de la psychologie. Jérôme Millon, Paris

Trakas G, Corillon P, Weinberger LaF, Clément G, Provoost I, Denicolai S, Suzuki A, Berclaz J-D, Ristelhueber S, Huws B (2008) Paysages: regards croisés sur les paysages, l'espace d'un parc, le temps d'un artiste. Saint-Julien Molin-Molette, Jean-Pierre Huguet éditeur, Valence

Chapter 9
The Prospective Vision: Integrating the Farmers' Point of View into French and Belgian Local Planning

Claire Planchat-Héry

Abstract The integration of the farmers' point of view is rarely considered in planning procedures. This chapter presents a participatory method of landscape mediation termed the Prospective Vision, involving graphic and social landscape representations as collaborative learning processes. It is applied in two French-speaking contexts: Urban Local Planning in Billom, France, and the Landscape Charter of Attert, Belgium. The study's two main findings were that the use of landscape representations reveals specific landscape and territorial features at different scales of observation, and it facilitates the expression of farmers' points of view and

C. Planchat-Héry (✉)
Clermont University and UMR Metafort Agro Paris Tech, BP50085, F-63172, Aubière, France
e-mail: metafort.planchat@gmail.com

M. Jones, M. Stenseke (eds.), *The European Landscape Convention*,
Landscape Series 13, DOI 10.1007/978-90-481-9932-7_9,

their involvement in planning operations. This heuristic and constructivist approach to sharing local knowledge highlights the challenges of finding new methods for applying the participatory objectives of the European Landscape Convention.

Keywords Participatory method · Landscape mediation · Territorial features

9.1 Introduction

The European Landscape Convention (ELC) requires Parties to integrate landscape into policies concerning regional and town planning, agriculture, environment, and any other policy that has a direct or indirect impact on the landscape. Parties are further obliged to establish procedures for the participation of the general public, local and regional authorities, and other parties with an interest in landscape policies (Council of Europe, 2000: Articles 5c, d). The Convention hence requires the mobilization of stakeholders in order better to consider landscapes. The European Landscape Convention is a step forward in relation to the landscape and urban planning law in France. Belgium does not have a landscape law; the Convention is therefore one of the most important landscape guidelines. In both cases, the local application of the objectives of Article 5 of the ELC is not simple: at the municipal level, planners and elected officials are searching for methods of implementation in order to meet the objectives of the Convention.

Parties to the Convention also undertake to increase awareness in civil society, private organizations, and public authorities of the value of landscapes (Council of Europe, 2000: Article 6A). The guidelines for the implementation of the ELC point out that awareness-raising 'is made up of the experience gained from exchanges between local people affected by the planning decisions to be taken and persons possessing scientific and technical knowledge, that is, scientists and experts' (Council of Europe, 2008: II.2.3.B). Against this background, this chapter presents a methodological approach that was developed in a French-speaking context, based on the use of landscape representations with participative methods. This methodological process was used to help 'interested parties' (Council of Europe, 2000: Article 6C) to think about agrarian landscapes in the development of a planning project and to integrate agricultural issues into local planning processes. Two projects were studied: Urban Local Planning in the city of Billom in Auvergne, France, and the Landscape Charter for the Attert Valley Municipal Natural Park in Wallonia, Belgium (Fig. 9.1). In the past, agriculture was the driving force of development in both areas. Today agriculture is spatially, economically, and socially threatened by urban sprawl.

9.1.1 The Farming Issue in Planning Processes

In Europe, two major types of planning are regional strategies and spatial planning. These approaches are influenced by national rules for urban and local planning,

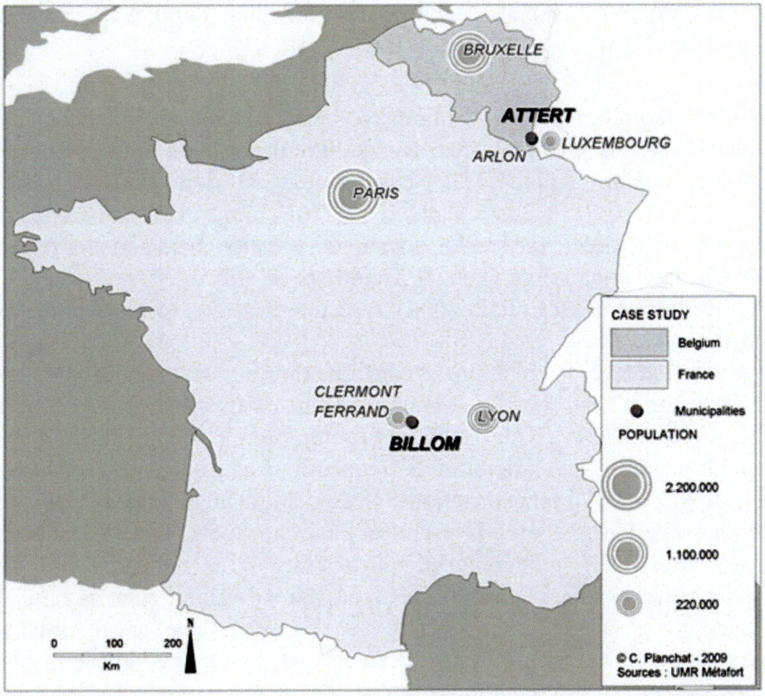

Fig. 9.1 Location of Billom in France and Attert in Belgium in relation to major urban centres

which integrate general or sectorial policies according to local territorial expectations (Lebreton, 2009). Regulatory spatial planning involves restrictions in terms of property and urban law. Regional strategies have a prospective use, defining issues for implementation. Sectorial policies, particularly related to landscape and environment, have appeared in France in the last 20 years and are just starting in Belgium.

Whatever the type of planning, the procedures come from planners who try to avoid the 'heterogeneity of knowledge' (Brand and Gaffikin, 2007: 290). The aims and content of stakeholders' projects, and current processes on the territory in question, are often not clearly considered. This observation is even more manifest on territories halfway between the city and countryside, where the consideration of farming development projects is often forgotten.

Farming can be considered to be a 'trilogy' interlinking stakeholders, activities, and areas (Benoît et al., 2006). For the geographer Jean-Paul Diry (2008), this system is composed of:

- Agrarian landscapes: multifunctional farming systems and land cover, which can be seen as landscape features
- Economic systems of production: local or global laws and markets in relation to modes of production

- Agrarian structures: social structures (farmers, their families, their employees, etc.) and material structures (farms, tractors, etc.).

Local elected officials often continue to consider territories influenced by urban areas as rural territories, but it often seems difficult for them to formulate expectations concerning farming areas. They have trouble considering balance, changes, or continuity (Antrop, 2003; Busck et al., 2009). The contrast between urban and rural no longer helps to define the spatial, social, and identity characteristics of these territories. Through expressions such as 'Tiers-Espace' (Third Area) (Vanier, 2007) to 'rurban' territories (Busck et al., 2009), research continues to try to come to terms with this.

However, whatever the planning project, the farming issue tends to be thought of only in a sectorial way, too often simply in terms of agrarian areas and production systems (Laurent, 2005). Thinking about farming in a more holistic way seems to be difficult for officials. Agriculture is frequently used as a generic term, based on decontextualized social representations, or even stigmatized, and as a consequence provides conceptions that are irrelevant for landscape management. Planning documents in France and Belgium relay an image of farmers as 'industrial' or 'polluters'. On the other hand, social values that reflect positive values, such as rural solidarity, the culture of farming communities, and the countryside, are reported without considering actual agrarian structures. There appears to be a dichotomy between practices and social representations (Table 9.1).

Those most able to speak about the evolution and current dynamics of farming are the farmers themselves. Consequently, it seems important to integrate the farmers' point of view in regional strategy and spatial planning procedures. Spatial and social landscapes representations are used in the research presented in this chapter as a means of landscape mediation, as developed by Joliveau and Michelin (2001). These instruments, used in participative workshops, facilitate the involvement of stakeholders who are seldom integrated in the decision process. Co-constructing representations of the agrarian landscape with farmers is a means whereby elected officials and planners can become aware of and emphasize agrarian needs and structures.

Table 9.1 Farming dimensions in planning documents for territories under urban influence (based on Benoît et al., 2006; Diry, 2008)

	Agrarian landscape	Agrarian and farming structures	Agrarian system	Themes in planning
Practices	Real estate and land occupancy	Farmers, polluters	Harvests, production	Economic development
Social representations	Country landscape, rural	Farmers, people from countryside	Nature, leisure activities, harvesting	Environmental protection

9.1.2 Participative Landscape Mediation

For more than a decade, much research has been undertaken investigating participative tools that help planners by means of stakeholders' landscape representations, such as landscape models used to construct prospective scenarios of landscape change (Joliveau and Michelin, 2001; Tress and Tress, 2001; Caspersen, 2009). This chapter discusses the use of landscape tools for mediation and participation specifically for farmers involved in regulatory planning procedures. The landscape is considered as a tool of mediation in the sense that, from the point of view of cognitive psychology and semiotics (MacEachren, 1991), sensorial knowledge is transformed into data. According to Augustin Berque's theory (2000: 159), the landscape is *médiance*, i.e. a medium between the material (objective) and the ideal (subjective). Spatial and social landscape representations are two interconnected dimensions:

- The landscape is a spatial representation of a territory in the sense that it shows spatial organization, dynamics, and material images (e.g. maps, photographs, drawings, or block diagrams).
- The landscape is a mental and social representation of an inhabited area, from which a dialogue between stakeholders can be started.

To summarize, the landscape is founded on a material geographic reality, which exists independently of the observer, but only has meaning through the view of the observer (Michelin, 2000). The geographical elements taken in by this view are appreciated and interpreted through cultural and social filters. The observer perceives the landscape by mobilizing his or her knowledge. Landscape mediation consists of using visual aids to reveal the knowledge of the observer, his comprehension of the area, his intentions, and his development projects regarding landscape elements. The sharing of this knowledge and these management intentions during participative workshops in the planning process becomes pertinent for designing a shared project.

Therefore, when the stakeholders are asked individually and collectively to transfer what they consider to be important landscape elements on to spatial models, they can be helped to explain, at different scales, the ways in which they perceive and live in their area and to position these perceptions in space. The participative approach presented in this chapter aims to enable the stakeholders (elected officials, planners, and farmers) to specify realistically their territorial and personal ambitions in the planning process. Using landscape mediation helps to understand the modalities of technical and political management of agrarian landscapes. To attain these objectives, various modes of participation (individual and collective) are set out, using a similar methodological approach, but which can be adapted to key stages of different types of planning procedure. The researcher has a dual role, both as facilitator of the participatory process and in its evaluation. The facilitator helps the stakeholders to trust in the participatory process through all its stages. The evaluation concerns two aspects:

1. The production by stakeholders of different spatial and social representations as a means to express their points of view concerning future landscapes relevant for the planning project.
2. The way in which the participation of the farmers was encouraged or not by these means, in relation to the ELC's objective to raise the awareness of local people affected by planning decisions.

In the following, the case areas and their territorial context are first briefly described. A participative process termed the Methodological Itinerary of Prospective Vision is then presented. Finally, the uses and the outputs of the Prospective Vision as a means of taking the farmers' point of view into account in planning procedure are discussed.

9.2 Case Areas and Method

The point of departure for the study was in questions from elected officials about the farming issue in two types of planning: a general spatial plan, represented by the Urban Local Plan (*Plan Local d'Urbanisme*) of the municipality of Billom in France; and a sectorial regional strategic plan, represented by the Landscape Charter of the Natural Park of the Attert Valley, located on the perimeter of the municipality of Attert in Belgium.

9.2.1 Billom and Attert Valley

The study areas were chosen for two reasons. First, they present certain similarities in terms of territorial development under urban influence (Table 9.2). They are both characterized by:

- Political willingness to maintain a rural and ecological living environment while at the same time welcoming new inhabitants
- Farmers forming a social and electoral minority
- Threats to farming areas from urbanization.

Second, the methodological approach could be tested on two types of planning in order to assess the functioning and non-functioning outputs, and to appraise whether or not they can be reproduced.

The municipality of Billom formulated its Urban Local Plan during the period 2005–2008. For a municipality, this procedure determines both the development strategy in a report called the Programme for Sustainable Development and the spatial plan in a separate report. It defines four types of operational zones: (1) urbanized zones [U]; (2) future zones of urbanization [AU], which are predominately farming areas under urban influence; (3) natural areas [N]; and zones strictly reserved for

Table 9.2 Main characteristics of the case study areas

	Billom (Auvergne, France)	Attert (Wallonia, Belgium)
Area (ha)	1696	4893
Urban influence	20 km from the city of Clermont-Ferrand (137,154 inhabitants, density 3214 persons /km^2)	30 km from the city of Luxemburg (90,000 inhabitants, density 1485 persons/km^2)
Population	1999: 4246 2007: 4575	1992: 3328 2007: 4893
No. of farmers having at least part of their farm in the municipality	1988: 48 2000: 21	1987: 131 2009: 47
Farming areas	75% of the farming areas are crops	80% of the farming areas are livestock-breeding
Territorial project	Urban Local Plan: general regulatory planning	Landscape Charter: sectorial regional strategic plan
Question asked by elected officials	How to define a development project and adequate zoning for consideration of the farming issue?	How to integrate the farming issue (practices and values) in the consideration of landscape, environmental, and territorial transformations?
Period of participation	2006: With farmers in the middle of the procedure of finalizing the territorial and landscape diagnosis 2007: With elected officials just before the final zoning plan	2007: With farmers and officials of the Municipality and the Park Council during monthly orientations on the future park charter

agriculture [A]. The question faced by the elected officials of Billom concerned the rules to apply to the farming zones to maintain a balance between production, protection of natural areas and landscapes, and urban development. Faced with such questions, the municipal officials traditionally do not have the financial and technical resources for a farming survey. They leave the task to planners who simplify or even homogenize the area and farming activity without taking in account the expectations and impacts of present and future farming activity on the landscape. In most cases, we find in the Urban Local Plans the same phenomenon described by Busck et al. (2009) in the Netherlands, where zoning is negotiated in a one-way direction, focusing mainly on positioning urban elements in rural areas.

For the Attert Valley Charter, the objectives were different. Here, the urban plan (*Plan de Secteur*) has not been revised since 1974. In 2008, there were no longer any zones that could legally be urbanized. Farming zones are still not designated as such and are sometimes illegally converted into urbanization zones by individuals. The Natural Park must write its Landscape Charter before 2012 to comply with the new decree of 3 July 2008 for natural parks (Ministère de la Région Wallonne, 2008: Article 9), as well as with the objectives the European Landscape Convention

(Christians and Schmitz, 2009). According to this decree, the charter should be written in accordance with a comprehensive local plan (*Plan de Gestion*). The local plan should determine the direction of landscape development and list specific features of the landscape in respect of the local heritage. A risk of this type of document is that it would propose only the preservation of rural landscape elements considered 'emblematic'. The charter aims to be a strategic planning document that intends to have a certain effectiveness but for which the degree of application and the requirements are limited. As in the case of the urban plan of 1974, park officials considered the question of the protection of farming areas against intensive agricultural production, as well as against urban sprawl, with the aim of avoiding deterioration of landscapes and natural areas. However, the expectations of the farmers were initially not included even though they are the main actors in the landscaping process. For this reason, the charter risked presenting only an expert point of view.

Therefore, the local officials accepted a partnership with the researcher to design and test a method that would help them to manage actions that took better into account farming and landscapes. The developed method should take up the following challenges:

- To respect the framework of the procedure and remain a motivating factor
- To be used simply and accessibly, allowing individual and collective operation of different tools
- To design cheap tools, which are quick to build and efficient
- To conceive tools that do not replace those of the planners, but complement them
- To develop holistic thinking for the territory
- To contribute to a heuristic and constructivist approach to knowledge.

9.2.2 *Methodological Itinerary of Prospective Vision*

The Methodological Itinerary is a concept developed by S. Lardon and J.-P. Deffontaines in 1994 (Benoît et al., 2006). Its structure is built around different methodological modules (Fig. 9.2), which, when linked, lead to a thought process, necessary steps, and possible alternatives. In an iterative progression for a continual improvement of sharing knowledge, it is designed as a guide to anticipate different situations. Each module presents the results of previous stages, tools for implementation according to the objectives and knowledge produced, and individual and collective participation of stakeholders in the construction and use of these modules. In short, its implementation enables the specification of different methods and tools for dealing with data, and allows the stakeholders to participate in producing spatial representations in order to build progressively a shared and strategic vision of the territory (Benoît et al., 2006).

The concept of Prospective Vision is the orientation that was given to the Methodological Itinerary designed for Billom and Attert. It consists in building and specifying present and future spatial and landscape visions at key stages of the

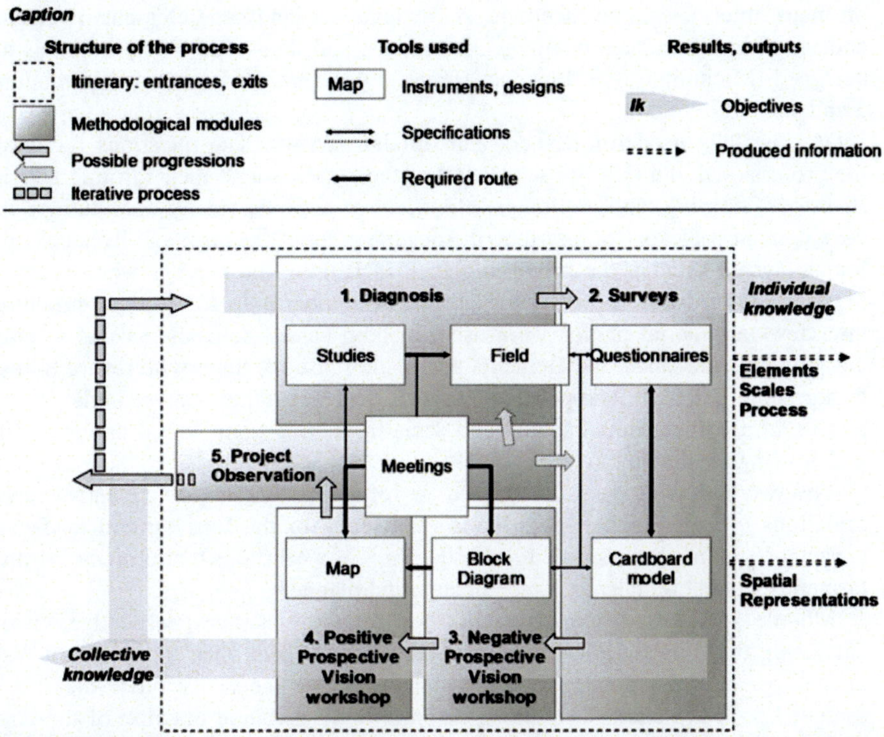

Fig. 9.2 Methodological itinerary of prospective vision (Benoît et al., 2006)

planning process. Rather than constructing landscape scenarios based on simulations that answered the question: 'What will happen?' (Carberry et al., 2002; Benoît et al., 2006), the question was refined to 'What will happen if?' (Tress and Tress, 2003). The scenarios resulting from the Prospective Vision approach included both a negative vision and a positive vision. The questions were posed in the following order: 'What do we not want to see appear or disappear in the landscape?' and then 'What do we have to do to get what we want to see?' The Methodological Itinerary of Prospective Vision provides inputs to the decision-making process before or during the planning project.

The researcher should build the itinerary to improve involvement in the participative approach by facilitating the input of stakeholders and deciphering the representations of the landscape elements made by the participants during the experiment and in the procedure.

The Methodological Itinerary of Prospective Vision comprises five modules. They integrate participative phases that can be renewed at different times in the procedure and with different stakeholders.

Module 1 enables the researcher to report the territorial and landscape diagnosis. It is based on observation and analysis of landscapes in the field, data gathered on maps, interviews, and meetings. A catalogue of landscape elements is created and visual graphics such as maps, 3D models, and landscape block diagrams are designed (Michelin, 2000; Planchat-Héry, 2008). These aids are used in modules 3 and 4.

Module 2 is used for farmers and official surveys. The questions focus on: the professional and family background of the stakeholders; their farming activity (livestock, farming land, buildings, etc.); leisure activities; daily commuting; and development projects (for a farmer, the development of the farm and livestock, and for an elected official, the urban projects). In Billom, the surveys were based on 3D cardboard models of the territory and a questionnaire. In Attert, semi-structured interviews and aerial photos were used. In both cases, stakeholders had to point out with pins the landscape elements that made sense for them with regard to their background, practices, and political projects. The researcher analysed this production in terms of individual spatial and social representations, based on recordings and drawings made on the models.

The two following modules were used for the production of collective representations in order to derive a shared knowledge of the farming and landscape expectations. They were carried out in the form of workshops that brought together farmers, elected officials, park technicians, and planners.

Module 3 is the participative workshop for the Negative Prospective Vision, answering the question: 'What are the landscape objects that one does not want to see appear or disappear in the next ten years?' A landscape block diagram is screened on a flip-chart. This block diagram is a schematic drawing of the landscape elements extracted from the catalogue and surveys. The block diagram is first used to depict the territory in order to discuss the place as it is at present or to locate missing objects. The participants are prompted to write on a piece of paper the elements they do not want to see. The block diagram is then used as a basis for discussion. The participants draw the landscape elements that they do not want to see on the block diagram. They contribute in this way to formalizing a worst-case scenario from the visual model. The researcher, as facilitator, asks the stakeholders to explain their choices concerning the elements drawn. These explanations are based on the individual or collective expectations of what to implement in the territory. For example, a tree in the middle of a field can be considered as starting point for a hedge. Inversely, it could be eliminated because it is considered as an obstacle for the passage of farming machines. The judgement on the necessity of keeping elements is also discussed as an action to be managed collectively. Certain elements may seem disconnected from the values of the researcher. However, the researcher can only intervene by helping the participant to develop a sound argument in relation to certain values that are the result of technical facts or an aesthetic point of view.

Module 4 is the participative workshop for the Positive Prospective Vision to define 'what we want to see'. The researcher asks the stakeholders to convert their negative vision into a positive vision. They were given the following instruction: 'Now that you know what you do not want, what should be done so that the elements evolve to what you do want to see?' As in module 3, the stakeholders are led to prioritize their expectations, and to position them spatially on a simplified map (showing municipal boundaries, roads, buildings, etc.). This spatial positioning can lead to the establishment of prototype zones. The elements identified in workshops 3 and 4 can be used to argue for development proposals to be integrated in the planning. The description based on the knowledge and practices of the stakeholders and their analysis of different spatial processes ends up in the identification of territorial stakes.

Module 5 corresponds to what the decision-makers have, or have not, incorporated into their planning procedure. The researcher analysed in which way the graphic representations and the elements of dialogue produced during participative workshops had been integrated. Two types of integration were distinguished: direct integration, legible in project documents, and indirect integration, through arguments cited by stakeholders during official meetings.

9.3 Representations of Landscape Futures, Desired or Not

Landscape mediation through different spatial representations is used to reveal landscape features specific to the territory at different scales of observation, and to facilitate the expression of points of view, shared or contradictory, about landscapes that are desired or not. The link between modules 2, 3, and 4 reveals shared knowledge and management orientations at different scales of agrarian landscape observation. There are three scales of representation: the scale of the element; the scale of a combination of elements (operation); and the scale usually called panorama (development), considered as an entity composed of a group of elements (Table 9.3).

On the *scale of the landscape element* each individual pays attention to elements he can identify and recognize in reference to his cultural system. These elements are the elementary particles in the spatial composition of landscapes. They maintain the link with landscape processes. By choosing to position an element in relation to another (module 3), the stakeholders reveal what stands out for them in relation to the places. The Prospective Vision approach enables stakeholders to incorporate an expectation for each element.

Starting with the element (Table 9.3), the stakeholders of the Attert Valley recognized visible spatial elements that they could directly influence. For example, they cited 'the fence' around the plot of land that is on the river bank. This fence should not disappear so that the cattle do not damage the riparian area. In the cases of representations linked to buildings and to the phenomenon of urban sprawl, the scale of the element stands out because they readily mention elements that are visible at the human scale – what is in front of them and makes sense for everyone (for example, the church bell tower). For objects that they do not want to see appear, the elements mentioned are specific and precisely located, for example electric power lines, and urbanization along main roads (urban sprawl, and the industrial zone project).

Through the use of block diagrams (Figs. 9.3 and 9.4), the scale of the element was very useful to describe or model what they did not want to see. The block

Table 9.3 Three scales of perception of the Attert Valley

	Elements		Operation		Development		Thoughts, arguments	
Elements	Not desirable	Desirable	Not desirable	Desirable	Not desirable	Desirable	Not desirable	Desirable
Urbanization, infrastructure	High voltage transmission lines Buildings GSM antenna Highways Urban development zone(s) near villages	Churches Nobressart hamlet Mill Old walls	Traffic lights Hedges 50 m from intersections Housing estates	Integrated housing Rural housing Public buildings	Urban development of Heinstert Development sprawl of buildings Housing in interesting landscape zones		Land rental from Luxemburg	
Farming practices	Rural paths Isolated spruce trees Speed bumps Fences or hedges along farming roads	Fences on river banks Family farms Christmas trees Paths Short cuts Hedges	Development near farms Extension of tree farms Golf courses	Farming fallow	Extended monoculture	Large plots of grassland Farming landscapes	Farming of bio-fuels Intensive farming	Farmers Cultivators Herds of cows crossing the road Buried wet zones and farming grasslands

Table 9.3 (continued)

	Elements		Operation		Development		Thoughts, arguments	
Elements	Not desirable	Desirable	Not desirable	Desirable	Not desirable	Desirable	Not desirable	Desirable
Economic and territorial stakes	Polluting factories Airport, airfield Enlargement of networks Malls Loading docks for wood		Closing of quarries				Increase of road traffic (due to residences)	
Natural elements		Remarkable trees Ponds	Reforestation New quarry	Forest		River bends Ecological zones of high value		Wild zones
Social relations							Farmers from Luxemburg Lack of responsibility of local residents	Natural Park
Concepts							Road safety	Countryside Rural character of villages

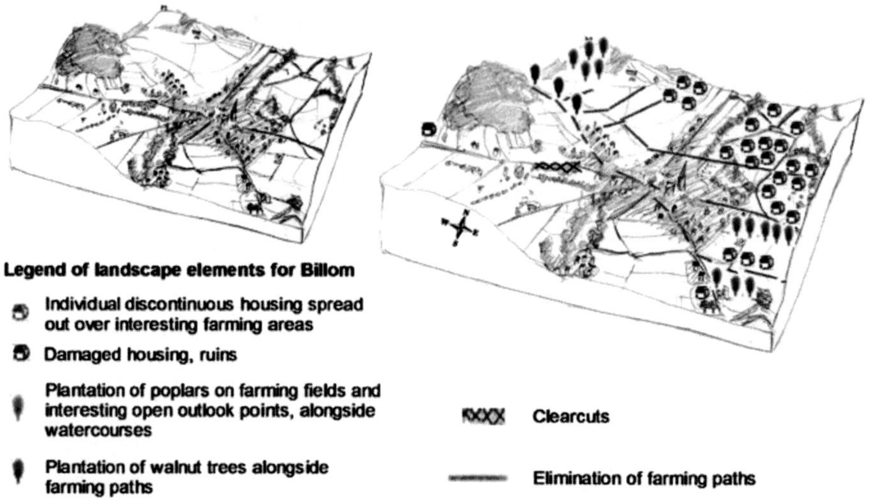

Fig. 9.3 Landscape block diagram showing a negative prospective vision for Billom

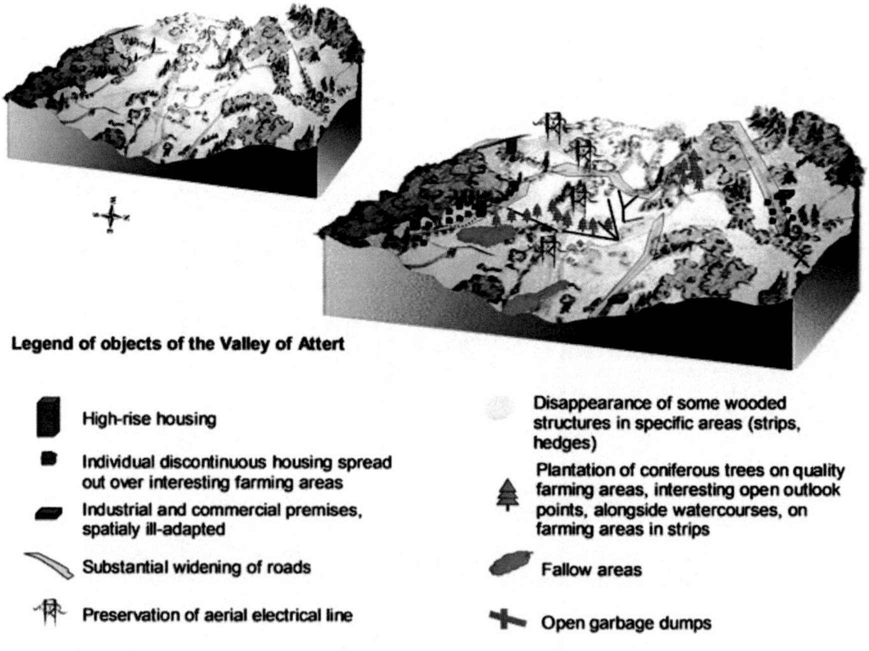

Fig. 9.4 Landscape block diagram showing a negative prospective vision for Attert

diagram was used to place, modify, and prolong features in order to make a rejected landscape more perceptible and concrete.

The *scale of groups of elements,* termed the 'operation scale', relates to the spatial approach in the planning procedure. It enabled the specification of landscape dynamics in relation to socio-economic mechanisms in the territories. For example in Billom, the farmers compiled a list of buildings that were too dilapidated for farming activities, but which were located in a hamlet and could be converted into residences. Along with the elected officials and urban planners, everyone agreed to propose both the reclassification of the central area of the hamlet in order to give it the appearance of a rural village, and the preservation of farming activity on plots of land close to the village in order to keep the traditional landscape as a backdrop to the buildings. This proposal accorded well with the territorial policy for reception of newcomers as well as for the preservation of farming areas on the outskirts of villages.

The choice of this scale gave the stakeholders an understanding of socio-political, economic, and ecologic mechanisms causing landscape changes that they considered non-desired. In the example of Billom, this scale revealed the impact on the landscape of the individual strategies of those owning property or farm buildings. The buildings were likely to be sold to the highest bidder because of financial pressure exerted by newcomers. On the other hand, the municipality was trying to defend the 'rural' coherence of the village.

The *scale of the panorama*, using the planners' term, refers to territorial development. The landscapes on this scale integrate multiple layers and forms of intervention in administrative and geographical territories. This not only concerns individual interests but also especially collective stakeholders and the general interest. This can be termed the 'development scale' in reference to the concept of governance and development of these territories. This scale is the one favoured in the procedures for talking about landscape quality. Here, the landscape element is often considered either as a useful element to characterize a panorama and make it emblematic (e.g. 'river banks'), or as an obstacle that will hide this panorama (e.g. 'residences in interesting landscape zones') (Table 9.3).

The Prospective Vision was used in the two studies as a way to bring the stakeholders together around what they did or did not want to see. Therefore, even people who were in direct opposition to one another concerning their expectations realized that they shared certain common values and that they rejected the same things. It then became easier to move towards the construction of a common project and to map it. Elected officials and developers greatly appreciated co-producing the map in module 4 after having been able to represent their territory in a less abstract way.

Going from one scale to another is crucial in the eyes of elected officials and planners. They are then able to suggest management operations that are adapted to the areas of their daily life and to the scales of perception of landscapes. This enables the improvement of the knowledge and social representations to be shared and integrated into the planning.

9.4 Integration in Planning

The results obtained with the participative approach were validated by indirect and direct integration of farmers' points of view into the planning procedures. Indirect integration was expressed by the participation of a large number of farmers in the participative phases. They explained their own farming operations and their commitment to develop them. Consequently, they indirectly instigated their personal long-term involvement in the planning project. Direct integration was assessed on the basis of results from the participation modules, for example the maps produced in module 4 that were reproduced in the final planning document.

9.4.1 Advantages of the Participative Approach for Farmers: Indirect Integration in the Project

The approaches that mobilized landscape representations paid consideration to farmers. With their agronomic arguments, they could deconstruct the views of elected officials who considered them, at the beginning of the process, as only 'landscape gardeners'.

The social representations of their farms helped the different stakeholders to specify the roles and responsibilities of each person in the local management of landscapes. For example, concerning Attert Valley, the results of modules 3 and 4 highlighted the role of cross-border farming practices by farmers from neighbouring Luxemburg, perceived negatively by many, as well as the ecological sensitivity of farmers. Farmers considered the management of meadows in an ecological way as an important part of their practices. During the individual surveys, many spoke of the difficulty of preserving their property against competition from Luxemburg farmers who have a need for cultivable land, and against the real-estate competition due to the development of urbanization. On the Luxembourg side, crops are favoured, while on the Belgian side, it is more mixed farming and breeding in a grazing system. This theme was taken up collectively during the participative workshop where management suggestions for these plots were thought through. By tackling the landscape question at the scale of positive landscape elements (e.g. 'large plots of grassland'), the stakeholders were able to undertake its management more easily. The 'development scale' revealed the impact of regional farming and forestry policies that must be integrated, as well as the interest in collaborating with cross-border natural parks in Luxemburg in order to contemplate a more appropriate management of farming areas.

Negative Prospective Visions are an important part of the participative approach. This module does not eliminate conflicts, but it reduces them substantially because divergences are not merely mentioned but also explained and, most importantly, spatialized. The presence of both young farmers and retirees in the workshops helped to induce friendly exchanges intended to favour the continuity of farming property. With the presence of elected officials and park technicians, the agronomic interest of these plots could be compared with their ecological value within a complex political interaction. Therefore, the charter was in a position to take into account the objective of preserving open agrarian landscapes, while also integrating a well-thought-out management of the property demands of young Luxemburg farmers.

In Attert, there was particular interest in module 2, the individual survey. Even if many technical-economic surveys are carried out among farmers (related to the EU's Common Agricultural Policy, property market surveys, etc.), they are rarely published in the planning documents. In the workshop, the farmers whose operations were confronted with other interests could better understand their role and the power they had concerning urban or environmental management. This is because 70% of areas left 'blank' on the map and traditionally excluded from the planning procedure belonged to them.

9.4.2 Benefits for Planning: Direct Integration

Billom, where the project lasted from 2006 to 2008, best illustrates direct integration of the results obtained during the participative workshops into the planning document (Fig. 9.5). The Prospective Vision and the participatory approach went beyond the objectives of obligatory consultation in the official procedure. Farming landscape issues were included in the sustainable development and spatial plan.

Officials considered farming in its present situation as the best way to ensure the conservation of open landscapes, balancing cereal crop production with the limitations of urbanization. The officials were especially concerned with maintaining the rural way of life. Parts of the farming areas continued to be accepted as scenery or as leisure areas in the final planning document. Here, areas of arable land are classified in zones designated as 'natural areas and landscape interest' [N] (Fig 9.5c) (Feydefont et al., 2007). Farmers are aware of the urgency of protecting natural areas and landscapes as a new resource for quality farming. However, although this image of 'rural heritage landscapes' is strongly favoured by officials, both farmers and planners realize that they will not be able to reproduce them. It is more a matter of targeting some features to conserve (e.g. vineyard cabins, or small forested areas) as a testimony to the past and that need to be integrated into new landscapes.

So as to encourage farming practices and not damage cereal activities, individual farming operations and results from the farming surveys were taken into account. Officials asked the planner to designate specific zones for the development of farm buildings [As] (Fig 9.5b). These buildings were to be developed to take into account the evolution of farming activities, the farmer's family, and the management of farm pollution, which had a negative impact on the neighbouring natural areas. 'Green sectors' on the outskirts of the urban areas were to be maintained. In the official zoning, the map presents only one agricultural zone [A]. However, where agricultural areas were very small and occupied by a farmhouse, the planner wrote a specific regulation in the official document with the aim of preserving them from urbanization [As](Fig 9.5c).

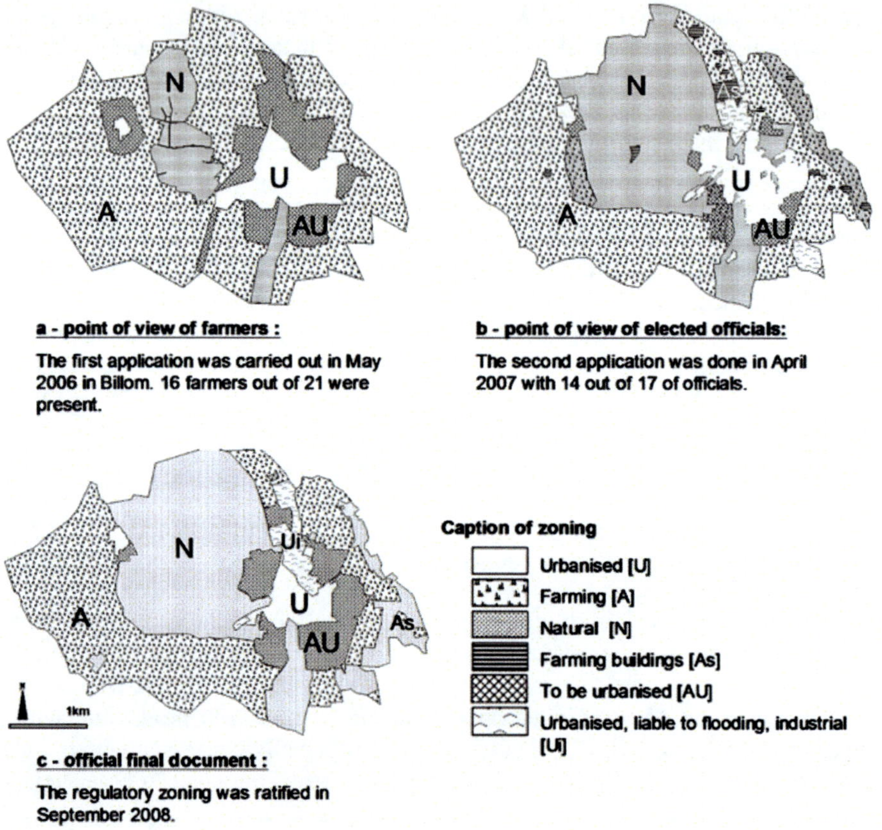

Fig. 9.5 Two maps of positive prospective visions and the official urban local plan (by Planchat Surveys) (Freydefont et al., 2007)

By taking components of landscapes and their systems into account, the regulatory approach adopted in Billom gave the officials the opportunity to limit what Marc Antrop calls the random effects of individual decisions that lead to a loss of landscape diversity (Antrop, 2003: 12). The balance between zones [A] and [N] could be renegotiated. However, even if current farming activity is maintained, the result will not be the 'picture postcard' landscape the officials initially wanted, but also not a purely functional landscape of cereal farming. The Urban Local Plan is the result of a compromise between the individual interests of the owners and farmers and the collective expectations expressed by the officials.

The Prospective Vision approach awakened the stakeholders' awareness of the political choices and the willing or unwilling changes that they bring to landscapes, to property, and to agrarian systems. Table 9.4 summarizes for the two case studies 'the experience gained from exchanges between local people affected by the planning decisions to be taken and persons possessing scientific and technical knowledge' (Council of Europe, 2008: II.2.3.B).

Table 9.4 Indicators of 'awareness raised' concerning farming issues as a result of the experience gained from exchanges between local people affected by the planning decisions to be taken and persons possessing scientific and technical knowledge

		Awareness raised Billom (Auvergne, France)	Both municipalities	Attert (Wallonia, Belgium)
Knowledge of practices	Farmers	Landscaping elements linked to harvesting: hedges, open fields, drains	Farming and land-use system as a social system	Ecological elements linked to farming (wetlands, cultivated banks, nesting in harvests)
	Elected officials	Taking into account individual farming projects for maintaining agriculture in the municipality	Farming between the global and the local	Affected farmers' practices and perceptions during the entire Landscape Charter process
	Planners	Participatory tools to involve farmers in decisions	Different scales for managing landscapes: elements, systems, scenery	Concept of rural landscape without overemphasis on remarkable architecture or scenery
Knowledge of social representations	Farmers	To give importance to the land as an outcome for the territory	Farmers as a guarantee for preserving country values	Better understanding of demand for urbanization.
	Elected officials	Thinking from viewpoint of farmlands rather than thinking of 'where to urbanize'	Multi-functionality of farming as a collective approach	To enhance the 'contemporary' way of life in rural landscape areas
	Planners	Local farming as a lever to define specificity of local landscape	Localize 'everyday' landscapes as well as remarkable landscapes	To integrate into planning both local and generic perceptions of the landscape

9.5 Interest and Limits of the Method

To a certain degree, the objective of enriching two planning procedures by participative approaches was reached. Ahern (2005) emphasizes that the development of new methodological contributions considers participation not as an objective in itself but as one of the tools to integrate in a process to bring out elements that raise awareness, and which are pertinent for the integration of the farmers' point of view in the procedures (Table 9.4).

The Prospective Vision approach helps shed new light on the use of landscape representations in planning as developed by Joliveau and Michelin (2001). In the same way that adding graphs to a text facilitates comprehension and memorization of information, the use of spatial representations in a discussion encourages debate, based on knowledge that is made accessible to the entire community, who can thus consider it and re-examine it.

The Methodological Itinerary of Prospective Vision is influenced more by agronomic than planning research, and validates the mobilization of limited technical instruments while aiming for a certain efficiency in the production of knowledge (Carberry, 2002; Benoît et al., 2006). The Methodological Itinerary initiated provides a 'tool kit' that helps both practitioners and elected officials (Brand and Gaffikin, 2007).

However, this approach depends not only on its immediate local application, but also on its appropriation by the local stakeholders in the long term. In Attert, the stakeholders described the block diagram drawings as 'not very pretty'. However, they appreciated redrawing, completing, and transforming the drawings, and discussing technical arguments, not only aesthetic ones. Naive forms of landscape representation are liberating in comparison with digital visualization designs with an attractive interface that have been developed recently. Often the vision presented by digital designs imposes the aesthetic view of the researcher or planner, which stakeholders are seldom able to challenge (MacEachren, 1991; Joliveau and Michelin, 2001).

The method of Prospective Vision contributes to 'developing a local policy of awareness and explanation for production systems in order to integrate practices (of production or management) for landscape functions' (Schmitz, 2004: 117). The farmers' involvement helps to present practices and management proposals on different scales, from the element to the panorama. Finally, it aims for 'public participation in order to integrate the invisible part that is too often neglected but which gives meaning to our landscapes' (Schmitz, 2004: 117).

In Billom, certain categories of inhabitants did not participate because the approach was exploratory and concerned particularly the farmers. The fact that 16 farmers out of 21 participated and that, in the end, the planning document took into account some of their needs represents in itself significant progress. Even if it is interesting to experiment with the sharing of local knowledge (Caspersen, 2009), the participation of all the actors of a territory needs other methodological devices. It is more complicated, needing a bigger team for facilitation and evaluation, with several workshops respecting the key stages of the regulatory procedure.

The adoption of research tools by practitioners is rare, in spite of the European Landscape Convention and the increasing number of recommendations proposed by national policies to encourage participation in public decisions. The researcher as facilitator is in itself an important research topic (Planchat-Héry, 2008). From a methodological point of view, 'the contemporary planner would need to be not only gifted in facilitation and arbitration for proactive engagement, but also skilled in semiotics and hermeneutics for analysis' (Brand and Gaffikin, 2007: 308). In order to respect the provisions of the European Landscape Convention on local participation, a pre-established solution, made from a combination of 'good tools' used by 'good planners', cannot exist. It is necessary to develop a reflexive process in accordance with the questions that stakeholders have and the questions the researcher wants to ask the stakeholders.

9.6 Conclusion

The position of the researcher in the experiments in Billom and Attert was not so much to provide expertise as to help the 'interested parties' to find their own answers to questions using the Prospective Vision. It is not the vision itself that is most important but the questions asked around the vision that will guide strategies. This method helped decipher the scales of landscape perception to which these strategies would apply. Using different steps and tools for spatial and landscape representations enabled the interpretation of territory both as a whole and through its components, which make up the singularity of each case study. If the stakeholders mention and debate certain elements without realizing the scale changes, the researcher can then help highlight these scales in order to facilitate the reading of representations produced by the stakeholders. In the end, 'the decisions still have to be made by those who have the responsibility and vote' (Tress and Tress, 2001: 13). The collaboration between researchers and stakeholders can nevertheless contribute to making this decision more specific to the territory in question and more easily acceptable to all parties.

Encouraging the sharing of knowledge, in association with a local and official partnership, will help the European Landscape Convention to gain in recognition and effectiveness. It is important to rebuild the social and material link with the territory, particularly in 'rurban' municipalities: if this link is created, the management of the landscape that will be produced from this knowledge will be all the more pertinent. While there will never be a universal method to create this link, this perspective can encourage the development of guidelines for an approach that on each occasion will be adapted to the specificities and evolution of the territories in question.

Acknowledgments This work is part of the author's Ph.D., supervised by Professor J. P. Diry of University Blaise Pascal, and Professor S. Lardon and Professor Y. Michelin, UMR Metafort at Clermont-Ferrand. For the partnership in Belgium, thanks are due to Professor C. Feltzt of the University of Agronomy of Gembloux, Professor C. Christians of the University of Liège, and the Attert Valley Park. For the French partnership, thanks are due to the City Hall of Billom, the Chamber Farming Council of the Puy-Dôme, and the Urban Planners Cabinet of Sycomore.

References

Ahern J (2005) Theories, methods and strategies for sustainable landscape planning. In: Tress B, Tress G, Fry G, Opdam P (eds) From landscape research to landscape planning: aspects of integration, education and application. Springer, Dordrecht, pp 119–131

Antrop M (2003) Continuity and change in landscapes. In: Mander Ü, Antrop M (eds) Multifunctional landscapes. WIT Press, Southampton, pp 1–14

Benoît M, Deffontaines J-P, Lardon S (2006) Acteurs et territoires locaux: Vers une géoagronomie de l'aménagement. INRA, Paris

Berque A (2000) Médiance de milieu en paysages. Reclus Géographiques, Belin, Paris

Brand R, Gaffikin F (2007) Collaborative planning in an uncollaborative world. Plan Theory 6(3):282–313

Busck AG, Hidding MC, Kristensen SBP, Persson C, Søren P (2009) Planning approaches for rurban areas: case studies from Denmark, Sweden and Netherlands. Geogr Tidsskr – Dan J Geog 109(1):15–32

Carberry PS, Hochman Z, McCown RL, Dalgliesh NP, Foale MA, Poulton PL, Hargreaves JNG, Hargreaves DMG, Cawthray S, Hillcoat N, Robertson MJ (2002) The FARMSCAPE approach to decision support: farmers', advisers', researchers' monitoring, simulation, communication and performance evaluation. Agric Syst 74(1):141–177

Caspersen OH (2009) Public participation in strengthening cultural heritage: the role of landscape character assessment in Denmark. Geogr Tidsskr – Dan J Geogr 109(1):33–45

Christians C, Schmitz S (2009) Conformités des paysages ruraux et parcs naturels: L'exemple wallon du parc naturel de la Vallée de l'Attert. In: Van Eetvelde V, Sevenant M, Van De Velde L (eds) Re-Marc-able Landscapes – Marc-ante Landschappen: Liber Amicorum Marc Antrop. Academia, Gent, pp 159–170

Council of Europe (2000) European Landscape Convention, Florence, 20.X.2000. ETS No. 176. http://conventions.coe.int/Treaty/EN/Treaties/Html/176.htm. Accessed 3 Jan 2007

Council of Europe (2008) Recommendation CM/Rec(2008)3 of the Committee of Ministers to Member States on the guidelines for the implementation of the European Landscape Convention. https://wcd.coe.int/ViewDoc.jsp?id=1246005. Accessed 20 Jul 2009

Diry J-P (2008) Les espaces ruraux. Armand Colin, Paris

Freydefont JM, Olivier M-A & Groupe Sycomore (2007) Plan Local d'Urbanisme de la commune de Billom: Rapport de présentation. Mairie de Billom, Billom

Joliveau T, Michelin Y (2001) Modèles d'analyse et de représentation pour la prospective paysagère concertée: Deux exemples en zone rurale. In: Lardon S, Maurel P, Piveteau V (eds) Représentations spatiales et développement territorial. Hermes, Paris, pp 239–266

Laurent C (2005) Les agriculteurs de l'Union Européenne et leurs territoires. In: Laurent C, Thinon P (eds) Agricultures et territoires. Hermes, Paris, pp 31–55

Lebreton J-P (2009) La planification spatiale en Europe: introduction générale. In: Collège Aménagement, urbanisme, habitat et mobilité. Conseil Général de l'environnement et du développement durable. www.gridauh.fr/sites/fr/fichier/4a4e294eb0360.pdf. Accessed 20 Oct 2009

MacEachren AM (1991) The role of maps in spatial knowledge acquisition. Cartogr J 28(2): 152–162

Michelin Y (2000) Le bloc-diagramme: Une clé de compréhension des représentations du paysage chez les agriculteurs? CYBERGEO 118. www.cybergeo.presse.fr. Accessed 8 Feb 2006

Ministère de la Région Wallonne (2008) 3 juillet 2008 – Décret modifiant le décret du 16 Juillet 1985 relatif aux Parcs Naturels. Moniteur Belge, DRW/20080703/AG, 01/08/2008 – Ed. 2, 40662–40666. http://www.ejustice.just.fgov.be//mopdf/2008/08/01_2.pdf#Page32. Accessed 31 Jan 2010

Planchat-Héry C (2008) Les outils graphiques paysagers, révélateurs des enjeux agricoles dans un Plan Local d'Urbanisme. In: Lardon S, Chia E, Rey-Valette H (eds) Dispositifs et outils de gouvernance territoriale. Norois, PUR, Rennes, pp 109–125

Schmitz S (2004) Introduction aux paysages ordinaires. In: Conférence Permanente du Développement Territorial (ed) Territoires, urbanisation et paysages. Actes 4ème rencontre de la CPDT. Région Wallonne, Liège, pp. 116-117. chmi

Tress B, Tress G (2001) Scenario technique and participation: a new approach for rural planning in Denmark. In: Akademie fiir Raumforschung and Landesplanung (ed) Cities: engines in rural development? Arbeitsmaterial 268. Akademie fur Raumforschung und Landesplanung, Hannover, pp 72–77

Tress B, Tress G (2003) Communicating landscape development plans through scenario vizualization techniques. In: Palang H, Fry G (eds) Landscape Interfaces: cultural heritage in changing landscapes. Kluwer Academic Publishers, Boston, pp 185–220

Vanier M (2007) Métropolisation et tiers espace. Rencontres de l'innovation territoriale www.pacte.cnrs.fr/IMG/pdf_41_Vanier_Tiers_espace.pdf. Accessed 21 Jun 2009

Chapter 10
'Landscape Quality Objectives' for Remote Rural Landscapes in Portugal: Addressing Experts' and Stakeholders' Perspectives on Future Developments

Isabel Loupa Ramos

Abstract The European Landscape Convention calls for the definition of 'Landscape Quality Objectives' (LQOs) as 'the formulation by the competent public authorities of the aspirations of the public with regard to the landscape features of their surroundings'. The Convention itself does not propose any specific approach to identify and integrate the aspirations of the public in the formulation of LQOs. There is thus a need to develop and test specifically adapted methodological approaches. As the identification of aspirations cannot be dissociated from an

I.L. Ramos (✉)
Centre for Studies of Regional and Urban Systems (CESUR), Technical University of Lisbon, 1049-001 Lisbon, Portugal
e-mail: isa.ramos@civil.ist.utl.pt

M. Jones, M. Stenseke (eds.), *The European Landscape Convention*,
Landscape Series 13, DOI 10.1007/978-90-481-9932-7_10,

informed discussion about a desired future, it does not seem a straightforward task in remote rural landscapes that are being abandoned by people and consequently by their activities. This chapter aims at presenting the perspectives of stakeholders and experts towards the future of the landscape of Mértola (southeast Portugal) as it surfaced through the use of 'exploratory landscape scenarios' in the context of the methodological approach developed to address the formulation of LQOs.

Keywords Landscape Quality Objectives · Stakeholders · Exploratory Landscape Scenarios · Rural Landscapes · Portugal

10.1 Introduction

In line with other policy documents such as the Aarhus Convention (UNECE, 1998), the public and its involvement is a major pillar of the European Landscape Convention (ELC). This aspect is first presented in the definition of the landscape concept itself, which builds on the 'perception of the public'; second, the ELC calls for the formulation of 'Landscape Quality Objectives' (LQOs) derived from 'the aspirations of the public'; and third, it stresses the need to establish procedures for 'participation of the public' (Council of Europe, 2000: Articles 1a, 1c, 5c). However, it is neither clearly defined which public should be addressed nor to what extent it is expected to be involved, nor how the aspirations should be identified. To the criticism concerning the lack of clear directives towards an effective implementation of the ELC in all signatory states, the Council of Europe states that the subsidiarity principle must prevail and, therefore, the adequate 'formula' should be developed for each landscape, encouraging 'creativity in the research and experimentation' (Council of Europe, 2008: II.2).

The definition of 'Landscape Quality Objectives' is a centrepiece of the ELC as it aims at setting the basis for further actions. According to the ELC, LQOs are 'the formulation by the competent authorities of the aspirations of the public in their surrounding landscape' (Council of Europe, 2000: Article 1c). This definition calls for the integration of landscape characteristics and the views of the public towards the landscape, serving as the basis for landscape policy formulation, landscape planning and landscape monitoring. It is this pivotal role of LQOs in the context of the ELC that makes its formulation so sensitive.

The present chapter aims to present an approach for identifying aspirations of stakeholders and experts towards the landscape in remote rural areas as it surfaced through the use of 'exploratory landscape scenarios', and thereby to show how this type of scenario can be a useful tool in the formulation of LQOs. Remote rural areas pose particular challenges in ascertaining the aspirations of stakeholders because of out-migration and abandonment of landscapes. The first set of questions concerns the 'public': 'Who is the public?' and 'How to identify the public of a specific landscape?' The second set of questions relates to 'aspirations': 'How to conceptualize aspiration in landscape studies?'; 'How do aspirations relate to preference?'; and 'How to address their future-oriented nature?'

10.2 The 'Public' and Its Involvement

The focus of the ELC on the 'public' has been extended from the initial 'local population' to include 'visitors' to these landscapes in later documents (Council of Europe, 2008). The early focus on the local public is not surprising taking into account the role played by the Congress of Local and Regional Authorities (CLRAE) in the preparatory work leading to the ELC's adoption (Phillips and Clark, 2004).

There has been a rising awareness of the role of 'outsiders' in the construction of the rural landscape (e.g. Halfacree and Boyle, 1998). Urban population is being recognized as one of the main drivers of rural landscape change through its demands and uses (e.g. Ferrão, 2000; Antrop, 2004; Klijn, 2004). In addition to the traditional functions related to food and fibre, new demands that value other landscape functions are arising (Pinto-Correia and Breman, 2009; Wiggering et al., 2006).

It becomes evident that the formulation of LQOs that accommodate 'the wide range of social perceptions, which reflect the population's diversity' (Council of Europe, 2008: II.2.2) must comprise all the public. However, this is challenging. As Jones (2007) points out, the public to be taken into consideration by the ELC is manifold. As you move away from the local public, the identification and location of other users of the local landscape become less clear. The blurring of the urban-rural divide (Baptista, 2001; Roberts, 2002; Mormont, 1996) does not make it easier to identify and classify the public as far as its relation to a specific landscape is concerned.

To embrace the public in a broad sense may mean not only the actual users but also potential users. Literature provides little experience on methodological approaches for addressing such a vast spectrum of public in landscape studies. These variously focus on: local population, possibly differentiated according to socio-demographic characteristics (e.g. age, gender, occupation, period of residency); direct users of a specific landscape, for instance visitors; users in relation to specific activities, for example managing, working, hunting, beekeeping, mushroom picking, etc. (e.g. Surova and Pinto-Correia 2008); or decision-makers, public administrators and planners (Ryan, 2006). Jones (2007) refers explicitly to local communities, landholders, visitors, immigrants and groups with special needs as among those that must be addressed in the context of the ELC.

Another question concerns the geographical distribution of the public and consequently the need to define a 'sphere of influence' or 'attraction' of a specific landscape. This issue has not received much consideration in the literature either. In remote rural areas it is crucial in the definition of the public to be considered. Extremely low population density together with peripheral location increases the relevance of the role played by the external public in the construction of the landscape.

Nevertheless, there is little doubt about the crucial role of the local population in the landscape, both as actors and as observers (van der Leeuw, 1999), taking actions in physical construction and re-creating landscape as a mental construction; thus it becomes part of their individual identity (Duby, 1991; Groth, 1997). Landscape changes at the local level mostly affect the local public's personal living conditions.

As already experienced by Patel et al. (2007), the local public seems to face some constraints in envisioning alternative futures. Consequently, exercises that involve dealing with the future might benefit from the integration of multiple visions and notably 'outside' views, which bring in new information that can be used to develop new future landscapes.

The local public is diverse in itself. There are significant social and economic differences among individuals, as well as concerning their role within the community. Typically, more analytical approaches may prefer to address a representative sample of the local public to gather the whole diversity of landscape perceptions, whereas approaches that are more focused on landscape planning and management may give preference to individuals that hold a special position within the community and thereby may be better equipped to motivate the whole community towards action. Even though the concept of stakeholder has emerged from a corporate context as 'those groups without whose support the organization would cease to exist' (Friedman and Miles, 2006: 5), it has been extended to other organizational structures, such as communities, mainly due to the richness of their local and contextual knowledge (Marjolein and Rijkens-Klomp, 2002).

For the purpose of implementing landscape policies, the integration of LQOs into local spatial planning documents appears to be crucial (Zoido-Naranjo, 2005; Ramos and Saraiva, 2008), and the success of their implementation depends strongly on the mobilization of the local public. This calls for a level of public participation in the definitions of LQOs that might have to surpass 'public consultation'. The latter level may be insufficient, as it means that the public may hear and be heard but lacks the power to ensure that its views will be taken into account in policy making (Arnstein, 1969). Shortcomings in integrating the public in the process may hinder future acceptance and implementation of landscape plans (Luz, 2000) that build on LQOs.

10.3 Framing the 'Aspirations' of the Public

Formulating LQOs is about creating a vision for the future of the landscape based on the 'aspirations of the public'. Merriam-Webster's dictionary (2009) defines 'aspiration' as (a) 'strong desire to achieve something high or great' (b) 'an object of such desire'. This concept has not been much theorized in literature and in the scope of landscape studies, where 'public preference' is traditionally explored. Returning to Merriam-Webster's dictionary, 'preference' calls for the (a) 'the power or opportunity of choosing' and (b) 'giving advantages to some over others', and is synonymous with 'choice'. Visions can be seen as preferred futures in comparison to all those futures available (van der Helm, 2009). Thus, the 'aspirations of the public' in the context of the ELC can be defined as a preferred or desired future state of the landscape chosen among a set of plausible futures.

The *Guidelines for the Implementation of the European Landscape Convention* state that the definition of LQOs should be based on the 'knowledge of the specific characteristics and qualities of the places concerned, and identification of their dynamics and of potential as well as of how landscape is perceived by the public' (Council of Europe, 2008: II.2.2), meaning that the choice is limited to those aspirations identified as meeting the sustainability criteria (Lugenbühl, 2006). This requirement places LQOs very close to the *Leitbild* concept where, besides the aspirations for desired futures, social, cultural, political and environmental perspectives also have to be integrated (Klug, 2007).

Creating this vision (van der Helm, 2009) is about combining an idealized future (aspirations) and deliberate change (planning), based on the assumption that there are no futures out there to be discovered (Schwartz, 1996; Godet, 2001; van der Heijden, 1996) and that there is collective responsibility in the construction of futures (Healey, 1997, 2008) which are more sustainable and improve people's quality of life. Hence, LQOs are expected to be forward-looking, developing visions for landscapes that are capable of converging for the development of landscape policies and planning into the desired direction. As Ackoff (2006) stresses, one has to be aware that a discussion on a desired future draws foremost from consciousness of present conditions and therefore it is a discussion on reducing the gap between where we are and where we want to be. Therefore discussing the future can be less compromising than discussing the present as it provides stakeholders with more space to envision a desired future. This vision acts then as a joint platform or common ground for taking preparatory steps towards action.

The creation of a vision in remote rural areas may be notably compromised by their demographic characteristics. As the population with the most initiative tends to be the first to abandon these landscapes (Correia et al., 2004), the remaining population shows difficulties in creating a vision and in motivating themselves towards action.

10.4 Scenarios as a Communication Tool

Landscape scenarios are commonly used as a way to communicate future landscape change and thereby play a relevant role in fostering public participation (Jones et al., 2007). Their communicative capabilities rely strongly on a wide spectrum of visualization techniques that may include handmade drawings and paintings (e.g. Jones and Emmelin, 1995; Palang et al., 2000), collages (e.g. Artner et al., 2006), and systematic use of computer techniques, such as photorealist representations (e.g. Nassauer et al., 2002; Tress and Tress, 2003) or digital simulation using GIS that enables more interactive 'walk-throughs' and 'fly-throughs' (e.g. Dolman et al., 2001; Dockerty et al., 2005). Sheppard (2005) draws attention to the need of carefully choosing the most adequate visualization techniques according to the characteristics of the landscape and of the public involved.

Nassauer and Corry (2004) point out the wide array of landscape scenario approaches that have been used by landscape ecologists in order to integrate ecological values in landscape planning practice by making policy options visible (Nassauer et al., 2002). Landscape scenarios can be based on different policy options (Jones and Emmelin, 1995; Palang et al., 2000), alternative development models (Hulse et al., 2002; Steinitz et al., 2003), current conflicting interests on the landscape (Tress and Tress, 2003) or implications of present decisions or societal behaviour in the future as, for instance, the impacts of climate change on future landscapes (Dockerty et al., 2005; Sheppard, 2005). While landscape scenario approaches tend to be normative, showing what the future should or should not look like, 'exploratory landscape scenarios' (Ramos, 2008) aim at finding plausible landscape futures. Their development is based on a procedure that builds on the 'intuitive logics' approach (Schwartz, 1996) by focusing on the production of a variety of scenarios as a starting point for a discussion about the future rather than on finding an optimal or likely one (Ramos, 2008). Scenarios can be considered as tools for 'ordering one's perceptions about alternative futures' through constructions of 'internally consistent views of what the future might turn out to be' (Porter, 1985; Schwartz, 1996). Their intuitive and qualitative nature makes scenarios well suited to deal with 'discontinuity'.

'Exploratory landscape scenarios' arise from the need to envision futures in remote rural landscapes, proving to be a tool able to accommodate the high level of uncertainty they face. These landscapes are at the threshold of major change throughout Europe. Some of these remote rural areas are struggling with the role of agriculture which is entering a downward spiral of rural decline (OECD, 2006; Dhubháin et al., 2009). It is widely accepted that agriculture is no longer the main economic force nor does it provide the rationale for the organization of rural societies (Mormont, 1996; Roberts, 2002; Antrop, 2006). The role of agriculture is being altered under the influence of the new rationale of the Common Agriculture Policy (CAP), which has been the subject of incremental reform largely in response to budgetary pressures, the liberalization of agricultural trade and also societal expectations towards rural landscape (e.g. Roberts, 2002; Potter and Tilzey, 2005).

In this context, driving forces of future landscape change cannot be expected to be those of the past. As Naveh (2005: 353) points out, it is not possible to 'predict the future of our landscapes and their rapid sometimes even chaotic changes by simply extrapolating from the past and present into an uncertain future'.

Even though agriculture may gain a new 'productivist realm', particularly in light of the demand for bio-fuels, other forces located in different sectors and at different levels of decision-making have undoubtedly to be taken into account (EEA, 1999; Potschin and Haines-Young, 2006; Hersperger and Burgi, 2009), such as energy, transport, demography, communications, equity, employment, consumption behaviour, environmental degradation, technological changes, changes in social values or aspects of governance (e.g. Busch, 2006; Biggs et al., 2007). Although the presence of internal driving forces has to be acknowledged, decisions concerning the transformation of the local landscape are progressively located further away from the local landscape itself. Hence it is being recognized that local driving forces depend on or even coincide with those located at a global level (e.g. Schmitz et al., 2003; Pereira et al., 2003; Soliva et al., 2008). Global driving forces are capable of acting directly on the landscape but can also be filtered and reoriented by regional or national policies that are ultimately put in place by the local public (Fig. 10.1).

Fig. 10.1 The flow of global driving forces to the local landscape

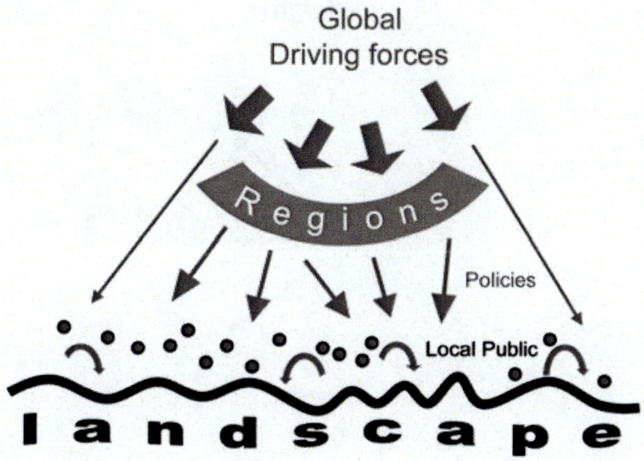

When complex issues with a high level of uncertainly are at stake, such as those related to future remote rural landscapes, expert panels can be useful in helping to explore the future. Their special knowledge on each driving force that may shape landscapes in the future enables them to express intuitions about future developments in a structured way. Turoff (1975: 96) developed Policy-Delphi techniques as a 'forum of ideas' that aim at generating broad and contrasting perspectives about the future (Loe, 1995). The capabilities of expert panels are notably appealing in situations where open-minded discussions attempt to challenge the validity of present mental models in the observation of the future (Turoff, 1975). Therefore, the contribution of experts for the development of 'exploratory landscape scenarios' can be twofold: on the one hand, they can set the basis for the formulation of a set of scenarios based on the articulation of alternative driving forces of development and, on the other hand, experts can provide the needed 'outside view' that may stimulate the imagination of local stakeholders and trigger discussion concerning aspirations for their landscape.

Landscape scenarios have predominantly found their purpose in communicating with stakeholders at a landscape scale. Whereas major scenario work done recently, such as PRELUDE (EEA, 2007), EURURALIS (Klijn et al., 2005) or SCENAR 2020 (EC, 2007), has put much effort in developing scenarios, frequently taking advantage of expert knowledge, it has generally failed to engage local stakeholders in the assessment of these scenarios (Soliva et al., 2008). 'Exploratory landscape scenarios', by combining these two approaches, rely on the experts' contribution for the formulation of plausible scenarios that are subsequently assessed by stakeholders, bringing out major concerns and desires regarding the future of their landscape. Derived from the categorization of participatory methods proposed by Marjolein and Rijkens-Klomp (2002), the role of experts in broadening the perspectives ('mapping out diversity') about the future in an exploratory way is followed by the definition of a desired future by stakeholders based on 'reaching consensus' (Fig. 10.2).

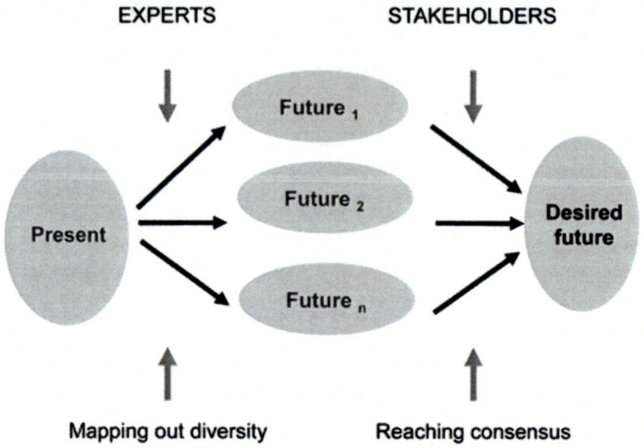

Fig. 10.2 The role of experts and stakeholders in the development of LQOs based on 'exploratory landscape scenarios'

10.5 Selection of the Case Study

The municipality of Mértola, located in the Alentejo region in southeast Portugal on the border with Spain (Fig. 10.3), was chosen for a case study mainly because in recent decades it has become the icon of remote rural areas in Portugal, as multiple dynamics leading to the decline of rural livelihoods occur simultaneously in the same landscape: drought (Rodrigo, 2009), soil erosion (Casimiro, 2002), desertification (Correia et al., 2004), abandonment of agriculture (van Doorn and Pinto-Correia, 2006), and deterioration of social and economic relations (Ferrão et al., 2000; Carolino, 2006; Oliveira, 2007).

Owing to these dynamics, the municipality of Mértola has been considered 'fragile' in terms of rural development (van Doorn and Pinto-Correia, 2006) and classified as 'critical' in the context of the national spatial planning document, mainly due to its decreasing population (8712 inhabitants in 2001 and 7514 in 2007) distributed over an area of 1279.4 km^2. Mértola, like other remote rural areas, also presents a very low level of activity (71% of the population depends on the activity of others). Its geographical location is peripheral in relation to the main urban centres and infrastructure. Agricultural productivity suffers from poor soils and little rainfall. Since the early 1990s cereal growing has been shifting progressively to extensive livestock breeding (Oliveira, 2007). These marginal conditions mean that the maintenance of agriculture depends heavily on CAP subsidies. Nevertheless, agriculture is still a relatively important source of income for the sparsely distributed population (Pinto-Correia, 2000; van Doorn and Bakker, 2007), contributing to 16.5% of the employment. Simultaneously, the municipality of Mértola embodies significant natural and cultural values, associated with the presence of areas designated for nature conservation integrated in the European Union's Natura 2000 network (EC, 2009) and a remarkable heritage from the Moorish era.

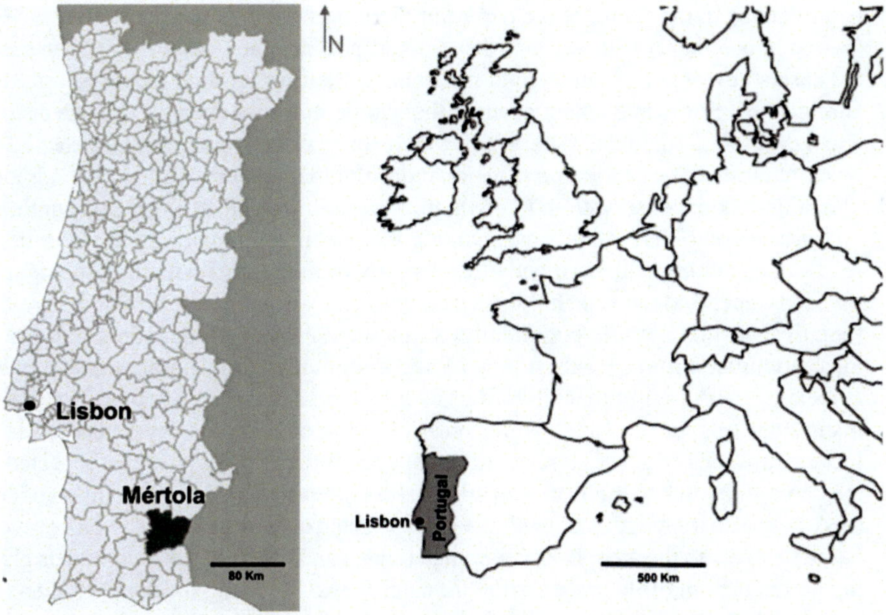

Fig. 10.3 Geographical location of Mértola in Portugal and Western Europe

10.6 Selection of Expert and Local Stakeholder Panels

Since the early development of Delphi techniques, one major issue has been the definition of who is an expert (Helmer, 1967). This issue, in the absence of clear criteria, needs to be redefined for each case. Alternative criteria, such as the number of scientific publications (Tapio, 2002) or the position one holds in an organization and the level of responsibility concerning the topic (Rauch and Randolph, 1979) have been used instead. In order to be able to produce a wide spectrum of futures, Tapio (2002) and Scapolo and Miles (2006) argue that one must avoid inviting only well-known or acquainted experts because they might share the same world-views. Therefore, whenever possible, two experts covering one issue should have contrasting worldviews.

In line with these recommendations experts were selected according to their nationwide recognized expertise concerning the main potential driving forces that may shape the future of Mértola's landscape. Most have a sound academic background presenting a wide array of publications in their field of expertise. Some hold or have held leading management positions, either in private companies or public administration – three experts had been government ministers in the past and one became secretary of state after participating in this study. Besides their recognized knowledge, experts had to meet two additional criteria related to the case study area to prevent responses biased by personal or professional interests: first, they should have no specific emotional relationship to Mértola (i.e. never have lived there, never had relatives living there, never had been there on holidays, nor could they have a second house there) and, second, they should not have any responsibility towards the case study area (i.e. they could not hold, at the time of their participation, any administrative or management position that would make them able to take decisions about Mértola). Eighteen experts (of which three were women) were selected to cover ten issues (i.e. agriculture, regional development, energy, forestry, landscape planning, nature conservation, tourism, transportation, social change, and hunting).

Local stakeholders were chosen among those living and working in the municipality, contributing with their activities, directly or indirectly, to the management of the landscape. Sixteen stakeholders (including one woman) were selected as representatives of this specific community. Some were elected by the community (e.g. local administration) or sub-groups of the community (e.g. farmers and hunters associations or environmental NGOs), while others were responsible for public institutions (e.g. health care, nature conservation or education) or were the heads of local institutions (e.g. social services or culture). Individuals who were respected in the community due to innovative or successful entrepreneurship (e.g. tourism operators) were also included. Stakeholders were selected according to a snowball process based on visits to the area. Both relevant activities and the names of individuals playing an outstanding role arose during the interviews. Selecting local stakeholders in remote rural areas is quite challenging due to the extremely low population density.

10.7 Scenario Development

All four scenarios envisioned for 2030 feature major changes relating to alternative developments of the driving forces considered to be relevant for the future development of Mértola (Ramos, 2010): these forces were social cohesion, environmental challenges, energy, technological innovation, demography, and spatial planning model. Multiple combinations of these driving forces were organized along two axis which represented critical uncertainties (Fig. 10.4): (1) the level of protection of the State (i.e. levels of support by the state to the different sectors) and (2) the origin of demand on the landscape (i.e. development based on external or local values and knowledge), which leads to 'continuity' or 'discontinuity' under the current conditions.

Each of the four scenarios aimed at presenting different but equally plausible worlds which are not 'paradise' or 'doomsday' representations. In the following, brief descriptions of the scenarios are provided.

Scenario 1 – 'Fashion wilderness'. The first scenario corresponds to the continuing process of globalization. The CAP is assumed no longer to exist. Depopulation and agricultural abandonment are present in most marginal landscapes like Mértola. In a society of great heterogeneity, where the very rich fly in from the urban areas for very short periods of recreation – hunting, bird-watching and other outdoor activities – scrublands are managed in order to offer opportunities for both hunting and nature conservation. Since the state does not intervene in these areas, these

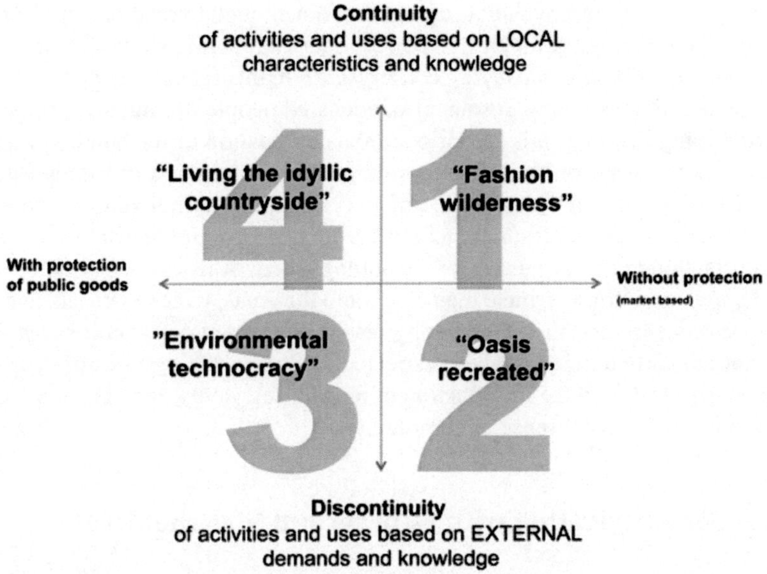

Fig. 10.4 Organization of scenarios according to critical uncertainties

activities depend on market forces to develop. Local communities disappear. There are only a few people remaining, and immigrants provide the small labour force needed for the private companies that have taken over the area. Small areas of agriculture contribute to biodiversity, as well as to the production of high-quality niche products sold in exclusive stores in metropolitan areas.

Scenario 2 – 'Oasis recreated'. The main driver of the second scenario is demography. The predictable demographic change of a European aging population with progressively higher life expectancy is assumed to be a key determinant for explaining socioeconomic changes. The natural conditions of Mértola, especially the sun and the river, as well as its rich heritage, attract population from Central and Northern European countries to live here permanently or stay over the winter months. The abandoned villages become transformed, being expanded in a traditional style while well equipped to meet the needs of an active senior population. New services create employment for specialists in, for instance, health care. In addition to the senior population, other new residents arrive with their families and create a new multi-cultural community. In this scenario, agriculture is transformed into a new form of landscape gardening, emphasizing scenic and recreational value.

Scenario 3 – 'Environmental technocracy'. Here agriculture and forestry play a leading role by responding to the main driving force of this scenario: energy crisis and compliance with the Kyoto protocol on reducing carbon emissions. Energy crops cover the most suitable soils, and in remaining areas forest produces biomass to serve as an alternative energy source. Even though these activities are financially supported by the state due to their strategic value, big private companies run the businesses. Some employment is created but not enough to retain local communities and avoid the concentration of manpower in the main cities of the region.

Scenario 4 – 'Living the idyllic countryside'. In this scenario depopulation trends change direction as young, resourceful, educated people immigrate to the area. The major driving forces of this scenario are the degradation of the quality of life in the cities, and the need of the state to protect natural and cultural values. Mértola is regarded as a better place to live in and to raise children. Individuals whose professional background enables them to work from home, taking advantage of advances in information and communication technology, benefit from this scenario. The new rural population brings critical mass back into the areas, which then can demand better access to services, thereby creating new employment opportunities for the local population. Although landscape is expected to change, losing some of its traditional characteristics due to the introduction of a new lifestyle by new inhabitants, mutual respect keeps alive the sense of community.

10.8 Communicating with Experts and Stakeholders

Although literature recommends scenario workshops as a way of using group dynamics to generate more ideas about the future (e.g. Schwartz, 1996; van der Heijden, 1996; van Notten et al., 2003; Shell, 2003), it was decided to use individual in-depth interviews because it has proved difficult to bring together high-profile experts for a workshop. Individual interviews give better insight into arguments in favour or against a specific scenario. For the same reason, individual interviews with local stakeholders were preferred. Closed questionnaires for rating the scenarios were not considered because it would not be possible to disentangle to what stimuli panelists were actually reacting when confronted with the images of the scenarios (Soliva et al., 2008). Kok et al. (2006) argue in favour of a workshop due to the added value of discussing matters in a group setting, which can stimulate social interaction and learning processes, enabling stakeholders to create a common vision. A possible drawback of a workshop is that not all individuals might feel comfortable expressing their opinions. This may be particularly valid for small rural communities, as in the case of Mértola, where there is a well-established social order.

Different media were used to communicate with experts and stakeholders: experts were confronted with a brief 150-word description of each scenario and local stakeholders with simulated photorealistic landscapes.

A pre-test with experts showed that this group seemed to be very critical concerning technical details of specific elements included in the images related to their field of expertise (such as if the right species of trees was used or if the solar panel had the right angle, etc.) and therefore showed difficulties in addressing the general content of the scenario and the driving forces behind it. A brief narrative proved to be more abstract and therefore more useful as a way of engaging experts in a discussion about the interpretation of the signs of change the world and how they might influence landscape in Mértola.

Fig. 10.5 Simulation of Scenario 4 – 'Living the idyllic countryside' through the introduction of new features (*right*) into the original image (*left*): dispersed housing for the new residents; horses and cars representing recreational needs of this new population; and a group of three Iberian pigs relating to a new lifestyle that values hobby farming based on traditional uses

Despite the photorealistic simulations that were used to communicate the scenarios to the stakeholders, it was decided not to make an effort to achieve a very realistically detailed image. Encouraged by the scenario work developed by Artner et al. (2006), who used iconic collages to emphasize the content of the landscape, the elements introduced into the simulations were made clearly visible at first glance. An example of this type of representation is illustrated in Fig. 10.5. This approach proved to be quite efficient for discussing the content and meaning of each image. The concerns of Soliva and Hunziker (2009) that stakeholders strongly judge landscape images by their visual quality were not supported.

Interviews with both experts and stakeholders were guided by the same set of three questions: (1) Which scenario is the most likely to happen? (2) Which scenario is the most desirable to happen? (3) Which scenario will be preferred by stakeholders at the landscape level? The interviews took about an hour and notes taken were subsequently organized for further analysis.

10.9 Results and Discussion

The content of the interviews was analysed using the answers to the set of three questions addressed to experts and stakeholders. It can be argued that the result from experts and stakeholders might not be directly compared since different means of communication were used (i.e. narratives for experts and images for stakeholders). Despite these limitations, the issues discussed were the same, and a joint analysis might provide a useful insight into the motivations of both groups interviewed. A summary of the results is presented in Table 10.1.

The results show that both groups had a similar understanding of the trend scenario (Scenario 1 – 'Fashion wilderness'). There was consensus that current drivers would continue to stimulate agricultural abandonment. Property structured in big estates would favour the use of landscapes for recreation by elites, especially linked to hunting. The resident population would disappear except for those employed in

Table 10.1 Responses of experts and local stakeholders

	Experts			Local stakeholders		
	Most likely	Most desired	Desired by others	Most likely	Most desired	Desired by others
Scenario 1 'Fashion wilderness'	●			●		
Scenario 2 'Oasis recreated'			●		●	
Scenario 3 'Environmental technocracy'						
Scenario 4 'Living the idyllic countryside'		●				●

this business. Some experts judged this future very negatively, associating it with a decaying rural community, with loss of identity and the presence of a society characterized by social and economic inequities. Others agreed with some local stakeholders who defended this scenario as the most appropriate and, therefore, the most sustainable, taking into account the existing natural conditions such as poor soils, steep slopes and reduced availability of water. Even though this scenario was not in general seen as a desirable option for the future, local stakeholders tended to perceive it as less dramatic than some experts, probably because they were used to observing decrease in population and increase in scrubland. To local stakeholders this scenario meant both 'continuity' and 'resignation'.

There was little consensus between expert and local stakeholder panels regarding the 'most desirable' future development for the landscape in Mértola. These differences seemed to be rooted in different sets of values and beliefs, but also in differences related to the identification of driving forces that might shape future landscapes. Experts shared the belief that a future based on exploring local knowledge and local values as natural and cultural heritage would be able to create a living environment. This, in turn, could attract new residents who, in the future, would be able to access basic services and to work taking advantage of evolving information and communication technologies (Scenario 4 – 'Living the idyllic countryside'). Experts believed that these new residents could play a central role in widening the future economic basis of the rural areas. To local stakeholders this scenario was considered rather 'unlikely' or 'undesirable'. A significant minority of the local stakeholders selected it as the 'most desirable' but, for a series of reasons that almost everyone made reference to, this scenario was considered 'not likely' to happen. The first reason was associated with dispersed housing. In addition to not being traditional in this landscape (and therefore not desirable), it was considered too expensive considering the costs involved in making basic utilities available in all houses (electricity, water, roads, sewage). Besides, the present regulation of the Natura 2000 area that covers 16% of the area of Mértola municipality restricts this kind of development. The second aspect is related to the change of property structure that would have to happen. It was perceived as unlikely that owners of big estates would want to split them into smaller areas, so that they could become more affordable for newcomers. Third, these stakeholders did not believe that an urban population would ever want to move out of the cities into rural areas.

Also for the majority of local stakeholders the 'most desired' future seemed 'unlikely'. The scenario in which abandoned villages might be transformed into residential tourism areas for Europe's retired population (Scenario 2 – 'Oasis recreated') was attractive because it mainly would 'bring life back'. Notably, local administration assessed this scenario as an opportunity to take advantage of existing infrastructure and the expertise of current institutions specialized in assisting the local senior population. Although most local stakeholders thought this scenario would be beneficial for them, in their opinion it was 'not likely' to happen as the landscape was not attractive enough for the senior population that it was expected to serve. They argued that visitors would prefer coastal areas where life and entertainment are more appealing. Experts also used this argument but since this scenario depended strongly on private investment they saw it happening in articulation with other landscapes (i.e. seaside, city, countryside). Nevertheless, there was a strong consensus among experts that this future was 'undesirable', referring explicitly to what they called 'loss of landscape authenticity'. A minority of local stakeholders also conveyed the same idea because too many people from abroad (both senior population and the employees) would be attracted and thus likely threaten the existing cultural values that should be preserved.

Curiously, both groups anticipated the other's preferences. Experts were quite unanimous in saying that the local population would mostly prefer the alternative most capable of attracting more people to the area (Scenario 2 – 'Oasis recreated'). Local stakeholders were also quite sure that the most preferred future for visitors would be the most traditional one (Scenario 4 – 'Living the idyllic countryside'). This suggests that experts themselves were unable to dissociate their position both as expert and, simultaneously, as part of an urban elite with a romanticized view of this landscape. This nostalgia for the countryside as a place of virtue, coming from an intellectualized urban society, seems neither novel nor surprising (Williams, 1973; Donadieu, 2002; Cavaco, 2005). Therefore, preferences of experts may be interpreted as those of potential users of this landscape, meaning that they respond according to their own aspirations towards this landscape. This positioning makes these experts also part of the public that would need to be addressed in the formulation of aspirations in terms of the ELC.

10.10 Conclusion

Lessons learned from this exploratory approach to experts' and stakeholders' perspectives on future developments are that it is quite clear in the minds of both groups which future is 'desirable', which is 'undesirable', and why. The main difference between the two groups is that experts believe that there are various plausible futures depending on the combination of exogenous driving forces but also on the policy decisions that might be taken, whereas local stakeholders did not see any possible future but the abandonment of the scenario they felt to be trapped in. This may be specifically valid for the way stakeholders in remote rural areas perceive their future.

Thus, it seems that a 'consultation process' based on the collection of opinions and views of all concerned groups or stakeholders, in the definition of LQOs as called upon by the ELC, might serve to help understanding what is the future desired by stakeholders, but it is not able to create a vision, especially in remote rural areas where present conditions do not seem to motivate the public for the future.

This study confirms that in the formulation of LQOs the 'policy formulation and action' that should follow 'public consultation', as recommended in the *Guidelines for Implementation of the European Landscape Convention* (Council of Europe, 2008), need to build on an adequate public participation process. Although public participation 'should not be seen as a substitute for official decision-making but as a complement to it' (Jones, 2007: 616), the development of LQOs needs to be addressed as a social process that demands the creation of a common vision to guide and engage the public into action by providing the empowerment capable of 'generating the energy and commitment needed for reaching out another tomorrow' (van der Helm, 2009:100).

Concerning the questions initially addressed, it has been shown that there is still need for research on how to approach the definition of who is the public of a specific landscape. It has also been shown that aspirations may vary among different segments of the public; for example, local stakeholders may differ from experts who ultimately represent a part of the urban population. These findings raise a new question on how to integrate and balance the 'aspirations' of different types of public in the formulation of LQOs at the landscape scale. The discussion on how to frame conceptually the 'aspirations of the public' shows that they strongly relate to landscape preferences. Hence the methodological approach presented, based on the development of exploratory landscape scenarios, may provide a useful tool for identifying preferred future states of the landscape chosen among a set of plausible futures. Despite these methodological developments, it has to be recognized that formulating LQOs is still a challenging task that requires social mobilization in order to create a 'shared vision of a desirable and sustainable future' (Costanza, 2003: 667).

References

Ackoff R (2006) Idealized design: how to dissolve tomorrow's crisis... today. Wharton School Publishing, New York, NY

Antrop M (2004) Rural–urban conflicts and opportunities. In: Jongman R (ed) The new dimensions of the European landscape. 91UR Frontis Series 4. Springer, Wageningen, pp 83–91

Antrop M (2006) Sustainable landscapes: contradiction, fiction or utopia? Landsc Urban Plan 75(3–4):187–197

Arnstein S (1969) A ladder of citizen participation. J R Town Plan Inst 35:216–224

Artner A, Frohnmeyer U, Matzdorf B, Rudolph I, Rother J, Stark G (2006) Future landscapes: Perspektiven der Kulturlandschaft. Bundesmnisterium für Verkehr, Bau und Stadtentwicklung, Berlin

Baptista F (2001) Agriculturas e Território. Celta Editora, Oeiras

Biggs R, Raudsepp-Hearne C, Atkinson-Palombo C, Bohensky E, Boyd E, Cundill G, Fox H, Ingram S, Kok K, Spehar S, Tengö M, Timmer D, Zurek M (2007) Linking futures across scales: a dialog on multiscale scenarios. Ecol Soc 12(1):17 http://www.ecologyandsociety.org/vol12/iss1/art17/. Accessed 2 Apr 2008

Busch G (2006) Future European agricultural landscapes – what can we learn from existing quantitative land use scenario studies? Agric, Ecosyst Environ 114(1):121–140

Carolino J (2006) Land as a manifestation of care: an anthropological approach to relations with the land in Southern Alentejo. Ph.D. dissertation in Social Anthropology, Brunel University, London

Casimiro PC (2002) Uso do solo, teledetecção e estrutura da paisagem, Tese de Doutoramento em Geografia Regional especialidade Novas Tecnologias em Geografia. Universidade Nova de Lisboa, Lisboa

Cavaco C (2005) As paisagens rurais: do 'determinismo natural' ao 'determinismo político'? Finisterra XXXIX(79):73–101

Correia FN, Landeiro C, Ramos IL, Saraiva MG, Silva FN, Monteiro F (2004) Desertificação em Portugal: Incidência no ordenamento do território e desenvolvimento urbano: Os agentes locais e os processos de desenvolvimento nas zonas sujeitas a desertificação. DGOTDU, Lisboa

Costanza R (2003) A vision of the future of science: reintegration of the study of humans and the rest of nature. Futures 35:651–671

Council of Europe (2000) European Landscape Convention, Florence, 20.X.2000. ETS No. 176. http://conventions.coe.int/Treaty/en/Treaties/Html/176.htm. Accessed 1 Nov 2009

Council of Europe (2008) Recommendation CM/Rec (2008)3 of the Committee of Ministers on the guidelines for the implementation of the European Landscape Convention. https://wcd.coe.int/ViewDoc.jsp?id=1246005. Accessed 1 Nov 2009

Dhubháin AN, Fléchard MC, Moloney R, O'Connor D (2009) Stakeholders' perceptions of forestry in rural areas – two case studies in Ireland. Land Use Policy 26:695–703

Dockerty T, Lovett A, Sunnenberg G, Appleton K, Perry M (2005) Visualizing potential impacts of climate change on rural landscapes. Comput, Environ Urban Syst 29:297–320

Dolman P, Lovett A, O'Riordan T, Cobb D (2001) Designing whole landscapes. Landsc Res 26(4):305–335

Donadieu P (2002) La société paysagiste. Actes SUD. École Nationale Supérieure du Paysage, Arles

Duby G (1991) Quelques notes pour une histoire de la sensibilité au paysage. Étud Rurales 121–124:11–14

EC (2007) SCENAR 2020 – scenario study on agriculture and the rural world. European Commission. http://ec.europa.eu/agriculture/publi/reports/scenar2020/index_en.htm. Accessed 25 May 2008

EC (2009) Natura 2000 network. European Commission. http://ec.europa.eu/environment/nature/natura2000/index_en.htm. Accessed 1 Nov 2009

EEA (1999) Environmental indicators: typology and overview. Report No 25. European Environmental Agency, Copenhagen. http://www.eea.europa.eu/publications/TEC25. Accessed 25 May 2008

EEA (2007) Land-use scenarios for Europe: qualitative and quantitative analysis on a European Scale (PRELUDE), Report No 9. European Environmental Agency, Copenhagen. http://www.eea.europa.eu/publications/technical_report_2007_9. Accessed 25 May 2008

Ferrão J (2000) Relações entre mundo rural e mundo urbano: evolução histórica, situação actual e pistas para o futuro. Sociologia, Problemas e Práticas 33:45–54

Ferrão J, Ferreiro F, Lopes R, Monteiro H (2000) Peripherality, rurality and rural areas in Europe: a literature survey. Project HPSE-CT-1999-00013 – The future of Europe's rural periphery: the

role of entrepreneurship in responding to employment problems and social marginalisation. ISCTE, Lisboa
Friedman A, Miles S (2006) Stakeholders: theory and practice. Oxford University Press, Oxford
Godet M (2001) Creating futures: scenario planning as a strategic management tool. Economica, London
Groth P (1997) Framework for cultural landscape studies. In: Groth P, Bressi T (eds) Understanding ordinary landscapes. Yale University Press, London, pp 1–21
Halfacree K, Boyle P (1998) Migration, rurality and the postproductivist countryside. In: Boyle P, Halfacree K (eds) Migration into rural areas: theories and issues. Wiley, Chichester, pp 1–20
Healey P (1997) Collaborative planning: shaping places in fragmented societies. Macmillan, Basingstoke
Healey P (2008) Civic engagement, spatial planning and democracy as a way of life. Plan Theory Prac 9(3):379–414
Helmer O (1967) Analysis of the future: the Delphi method. RAND, Santa Monica
Hersperger A, Burgi M (2009) Going beyond landscape change description: quantifying the importance of driving forces of landscape change in a Central Europe case study. Land Use Policy 26:640–648
Hulse D, Gregory S, Baker J (2002) Willamette River Basin planning atlas: trajectories of environmental and ecological change. In: Hulse D, Gregory S, Baker J (eds) Pacific northwest ecosystem research consortium. Oregon State University Press. http://www.fsl.orst.edu/pnwerc/wrb/Atlas_web_compressed/PDFtoc.html. Accessed 15 Nov 2010
Jones M (2007) The European Landscape Convention and the question of public participation. Landsc Res 32:613–633
Jones M, Emmelin L (1995) Scenario studies for the rural environment. In: Schoute JFT, Finke PA, Veeneklaas FR, Wolfert HP (eds) Scenario studies for the rural environment. Kluwer, Dordrecht, pp 405–413
Jones M, Howard P, Olwig K, Primdahl J, Sarlöv-Herlin I (2007) Multiple interfaces of the European Landscape Convention. Norsk Geogafisk Tidsskrift–Norwegian J Geogr 61:207–215
Klijn J (2004) Driving forces behind landscape transformation in Europe, from a conceptual approach to policy options. In: Jongman R (ed) The new dimensions of the European landscape, UR Frontis Series 4. Springer, Wageningen, pp 201–218
Klijn J, Vullings L, van der Berg M, van Meijl H, van Lammeren R, van Rheenen T, Tabeau A, Veldkamp A, Verburg P, Westhoek H, Eickhout B (2005) The EURURALIS study: technical document. Alterra-rapport 1196. Alterra, Wageningen, http://www.eea.europa.eupublications/technical_report_2007_9. Accessed 25 May 2008
Klug H (2007) Participative landscape planning using a GIS approach for facilitation. In: Mander U, Helming K, Wiggering H (eds) Multifunctional land use: meeting future demands for landscape goods and services. Springer, Berlin, pp 195–233
Kok K, Patel M, Rothman D, Quaranta G (2006) Multi-scale narratives from an IA perspective: Part II. Participatory local scenario development. Futures 38(3):285–311
Loe RC (1995) Exploring complex policy questions using the policy Delphi: a multi-round, interactive survey method. Appl Geogr 15(1):53–68
Lugenbühl Y (2006) Landscape and identification, assessment and quality objectives. In landscape and sustainable development: challenges of the European Landscape Convention. Council of Europe, Strasbourg, pp 97–115
Luz F (2000) Participatory landscape ecology – a basis for acceptance and implementation. Landsc Urban Plan 50:157–166
Marjolein BA, Rijkens-Klomp N (2002) A look in the mirror: reflection on participation in integrated assessment from a methodological perspective. Glob Environ Change 12: 167–184
Merriam-Webster (2009) Dictionary Online. http://www.merriam-webster.com/. Accessed 1 Nov 2009

Mormont M (1996) Le rural comme catégorie de lecture du social. In: Jollivet M, Eizner N (eds) L'Europe et ses Campagnes. Presses des Sciences Politiques, Paris, pp 161–176

Nassauer J, Corry R (2004) Using normative scenarios in landscape ecology. Landsc Ecol 19(4):343–356

Nassauer J, Corry R, Cruse R (2002) The landscape in 2025: alternative landscape future scenarios as a means to consider agricultural policy. J Soil Water Conserv 57(2):44A–53A

Naveh Z (2005) Towards a transdisciplinary landscape science. In: Wiens J, Moss M (eds) Studies in landscape ecology. Cambridge University Press, Cambridge, pp 346–354

OECD (2006) The new rural paradigm: policies and governance. OECD, Paris

Oliveira R (2007) Integrative landscape management: new parameters for the implementation of common agriculture policy: a case study in southern Portugal. Ph.D. Thesis, Universidade de Évora, Évora

Palang H, Alumäe H, Mander Ü (2000) Holistic aspects in landscape development: a scenario approach. Landsc Urban Plan 50(1–3):85–94

Patel M, Kok K, Rothman D (2007) Participatory planning in land use analysis: an insight into the experiences and opportunities created by stakeholder involvement in scenario construction in northern Mediterranean. Land Use Policy 24(3):546–561

Pereira H, Domingos T, Vicente L (2003) Report on the user needs and response options: Portuguese sub-global assessment of the millennium ecosystem assessment. http://www.ecossistemas.org. Accessed 20 Feb 2008

Phillips A, Clarke R (2004) Our landscape from a wider perspective. In: Bishop K, Phillips A (eds) Countryside planning – new approaches to management and conservation. Earthscan, London, pp 49–67

Pinto-Correia T (2000) Landscape identity, a key for integration. In: Pedroly B (ed) Landscape – our home: essays on the culture of the European landscape as a task. INDIGO, Stuttgart, pp 145–151

Pinto-Correia T, Breman B (2009) New roles for farming in a differentiated countryside: the Portuguese example. Reg Environ Change 9(3):143–152

Porter M (1985) Competitive advantage. The Free Press, New York, NY

Potschin M, Haines-Young R (2006) Rio+10, sustainability science and landscape ecology. Landsc Urban Plan 75(3–4):162–174

Potter C, Tilzey M (2005) Agricultural policy discourses in the European post-Fordist transition: neoliberalism, neomercantilism and multifuncionaltity. Prog Hum Geogr 29(5):581–600

Ramos IL (2008) As dinâmicas da paisagem rural: a formulação de cenários como instrumento de apoio ao planeamento. Ph.D. Technical University of Lisbon, Instituto Superior Técnico, Lisbon

Ramos IL, Saraiva MG (2008) 10 Anos de Lei de Bases da Política de Ordenamento do Território e de Urbanismo!... E qual o lugar da 'Paisagem' no Ordenamento do Território? Encontro anual da Ad Urbem.www.adurbem.pt/index.php?option=com_content&task=view&id=368&Itemid=343. Accessed 11 Sep 2009

Ramos IL (2010) 'Exploratory landscape scenarios' in the formulation of 'landscape quality objectives'. Futures 42:682–692

Rauch W, Randolph R (1979) Computer-assisted panel sessions (CAPS): review of an experiment in accelerated international teleconferencing. Technol Forecast Soc Change 13(3):235–255

Roberts S (2002) Key drivers of economic development and inclusion in rural areas: initial scoping study of the socio-economic evidence base for DEFRA. DEFRA, London. https://statistics.defra.gov.uk/esg/temp_rural/sion.pdf. Accessed 25 Apr 2008

Rodrigo I (2009) As seca: contextos, attitudes e comportamento: Baixo Alentejo Interior (1932–1995). IsaPress, Lisboa

Ryan RL (2006) Comparing the attitudes of local residents, planners, and developers about preserving rural character in New England. Landsc Urban Plan 75:5–22

Scapolo F, Miles I (2006) Eliciting experts' knowledge: a comparison of two methods. Technol Forecast Soc Change 70:83–101

Schmitz M, Aranzabal I, Aguilera P, Rescia A, Pineda F (2003) Relationship between landscape typology and socio-economical structure: scenarios of change in Spanish cultural landscapes. Ecol Mode 168:343–356

Schwartz P (1996) The art of the long view: planning for the future in an uncertain world. Currency Doubleday, New York, NY

Shell (2003) Scenarios: an explorer's guide. Global Business Environment Shell International, London. www.shell.com/scenarios. Accessed 29 Dec 2006

Sheppard S (2005) Landscape visualisation and climate change: the potential for influencing perceptions and behaviour. Environ Sci Policy 8(6):637–654

Soliva R, Hunziker M (2009) Beyond the visual dimension: using ideal type narratives to analyse people's assessments of landscape scenarios. Land Use Policy 26:284–294

Soliva R, Rønningen K, Bella I, Bezak P, Cooper T, Flø B, Marty P, Potter C (2008) Envisioning upland futures: stakeholder responses to scenarios for Europe's mountain landscapes. J Rural Stud 24:56–71

Steinitz C, Rojo H, Bassett S, Goode T, Maddock T, Mouat D, Peiser R, Shearer A (2003) Alternative futures for changing landscapes: the upper San Pedro River Basin in Arizona and Sonora. Island Press, Washington, DC

Surová D, Pinto-Correia T (2008) Landscape preferences in the cork oak Montado region of Alentejo, southern Portugal: searching for valuable landscape characteristics for different user groups. Landsc Res 33(3):311–330

Tapio P (2002) Disaggregative policy Delphi: using cluster analysis as a tool for systematic scenario formation. Technol Forecast Soc Change 70:83–101

Tress B, Tress G (2003) Scenarios visualisation for participatory landscape planning – a study from Denmark. Landsc Urban Plan 64:161–178

Turoff M (1975) The policy Delphi. In: Linstone HA, Turoff M (eds) The Delphi method: techniques and applications. Addison-Wesley Publishing Company, New York, NY, pp 80–96

UNECE (1998) Convention on Access to Information, Public Participation in Decision-making and Access to Justice in Environmental Matters done at Aarhus, Denmark on 25 June 1998. United Nations Economic Commission for Europe. http://www.unece.org/env/pp/treatytext.htm. Accessed 1 Nov 2009

Van der Heijden K (1996) Scenarios: the art of strategic conversation. Wiley, Chichester

Van der Helm R (2009) The vision phenomenon: towards a theoretical underpinning of visions of the future and the process of envisioning. Futures 41:96–104

Van der Leeuw SE (1999) Degradation and desertification: some lessons from the long-term perspective. In: Balabanis P, Peters D, Ghazi A, Tsogas M (eds) Mediterranean desertification: resource results and policy implications. European Commission, Brussels, pp 17–31

Van Doorn A, Bakker M (2007) The destination of arable land in a marginal agricultural landscape in south Portugal: an exploration of land use change determinants. Landsc Ecol 22(7): 1073–1087

Van Doorn A, Pinto-Correia T (2006) Refining the concept of land abandonment: experiences from southern Portugal. In: Bunce R, Jongman R (eds) Landscape ecology in the Mediterranean: inside and outside approaches. IALE Publication, Wageningen, pp 147–159

Van Notten P, Rotmans J, van Asselt M, Rothman D (2003) An updated scenario typology. Futures 35(5):423–443

Wiggering H, Dalchow C, Glemnitz M, Helming K, Muller K, Schultz A, Stachow U, Zander P (2006) Indicators for multifunctional land use – linking socio-economic requirements with landscape potentials. Ecol Indic 6(1):238–249

Williams R (1973) The country and the city. Oxford University Press, New York, NY

Zoido-Naranjo F (2005) Landscape and spatial planning policies. In: Landscape and sustainable development: challenges of the European Landscape Convention. Council of Europe, Strasbourg, pp 55–79

Chapter 11
Landscape Perception Through Participation: Developing New Tools for Landscape Analysis in Local Planning Processes in Norway

Morten Clemetsen, Erling Krogh, and Kine Halvorsen Thorén

Abstract The European Landscape Convention defines landscape as 'an area as perceived by people, whose character is the result of action and interaction of natural and/or human factors'. The definition can be viewed in several ways. This chapter presents a phenomenological perspective, which regards landscape as an integrated whole in which people create a place-related identity through meaningful dialogue based on active sensing and landscaping. Traditionally, landscape analysis

M. Clemetsen (✉)
Department of Landscape Architecture and Spatial Planning, Norwegian University of Life Sciences, NO-1432 Ås, Norway
e-mail: morten.clemetsen@umb.no

M. Jones, M. Stenseke (eds.), *The European Landscape Convention*, Landscape Series 13, DOI 10.1007/978-90-481-9932-7_11,

describes the landscape's natural and cultural features, in addition to its scenic qualities. This knowledge is then used as a basis for landscape planning and management. Based on a recently completed Norwegian research project on how recreational (and perceptual) values are dealt with in planning processes, we ask whether existing methods take fully into account knowledge of the comprehensive cultural and individual human-landscape relationships. This chapter presents an alternative and complementary approach, based on experience from pilot projects that used 'sense of place' methodology to identify and express people's own relationships and values linked to specific landscape character areas. This generates synergies by increasing awareness of the interaction between place/landscape and people. This again can lay the grounds for and mobilize broad public participation in the planning and development of future landscapes. The methodological aspects are illustrated by a landscape analysis from the fjord landscapes of western Norway.

Keywords Landscape resource analysis · Participation · Landscape perception · Landscape character · Landshaping

11.1 Introduction

The European Landscape Convention (ELC) defines landscape as an area as people perceive it. Moreover, the Convention emphasizes that landscape is formed in an active and reciprocal interplay with people. This underlines action in and experience of the landscape. According to a central objective of the ELC, people should be stimulated to actively participate in planning, shaping, and developing the landscapes in which they live.

In order to achieve this goal, it is necessary to develop models of landscape perception and methods for investigating relationships between people and landscapes that build on the best possible understanding of human landscape perception and experience. In this chapter we discuss the need for alternative models and methods, based on an evaluation of the existing traditions in professional practice. We also explain central characteristics of the relationship between people and landscape and argue for appropriate methods to understand this relationship.

Research in social science and the humanities shows that the perception and experience of landscape is complex and multifaceted. Perception is based on the use of multiple senses and physical skills. Experience is seldom verbalized, as it is generally rooted in implicit cultural values and interpretations, and is often linked to activities in the landscape. Only to a slight degree do existing landscape analysis and planning methods accommodate knowledge about how people perceive and experience landscape. Current methods are mainly based on mapping physical components and visual characteristics of a landscape, which in turn are used to analyse a landscape's scenic quality and character.

The gap between existing landscape planning methods and the basic characteristics of human perception and experience presents several challenges. First, discrepancies can easily occur between experts' assessments of landscape values and the perceptions and experiences of the landscape's stakeholders. Such stakeholders include all 'user groups' with links to and interests in a certain landscape, ranging from landowners and business interests to residents and various visitor categories. Second, it is necessary to develop methods that can bridge the gap between existing landscape planning methods and new insights into the human-landscape relationship.

In the following, we will first elaborate on the characteristics of human landscape perception, including a presentation of the concept of *landshaping* (Krogh, 1995). Hitherto there has been relatively little awareness of human-landscape relationships and insufficient integration of this knowledge into landscape planning methods. Such an elaboration provides a scientific basis for developing methods to explore 'sense of place'. To underline the need for improved methods with more emphasis on human perception, we will then present the results of a study of valuation methodology in Norwegian landscape planning practice. This study shows that the evaluation of an area by user groups can in part differ considerably from experts' priorities. To complicate matters further, the study also showed that the various user groups do not always agree among themselves, and in some cases have conflicting interests. How can landscape planning thus take people's landscape perception into consideration, when these perceptions show considerable variation? We will try to shed light upon this issue in the last part of this chapter, in which we introduce a method that integrates landscape assessments and sense of place within a concept we call *landscape resource analysis*. This is based on practical landscape work in Norway.

11.2 Landshaping Through Experience – Approaches to Human Interpretation of Landscape

11.2.1 Experiencing Landscape: From Visual Perception to Immediate, Reciprocal Participation

There is a wide range of studies of the perception and experience of nature and landscape from different professional fields, for example geography, ethnology, social anthropology, and psychology. However, the dominant research tradition is connected to environmental psychology. Empirical studies within environmental psychology focus mainly on measuring *visual* preferences, where cognition is assumed to occur on the basis of perception of visual stimuli, which then provide the basis for response and action. In this view, the human being is often considered to be a passive adapter to the landscape. Environmental-psychological studies of human visual preferences for landscapes are in keeping with the emphasis on scenic quality and landscape character in landscape architecture. As a result, other aspects of human-landscape relationships have received little focus in the methods developed for landscape planning.

New brain research challenges the stimuli–cognition–response chain as the predominant model for the connection between perception and action. In addition, transactional environmental psychology (Altman and Rogoff, 1987; Churchman,

2002; Gifford, 2007), which is basically a unifying position for contemporary environmental psychology, builds upon a *reciprocal* relationship between people and landscape as a basis for landscape perception and experience. However, the empirical tradition in environmental psychology builds on experimental studies, which isolate a few chosen variables, and the one-way stimulus–cognition–response chain. On the other hand, in the fields of geography, ethnology, social anthropology, and phenomenology the reciprocal relationship between people and their surroundings has been elaborated both theoretically and empirically (Relph, 1985; Feld, 1990, 1996; Tilley, 1994; Krogh, 1995; Basso, 1996; Casey, 1996; Gray, 2000; Ingold, 2000; Macnaghten and Urry, 2001). Our point of departure is in phenomenology, which elaborates on characteristics of the human-landscape relationship. However, we will first present new insights on human perception based on the discovery of mirror neurons.

The function of mirror neurons yields three basic insights concerning human perception (Rizzolatti and Sinigaglia, 2008). The first insight documents the human ability to immediate perception, cognition, and action as a unified whole. This immediate cognition is not connected to formation of concepts and reflections, but perception is nevertheless a coding and cognitive act. The second insight is that the designation 'mirror neurons' is misleading. Perception and processing of fractions of intended actions by others is not limited to mirroring, but implies that the 'I' immediately and creatively can connect to possible chains of action and results of actions. A person not only performs imitative and passive coding, but also has a creative potential. 'Aping' can be substituted with the potential of 'shaping'. The third insight widens understanding of perception from being egocentric to also being other-centric. The intentional 'I' is detached and conscious of its detachment from other people and perceives the landscape out of and through the body and the senses (Gallagher and Zahavi, 2007: 141–144), but also has the ability through perception to embrace other peoples' conditions, situations, and intentions immediately.

The scientific documentation of mirror neurons as *physical tools* for immediate, conscious cognition and intersubjective perception shows that mental representations build upon, and must therefore be related to, non-representational, immediate, continuous, and conscious perception, cognition, and action in a unified whole. To be able to understand people's primary and constituting view of the landscape, it is therefore necessary to use methods that enable the mapping of people's lives in and use of the landscape in situ, as a process in continuous change. This assumes a research method that builds upon participative observation through face-to-face encounters with the local population, and studies of their concrete use of the landscape over time.

11.2.2 The Physical Use of the Landscape as a Foundation for Experience and Meaning

This methodological approach is strengthened by an essential characteristic of the human-landscape relationship: that *meaning is constructed through activities*

(Krogh, 1995). Thus, meaning cannot possibly be derived directly from formal models disengaged from the activities themselves. This premise can be traced back to the phenomenologist Maurice Merleau-Ponty (1962), who claimed that, originally, consciousness does not imply 'I think that', but, rather, 'I am able to'. According to this way of thinking, reflection is founded upon previous intentional activities. Similarly, Martin Heidegger (1983) and Tim Ingold (2000) state that humans construct meaning through an interchange between dwelling and activities in space.

According to Merleau-Ponty, the relationship between the body and the world comes into being through meetings between the body and its surroundings. Thus, landscape is defined as the body's perceptible surroundings, which include what the bodily subject sees, smells, hears, feels, tastes, and senses in the meeting with her or his environment. Sight alone cannot represent or be the harbinger of the other senses and the specific qualities of other ways of sensing. It is therefore important to emphasize the specific qualities of the different senses:

- Sight can provide a general view, but also forms the foundation for participation and inclusion
- Sounds form layers which provide the rhythms and pulse of the place; they tell about movement, action, and creative activity and fill 'the space in between'
- Smells create associations and recall past events and experiences
- Touch organizes impressions and provides a physical framework for choice of action.

These sense qualities react together with the senses of balance, spatial dimensions, proprioception (perception of stimuli within an organism, especially related to position and movement of one's body), and alteroception (perception of body orientation and movements of others) to form primary perception. Through this interplay, the 'I' is provided with a continuous and unceasing tool for perception, cognition, and action in a unified whole. At the same time, bodily skills and sensibility will bear the mark of socialization in a cultural context, where different relationships as well as sense impressions from the landscape are given special meaning, while others are toned down or written off. Openness for the riches of the senses in encountering the diversity of the landscape is constantly in tension with socialized, controlled attention to the landscape.

The experiences which reveal themselves in the body's meeting with the world are always assigned meaning in relationship to earlier experiences. In this way, the phenomenological approach necessarily becomes also sociological and cultural-anthropological.

The mobile, sense-perceiving, and creative relationship between humans and landscapes is called 'landshaping' (Krogh, 1995). Landshaping always implies something new as it provides the possibility further to develop skills and extend the basis for experience and secondary reflection. Simultaneously, the landscape is shaped through human activities.

Human beings are immediately and always in the world – involved, perceiving, intuitively creating knowledge, and landshaping – that is to say, shaping landscape.

The mirror neurons provide the tool for experiencing the landscape through a unification of perception, cognition, and action. Whereas different landscapes can be assigned different characters and can represent qualities that are inductive for different activities, human character is a bodily-self which always and continuously encounters others and landscapes in actions that are intuitively knowledge-creating. Studies of the comprehension and experience of the landscape should occur near this point of encounter and therefore close to the landscape.

11.3 How are Different Perspectives on Landscape Taken into Account in Landscape Planning?

11.3.1 Landscape Analysis and Valuation in Practice: Presentation of a Norwegian Research Project

People's experience and perception of specific landscapes are generated in physical encounters with these landscapes, as actions and movements therein. Landshaping also bears the mark of socialization in a cultural context. Thus, cultural understanding, values, and practice will tend to vary a great deal between groups which relate to the same physical landscape. To what degree is this reflected in today's landscape analysis practice, which again forms the basis for further planning? To study how valuation was performed in practical landscape planning in Norway, a research project was conducted from 2001 to 2008. Outdoor recreation was chosen for study as a planning issue, because it clearly shows how the human-landscape relationship is taken into account. The project studied guidelines issued by national authorities on outdoor recreation (Vorkinn et al., 2002; Kleven et al., 2005) as well as local planning procedures (Thorén, 2008).

The results from the studies of local planning are especially of interest in this connection (Thorén, 2008). This subproject aimed to examine landscape values among experts and user groups in order to determine (a) which values were emphasized and (b) whether experts and users emphasized different values. The study was based on two cases in Norway located in different landscape types: Area A, a mountainous area of international importance; and Area B, which is close to densely populated urban areas south of Oslo. The experts' evaluations were studied through local plans and landscape assessments, while the perspectives of the general public were assessed by qualitative interviews and surveys. Since it cannot necessarily be assumed that different users have the same views on landscape, we selected informants from different user groups that are typical for the two areas. Local residents and second home owners were chosen from both areas. In Area B we included members of a local boating association, and in Area A random visitors such as hikers and people just driving through the valley were included.

As an analytical tool for understanding how valuation is performed at the national and local levels, we had initially established four pairs of value-types: (1) functional versus experiential values; (2) natural versus cultural values; (3) present-day versus future values; and (4) national versus regional or local values. In the present context, the first two of these are especially relevant.

11.3.2 Some Findings from the Project

The project confirmed our hypothesis that planning to a considerable degree is expert-oriented, and that the general public and experts have different views on landscape values – in this case for outdoor recreation purposes. The plans did not involve much participation, and planning methodologies varied, depending considerably on the planning goal and which experts had been involved. One of the big methodological differences was if and how much planning was based on the whole territory, as emphasized by the European Landscape Convention, or if the planning process had an attribute-oriented focus with emphasis on individual elements. The criteria on which experts based their evaluations were often not explicit. As a result, valuation was implicit and not very transparent.

Looking at the first value-type pair, *functional versus experiential values*, our study showed that the experts to a much greater degree than the users focused on quantifiable assets, such as the expanse of so-called 'untouched nature', or functional features of importance for outdoor recreation such as trails, beaches, etc. Nevertheless, planners and users more or less agreed on which functional features were important. Users, however, defined outdoor recreation much more broadly and had a much more comprehensive list of potential activities in each of the case areas.

Experiential values presented in the plans were mainly based on expert knowledge. Here, we identified two approaches, both focusing on physical qualities. One was based on experiencing landscape as a scenery, and can be traced back to a Norwegian methodology described in the work *Landskapsbilde i norsk naturforvaltning* ('Scenic quality in Norwegian nature management') (Bruun, 1996), while the other was based on spatial analysis and a division of the landscape into landscape subunits. Due to the one-sided use of expert assessments and the scenic approach to landscape, numerous aspects of the users' landscape experiences were not taken into consideration. This applied to experiential values linked to nature, traditions, childhood, social activities, recreational boating, adventure tourism, and similar dimensions that were not mentioned in the plans.

Users did not seem to distinguish as clearly between landscape use and experience as the plans did. For the users, landscape experience was part of their actions, which is fully in line with the concept of landshaping. When users are not involved in the planning process, the plans miss out important information about activities, i.e. actions in the landscape. Thus, the plans fail to include knowledge about people's connections to and experience of the landscape, which is contrary to the intentions of the European Landscape Convention.

The assessments of *nature value versus cultural value* showed even greater variation. To begin with, there were differences regarding the two landscape areas. Area A lies within a national park, and thus the main emphasis was placed on the protection of natural and cultural landscape assets. In this area, the wild, non-encroached areas and biological assets are unique. The experts who prepared the background material and plans in Area A mainly had scientific backgrounds and were primarily from regional or national authorities.

The users' evaluations of Area A were more diverse. The tourists using the landscape seemed to be on the same wavelength as the planners. They placed more emphasis than the local users on 'untouched' qualities and on what they considered nature values. For the tourists, the landscape formed by generations of mountain summer farming was thus 'living nature', and not an actively used cultural landscape. The area's residents looked at this differently. For them, the cultural landscape was a landscape that must be used to ensure their livelihoods. In contrast, the tourists would like to have seen the area protected, much more so than the residents. One could thus say that the planners are catering to the needs of the tourists, and not to those of the residents. However, it must be underlined that none of the user groups in the survey were in favour of extensive landscape changes.

In Area B, the planners did not specifically focus on distinguishing between nature and culture, conservation versus non-conservation, areas with so-called 'untouched nature', etc., but rather emphasized the area's 'green values' linked to the landscape's ecosystem services. In this urban fringe area, planners and users shared the same views; this was a 'green lung' and the distinction between natural and cultural landscape was meaningless – even though most respondents considered the area to be a cultural landscape. The amount of planning material for Area B is considerable. Municipal and regional authorities as well as a consulting firm were involved in the planning process, which included a diversity of experts, such as engineers, landscape architects, biologists, etc.

11.3.3 Lessons Learned

In sum, the project demonstrated that analysis and planning were expert-oriented and that the experts used a variety of methods, depending on their own professional background, the area's characteristics, etc. The experts' assessments differed from one to another. Further, experts and users did not always agree, but this may depend somewhat on the type of area in question. Last, but not least, the various user groups did not always agree among themselves on the landscape values.

Based on the results of this project, one could ask if analysis methods should perhaps be more standardized. Ratification of the Landscape Convention means that, pursuant the Specific Measures, one is committed to identifying and evaluating one's own landscapes (Council of Europe, 2000: Article 6C). What should such identification and evaluation thus include, and to what degree do we need standard methods, such as called for in the report on Nordic landscapes (Gaukstad and Sønstebø, 2003)? Our project showed that it could be difficult to implement the use of uniform methods. Rather, focus should be directed at the analysis process itself and at selecting the issues to be included in the planning process. There is also an obvious need to weaken the role of experts in landscape planning. In line with the ELC's landscape definition, one must take the 'landscapes as perceived by people' seriously, and develop strategies to involve all important stakeholders in landscape analysis and planning processes. The final section of this chapter shows how this can be practically achieved.

A study of the relationship between residents or visitors and the landscape is about understanding 'sense of place'. Through interviews, questionnaires, and observations, we can study how those who live in or visit a landscape comprehend it. The perspectives are characterized by human understanding of the landscape from a local position, whether the people are residents or guests in the landscape. These perspectives vary according to experientially based knowledge and skills, biography, and professional and other cultural background dimensions, including different interests in the landscape. This can lead to differing perspectives and conflicting practice between different governmental sectors with responsibility for landscape planning. It can also lead to value conflicts between the authorities and the general public, as well as different evaluations of the same landscape by different user groups and stakeholders. This issue will be elaborated in the presentation of professional practice in the next section. The same landscape or parts thereof can have different values for different people, and through a 'sense of place' study these values can be mapped, delineated, and systematized. This articulation of different values and relations can provide tools enabling those who live in the landscape to reflect over their own relationship with the landscape and to plan, shape, and develop the landscapes in which they live.

11.4 Landscape Character and Landscape Units

Landscape is a unifying concept aimed at expressing the complexity of different dimensions that a landscape contains. According to Phillips (2002), landscape is a meeting ground between:

- Nature and people – and how these have interacted to create a distinct *place*
- Past and present – and how therefore landscape provides a *record* of our natural and cultural history
- Tangible and intangible values – and how these come together in the landscape to give us a sense of *identity*.

In the light of this understanding, landscape can be considered a *social and cultural arena* that represents common resources and values. These resources and values need to be managed and at the same time they are a basis for value creation and development. It is therefore important to have methods for assessing the qualities and values linked to the landscape and for evaluating the importance of individual, social and cultural relations associated with it.

This multi-dimensional approach to understanding landscape was the basis for developing new landscape analysis methods in the 1990s (Swanwick, 2004). This also resulted in a clearer division between characterization of landscapes and evaluation of the landscape's meaning and value for people, or what is termed as 'sense of place'.

Landscape character is based on comprehensive interpretation of the landscape surrounding us, as perceived and experienced by people;

> Landscape character expresses the interaction between an area's natural conditions, land use, historical and cultural content, and spatial and other perceptible conditions that characterize the area and distinguish it from the surrounding landscape (Direktoratet for naturforvaltning & Riksantikvaren, 2010: 10).

Landscape character is here dependent on identification of landscape units, each with recognizable and individual characteristics that enable the separation of one landscape unit from the other. This provides a foundation for analysis which can be used at all scales from national to local level. This principle has been implemented in the English 'Landscape Character Assessment' method (Swanwick, 2002), which developed a stepless system for analysis based on the same criteria from the national down to the local community level. In Norway, the 'National Reference System for Landscape' (Puschmann, 2005) is an equivalent assessment system. It is based on a hierarchical structure of areas based on identification of landscape units at regional, sub-regional, and local levels. The main criterion in this process of identifying characteristic areas is based on spatial experience of dominating landforms. The approach is, however, the work of professionals, without any direct input from the general public.

Interpretation of landscape character within a defined land unit can provide a starting-point for developing a common platform of understanding between different stakeholder groups and local and regional administrative bodies, regarding planning and multifunctional use of the landscape.

11.5 Landscape Resource Analysis – A User-Oriented Mobilization Process Based on Landscape as a Public Good

The Landscape Convention emphasizes the importance of involving the public in analysis and planning processes. However, a user-oriented perspective and knowledge about the development of people's sense of belonging and identity appear only to a limited degree to be included in landscape analyses.

We need to understand more of how people are attached to and interact with landscapes. It seems that there is a need for new or complementary methodological procedures in landscape assessment – and there is also need to broaden the scope of current assessment tools in planning processes. In the following, we will discuss the possibilities for developing landscape analysis methodologies to become tools for mobilization, involvement, and value creation in local communities and landscape regions.

11.5.1 A Shift of Perspective on Landscape

Recent theoretical and empirical literature on landscape planning and landscape management documents show how the perspective on landscape within environmental management has changed in the past 20 years. There has been a shift of focus from considering the landscape as a sector-based category for conservation and management to treating the landscape as an arena for integrated planning, development, and management (Selman, 2006; Mose and Weixlbaumer, 2007). A strength of the European Landscape Convention lies in the way it has taken into account these changes by emphasizing the significance of paying attention to all landscapes, not just the most valuable ones. Sustainable planning, use, and administration of the landscape presupposes increased public awareness of the significance of the landscape in people's daily lives, thereby providing them the possibility of engaging in the development of the landscape. It is highly important to have good and easily available methods to collect knowledge and evaluate the landscape through broad public involvement and direct participation.

New methods and tools for assessment, local mobilization and strategic planning are being developed in Europe in order to release the versatile, value-generative potential within a sustainable management framework, e.g. the ECOVAST (European Council for the Village and Small Town) Landscape Identification guide to good practice (ECOVAST, 2006).

11.5.2 Participation in Analysis Processes – Involved Parties and Roles

Participation implies that all relevant parties make themselves heard. The landscape represents values for the population, and is a significant resource for local and regional development. A landscape analysis must be performed in a way that people can recognize that the landscape being described is *their* landscape. To ensure a successful analysis, it may often be desirable to involve actively a diversity of user groups in the process. For a complex issue such as landscape analysis, this represents a considerable professional challenge. In a landscape analysis, participation can involve seeking advice from professional communities and persons with specific expertise on issues that are important for understanding the landscape. However, participation can also imply cooperation with residents and the general public to ensure access to local knowledge and not least to gain insight into the relationships between an area and its residents. The Landscape Convention emphasizes the utilization of the entire scope of various knowledge forms.

'Local population' or 'users' are usually not a uniform group with a like-minded approach to bottom-up processes. According to Selman (2004), terms such as 'community at large' are too vague, and the management objectives and value creation potential associated with public goods is too wide-ranging, for a community to develop a common ownership to the analysis process. Communicating with the professional planner must therefore focus on involving people in concrete planning issues such as the development of recreation paths or access to and management of special landscape features.

Participation by specific interest groups, with the aim of influencing the planning process, is provided for by the Norwegian Planning and Building Act (Plan- og bygningsloven, 2008), and includes landowners, local residents, various user groups, and local and regional authorities. In analyses aimed at assessing and developing the landscape's resource potential for value creation, this influential role is often an integrated part of the terms of the project and its organizational structure.

11.5.3 Landscape Analysis as a Tool for Communication and Mobilization

The European Landscape Convention encourages the public to become involved beyond providing knowledge in the analysis process. The Convention encourages participation through actively taking part in the development of the landscape as a resource for local development and value creation (Council of Europe, 2000: Article 5c). Mobilization, involvement, and anchoring within the general public are prerequisites to developing new dynamic planning and administration strategies for the landscape.

Brunetta and Voghera (2008) point out three important aspects regarding landscape analysis as a tool for dynamic planning and conservation:

- Landscape analysis must be seen as a continuous learning process for all involved participants
- Landscape analysis should open for continuous dialogue between the general public and the authorities
- Landscape analysis is not static, but part of an ongoing process of development, with the goal to create new planning and management strategies.

These points indicate that Landscape Character Assessment must be complemented with more thorough knowledge about what landscape means to people – individuals or groups, at a given location, in order to be an active tool for place-based dynamic planning. Meaning and human-landscape relationships are to a considerable degree formed through individual activities, patterns of active use, movement, historical traditions, residence and ownership, etc. Acknowledging the value of these relationships is vital for enabling the development of a firm foundation for the planning process in the local community.

11.5.4 Sense of Place

Based on experience of 'sense of place' studies in England, such as the development of a Sense of Place Toolkit for the protected landscape of the Forest of Bowland Area of Outstanding Natural Beauty (2007), there have been efforts in Norway to test and develop tools for integration of knowledge concerning 'the landscape as perceived by people' in strategies for place-based development.

The concept of 'sense of place' is many-faceted and complex. It describes dimensions of the human-environmental relationship and can be defined as 'the collection of meanings, beliefs, symbols, values, and feelings that individuals or groups associate with a particular locality' (Williams and Stewart, 1998).

The use of the concept of 'sense of place' in landscape planning is aimed at collecting local and place-based knowledge which can give context and depth to knowledge generated through a landscape character analysis. Ideally, the many stakeholders involved in area planning processes will aim to 'make sense' together by building a common understanding of an area. Adding a 'sense of place' component in landscape assessments has the potential to amplify local inhabitant's current and historical place relationships in further planning and development efforts.

This approach also encourages involvement in developing knowledge of the landscape and people's relationship to it, and it contributes to making the value-generative potential of the landscape visible and understandable.

11.5.5 Landscape Resource Analysis

Linking an area-based landscape character analysis with systematic involvement of people through 'sense of place' studies has been termed 'landscape resource analysis'. Landscape resource analysis has the goal of laying the foundation for value-generative and developmental strategies within a certain area.

This enables two types of knowledge to be brought forth: expert-based analytical knowledge concerning natural factors, cultural factors (including land use and cultural history), and aesthetical and perceptual aspects of the landscape on one hand, and the knowledge linked to the individual and cultural importance of the area for its residents and visitors on the other. Implementing a landscape resource analysis in a specific landscape enables bringing forth complementary knowledge about the general characteristics of that landscape and its importance for people.

Whereas landscape character analyses are primarily performed by professionals according to a well-defined methodology, sense of place studies are carried out by the involved stakeholders themselves. In this context, 'involved stakeholders' includes many different user groups and roles. Their common denominator is having a link to or experience of the landscape area in question. Beyond that, the people involved can represent a wide variety of interests, including permanent residents, second-home owners, tourists, business interests, and other stakeholders.

The landscape resource analysis process encompasses four principal stages;

1. identifying the essential planning and value-creating issues at stake in the area
2. establishing a common understanding of the relevant landscape area, including meaningful sub-dividing boundaries
3. engaging in a community-based sense-of-place process involving a broad spectrum of stakeholders and user groups

4. extracting the experiential and entrepreneurial resource potential for value creation related to each landscape unit,

The process requires an open information flow and communication between experts, community groups, and other stakeholders in stages 2 and 3, and must be seen as a qualitative learning process.

Landscape resource analysis has been tested in a number of local communities, for example in the initial planning process for the Telemark Canal Regional Park project involving six municipalities, together with the County Council and the County Governor of Telemark (Clemetsen and Knagenhjelm, 2010). It has proven to be successful with regard to mobilization, involvement, and creativity when it comes to identifying a place's values and resource potentials. Such an analysis also enhances the ability of the authorities and the local population to gain a sound understanding of what the involved parties see as being important in their relationship to the landscape in question.

11.5.6 Use of Landscape Resource Analysis in Public Administration and Value Creation Processes

Landscape resource analysis has been tested and applied in several projects. One such project was carried out in the small farming community of Ornes in the municipality of Luster (Fig. 11.1). Here is found the protected Urnes stave church, which is a UNESCO World Heritage Site (Fig. 11.2). Based on the Cultural Heritage Act of 1978, a management plan for the landscape around the church was initiated by the County Council of Sogn og Fjordane in 2008 (Skjerdal, 2009). The preservation of the stave church and its surrounding landscape considerably limits the options of the local residents to make changes in the landscape. At the same time, tourists place a considerable strain on the landscape. To facilitate a good planning process, it was important to activate the local residents early in the process. An initial meeting with the local community action group, which represented the residents, was held, This meeting focused on what were considered the important issues and questions in the community of Ornes, and gave substantial input and legitimacy to setting up and carrying out meetings with focus groups, workshops, in-depth interviews, and questionnaires (for day tourists). Results from this work were directly used to compile a provisional analysis map of the area (Fig. 11.3). The landscape character map, together with descriptions of the area's management and value-creation potentials, were undertaken through a reciprocal process between local inhabitants and professional planners. For each area, its landscape character was described, its sensitivity to natural and man-made landscape changes regarding the qualities of the World Heritage Site was assessed, and a list of potentials for future development of the natural and cultural assets of the area was drawn up. This enabled the presentation of the residents' and tourists' values, wishes, and needs in a way that could be utilized directly in the further planning process. The results also provided a basis for compiling an interpretation plan for Ornes (Bjørnstad, 2009).

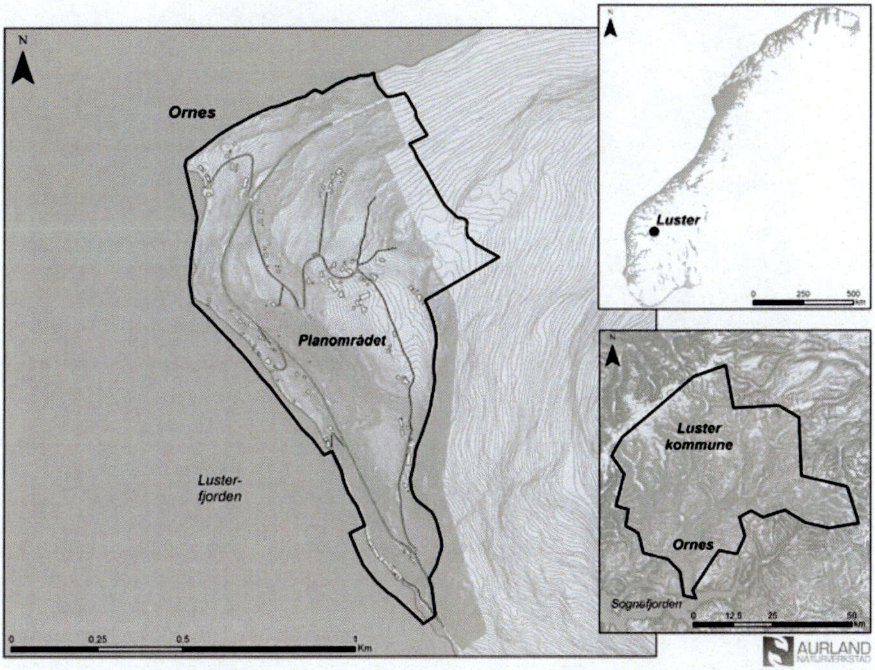

Fig. 11.1 Ornes plan area (*left*) and its location in Luster municipality in the fjord landscape of western Norway (Skjerdal, 2009) (Courtesy of Aurland Naturverkstad AS)

Fig. 11.2 The UNESCO World Heritage Site of Urnes, 2008. The 12th-century stave church is integrated in a rural landscape consisting of active farms, homes and small-scale businesses (Photo: Morten Clemetsen)

Fig. 11.3 Three-dimensional model of the farming community of Ornes based on aerial photography, showing identified landscape character areas (Skjerdal, 2009) (Courtesy of Aurland Naturverkstad AS)

What importance did the work on 'sense of place' have for the local population and the planning process? The concluding report from the sense of place study stated that:

> The local population is well aware that Ornes is a very scenic place, but the experience of one's own place of residence is generally not characterized by glamorous landscape descriptions. Landscape experience is generally based on the local residents' daily lives, in which farming and the use of natural resources play a major role. It seems as if the locals' experience of their place of residence is more closely linked to what could be called a 'functional dimension' than to the 'experiential dimension' typically associated with visitors/tourists to the area (Bjørnstad, 2009: 5).

The report also underlined that the continuous involvement of local stakeholders in the analysis and planning phases enhanced commitment to the project. Another important outcome of the active involvement from local residents in the initial phase of the project was the decision of the County Council of Sogn og Fjordane to prolong the planning process, This decision was essential in order to obtain a far better anchoring of the whole project in the local community.

11.6 Conclusion

Increased complexity of planning creates a distance between those who live in, work, and experience the landscape and the planners who determine the conditions for local enterprise and landscape use. Planning at the landscape scale often ends in compromises between different sector interests and perspectives. Drawing boundaries between different land-use categories divides up the landscape and complicates integrated landscape management in a way that is often not in accordance with how people themselves experience and interact with the landscape. This can result in alienating the local community from their daily living space.

The European Landscape Convention challenges this situation through the goal of stimulating the general public to participate actively in planning, forming, and developing the landscapes in which they live. Planning processes that maintain the distinctive character of a place enhance the population's relationship to their surroundings and their feeling of local identity. Mobilization and empowerment of a community are important for the democratic and sustainable development of local and regional communities. This requires in-depth knowledge of the landscape's physical, historical, and cultural features, and of the activities being performed therein. Experience from the landscape resource analysis model indicates that this can be a tool for increasing the sustainability of landscape planning that is rooted in both place and people.

A landscape resource analysis can be an appropriate tool for generating new knowledge and developing a basis for understanding and dialogue between top-down and bottom-up perspectives in planning. At the same time, this could contribute to releasing the mutual learning potential in a given landscape, which in turn could enhance a dynamic public administration process.

References

Altman I, Rogoff B (1987) World views in psychology: trait, interactional, organismic and transactional perspectives. In: Stokols D, Altman I (eds) Handbook of environmental psychology. Wiley, New York, NY, pp 7–40

Basso KH (1996) Wisdom sits in places: notes on a Western Apache landscape. In: Feld S, Basso KH (eds) Senses of place. School of American Research Press, Santa Fe, NM, pp 53–90

Bjørnstad K (2009) Stadkjensleprosjekt Ornes 2008/2009. Prosjektnotat 01/2009. Aurland Naturverkstad AS, Aurland

Brunetta G, Voghera A (2008) Evaluating landscape for shared values: tools, principles and methods. Landsc Res 33(1):71–87

Bruun M (1996) Landskapsbildet i norsk naturforvaltning. Institutt for landskapsplanlegging, Norges landbrukshøgskole, Ås

Casey ES (1996) How to get from space to place in a fairly short stretch of time. In: Feld S, Basso KH (eds) Senses of place. School of American Research Press, Santa Fe, NM, pp 13–52

Churchman A (2002) Environmental psychology and urban planning: where can the twain meet? In: Bechtel RB, Churchman A (eds) Handbook of environmental psychology. Wiley, New York, NY, pp 191–200

Clemetsen M, Knagenhjelm T (2010) Landskapsressursanalyse "Telemarkskanalen som regionalpark", Rapport 01/2010. Aurland Naturverkstad AS, Aurland

Council of Europe (2000) The European Landscape Convention. Florence, 20.X.2000. ETS No. 176. http://conventions.coe.int/Treaty/en/Treaties/Html/176.htm. Accessed 30 Mar 2010

Cultural Heritage Act (1978) Act of 9 June 1978 No. 50 concerning the cultural heritage. government.no. http://regjeringen.no/en/doc/Laws/Acts/cultural-heritage-act.html?id=173106. Accessed 23 Mar 2010

Direktoratet for naturforvaltning & Riksantikvaren (2010) Landskapsanalyse: Framgangsmåte for vurdering av landskapskarakter og landskapsverdi. Trondheim & Oslo. http://www.riksantkvaren.no/filestore/Framgangsmteforvurderingavlandskapskarakteroglandskapsverdi_24.2.2010..pdf. Accessed 22 Mar 2010

ECOVAST (2006) Landscape identification: a guide to good practice. http://ecovast.org/papers/good_guid_e.pdf. Accessed 30 Nov 2009

Feld S (1990) Sound and sentiment: birds, weeping, poetics and song in Kaluli expression. Pennsylvania University Press, Philadelphia, PA

Feld S (1996) Waterfall of songs: an acoustemology of place resounding in Bosavi, Papua New Guinea. In: Feld S, Basso KH (eds) Senses of place. School of American Research Press, Santa Fe, NM, pp 91–136

Forest of Bowland Area of Outstanding Natural Beauty (2007) Sense of place toolkit. http://www.forestof bowland.com/bn_sop. Accessed 2 Dec 2009

Gallagher S, Zahavi D (2007) The phenomenological mind: an introduction to philosophy of mind and cognitive science. Routledge, London

Gaukstad E, Sønstebø G (2003) Nordens landskap: Forprosjekt for oppfølging av den europeiske landskapskonvensjonen. Tema Nord 2003:550. Nordisk Ministerråd Miljø, København

Gifford R (2007) Environmental psychology: principles and practice. Optimal Books, Washington, DC

Gray JH (2000) At home in the hills: sense of place in the Scottish borders. Berghahn Books, Oxford

Heidegger M (1983) Hebel – friend of the house. In: Christensen DE, Redel M (eds) Contemporary German philosophy, Vol. 3 (trans: Foltz BV, Heim M). Pennsylvania State University Press, University Park, PA, pp 89–101

Ingold T (2000) The perception of the environment: essays in livelihood, dwelling and skill. Routledge, London

Kleven T, Thorén A-KH, Skogheim R (2005) Å sette pris på friluftsliv – bare for eksperter? NIBR-rapport 2005:02. Norsk institutt for by- og regionsforskning, Oslo

Krogh E (1995) Landskapets fenomenologi, Doctor scientarium theses 1995:15. Institutt for økonomi og samfunnsfag, Norges Landbrukshøgskole, Ås

Macnaghten P, Urry J (2001) Introduction. In: Macnaghten P, Urry J (eds) Bodies of nature. Sage, London, pp 1–11

Merleau-Ponty M (1962) Phenomenology of perception. Routledge & Kegan, London

Mose I, Weixlbaumer N (2007) A new paradigm for protected areas in Europe? In: Mose I (ed) Protected areas and regional development in Europe: towards a new model for the 21st century. Ashgate, Aldershot, pp 3–19

Phillips A (2002) Management guidelines for IUCN category V protected landscapes/seascapes. IUCN, Gland

Plan- og bygningsloven (2008) Lov om planlegging og byggesaksbehandling (plan- og bygningsloven), LOV-2008-06-27-71. Lovdata. http://www.lovdata.no/all/nl-20080627-071.html. Accessed 13 Apr 2010

Puschmann O (2005) Nasjonalt referansesystem for landskap: Beskrivelse av Norges 45 landskapsregioner. NIJOS rapport 10/2005. Norsk Institutt for Skog og Landskap, Ås

Relph E (1985) Geographical experiences and being-in-the-world: the phenomenological origins of geography. In: Seamon D, Mugerauer R (eds) Dwelling, place, and environment: towards a phenomenology of person and world. Martinus Nijhoff, Dordrecht, pp 15–32

Rizzolatti G, Sinigaglia C (2008) Mirrors in the brain: how our minds share actions and emotions. Oxford University Press, New York, NY

Selman P (2004) Community participation in the planning and management of cultural landscapes. J Environ Plan Manage 47(3):365–392

Selman P (2006) Planning at the landscape scale. Routledge, Taylor& Francis Group, London

Skjerdal I (2009) Fagrapport for områdeforvaltning av Ornes i Luster kommune. Rapport 01/2009. Aurland Naturverkstad AS, Aurland

Swanwick C (2002) Landscape character assessment: guidance for England and Scotland. Countryside Agency & Scottish Natural Heritage, Cheltenham & Edinburgh. http://www.landscape character.org.uk/lca/guidance. Accessed 23 Mar 2010

Swanwick C (2004) The assessment of countryside and landscape character in England: an overview. In: Bishop K, Phillips A (eds) Countryside planning: new approaches to management and conservation. Earthscan, London, pp 109–124

Thorén A-KH (2008) Verdsetting av friluftsliv: Empirisk uttesting av ulik verdsettingsmetodikk. Serie B, rapport B2/2008. Institutt for landskapsplanlegging, Universitetet for miljø- og biovitenskap, Ås

Tilley CY (1994) A phenomenology of landscape. Berg, Oxford

Vorkinn M, Clemetsen M, Kleven T, Thorén A-KH, Skogheim R (2002) Verdsetting av rekreasjonsområder: En kritisk gjennomgang av fem offentlige veiledere. ØF-Rapport nr. 14/2002. Østlandsforskning, NIBR, NLH, Lillehammer

Williams DR, Stewart SI (1998) Sense of place: an elusive concept that is finding a home in ecosystem management. J Forestry 96(5):18–23

Chapter 12
Participation Within the Landscape of the River Dart Catchment, Devon, England

Neil Spencer

Abstract This chapter presents a case study focused on the River Dart in south-west England, part of a transnational project focusing on the participatory management of river landscapes. It illustrates how the project brought together different interest groups and involved local communities in identifying shared values, priorities, and an action plan for the future management of the landscape. By reviewing the role of participation and the principles applied, it provides a rationale

N. Spencer (✉)
Sustainable Futures, Ashreigney Chulmleigh, Hayes, EX18 7NG Devon, UK
e-mail: neil.spencer@sustainablefutures.co.uk

M. Jones, M. Stenseke (eds.), *The European Landscape Convention*, Landscape Series 13, DOI 10.1007/978-90-481-9932-7_12,

for the approaches adopted. It describes the development process and the design of each event held to engage stakeholders and the public in the project and illustrates how the principles were applied in practice. The results of the study are examined with reference to the experience of the participants. The challenge of the long-term involvement and role of stakeholders after the participative decision-making activities have formally ended is considered with examples and recommendations.

Keywords Facilitating · Stakeholder · Engagement · River catchment

12.1 Introduction

The River Dart, located in south-west England, flows for 62 km from source to sea. Rising 550 m above sea level on Dartmoor, the river makes its way to the estuary at Dartmouth (Figs. 12.1 and 12.2). Its catchment area accounts for more than 10% of the land area of the county of Devon. The catchment of the river includes a wide variety of landscape types. On its journey to the sea the river traverses acid soils blanket bogs, and the granite rocks of Dartmoor. As it runs through steep-sided valleys into the estuary at Dartmouth, it has drained water from 475 km^2 of land.

The river, its tributaries, and the estuary are home to a diverse range of species and habitats. It is a highly valued landscape that underpins an important local tourist and recreational economy. The catchment is recognized as an area of high conservation value and incorporates a number of protected areas. These include a National

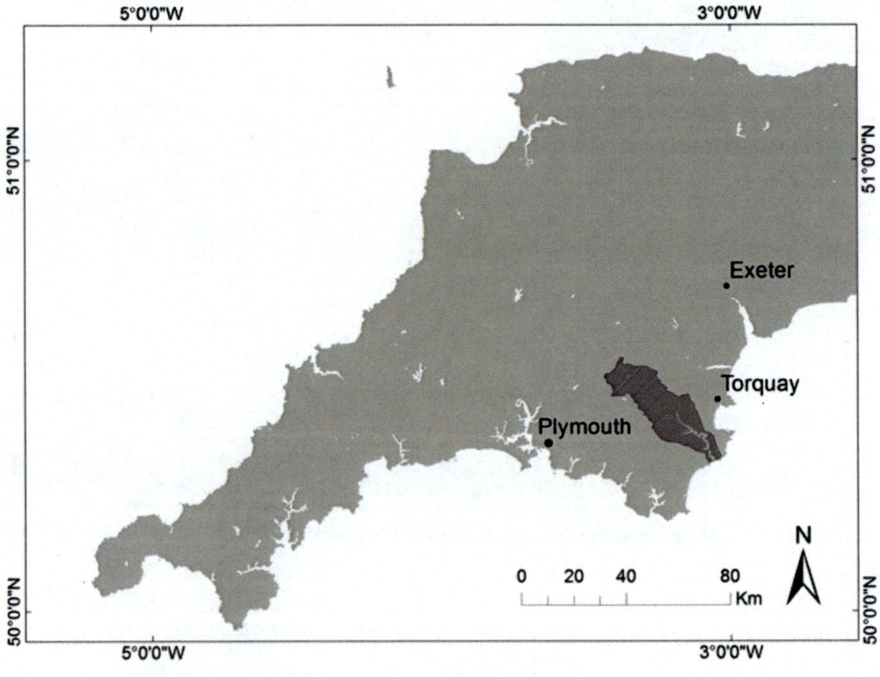

Fig. 12.1 Location of river Dart catchment in south-west England

Fig. 12.2 River Dart catchment showing estuary, tributaries, towns and other settlements (Courtesy of Devon Wildlife Trust)

Park, a Special Area of Conservation, an Area of Outstanding Natural Beauty, a Prime Biodiversity Area, and numerous Sites of Special Scientific Interest.

The landscape and the river have supported and sustained human life for centuries. Dartmoor itself has been inhabited and farmed for over 4,000 years. The continual human use of the land has shaped the catchment into a rich cultural landscape of green lanes, hedgerows, and archaeological remains.

In 2003, Devon Wildlife Trust joined a partnership of French, English, and Irish organizations who came together to develop, test, and exchange ideas for managing river-based catchments. The project, known as Cycleau, was funded through the European Regional Development Fund's INTERREG programme. Each of the partners represented predominantly rural areas with economies that are reliant on high quality coastal, estuary and river environments.

The focus of their collaboration was the European Water Framework Directive (WFD) (EC, 2000), which aims to overhaul the management of water throughout

Europe. Public participation is a core requirement of the Directive and its UK Regulation. The Directive requires environmental management on the scale of the river basin or catchment in order to achieve 'good status' standard for all surface waters by 2015 – incorporating elements of ecology, morphology, river continuity, and chemistry. To achieve this 'good status' standard, many issues, including pollution, land-use planning, land drainage, and flood defences, need to be addressed. In introducing the Directive the EU believes that it has proposed the best model for a single system of water management – the river basin. Under the Directive the landscape is categorized in terms of the natural geographical and hydrological unit instead of according to administrative or political boundaries.

The Directive is considered by the Foundation for Water Research to be the most substantial piece of European water legislation for over 20 years and includes the important principle of 'getting the citizen involved more closely'. This collaborative approach underlines the importance and the role of participation in engaging the citizen in the Directive's aims.

The Cycleau project used the WFD as a foundation to explore a number of objectives, including the objective of Article 14 that requires that 'Member states shall encourage the active involvement of all interested parties in the implementation of the Directive'.

Devon Wildlife Trust's role in the Cycleau project was to lead on the theme of 'Stakeholder and Public Participation'. Selecting the Dart and its catchment as a pilot project, the Wildlife Trust set out to engage the public and stakeholders in a process of awareness-raising and participative decision-making about the future of the river landscape. In the longer term the Wildlife Trust aimed to use the results of this pilot project to develop a framework that would be applicable to other river catchments in Devon. The pilot project provided an important opportunity to bring together different interest groups. Statutory planners worked alongside farmers, landowners, citizens, and the Wildlife Trust to identify shared values and priorities for the future management and the creation of a shared action plan for this highly valued landscape.

By using the framework of the river basin and the descriptive term 'catchment', the pilot project focused on the entire landscape connected by the river, its tributaries, and the estuary. In emphasizing the importance of the river catchment and the role of participatory techniques in the protection, planning, and management of the landscape, this pilot project offers a relevant case study for the implementation of the European Landscape Convention.

The Dart was chosen as a pilot project for its biodiversity and the sheer variety of its transitional habitats. This landscape area was also selected because it encompasses a variety of initiatives, projects, remits, and jurisdictions, which had two implications:

1. There was no comprehensive or coordinated approach across the catchment – this allowed the pilot project to test a methodology that 'would fill the gaps' between existing remits.

2. Where mechanisms were already in place, the pilot project would create partnerships to ensure it was delivered in an integrated and efficient manner.

As well as engaging the public and stakeholders in participative decision-making about the future of the catchment, the pilot project also included: a detailed profile of the catchment, looking at its economic, environmental and socio-economic balance; an audit of policy; the collection of data documenting the current state of the catchment in terms of environment, society and economics; an Action Plan that outlined environmental priorities for the river and its catchment; and an evaluation of the project development process.

To ensure the process would be facilitated independently, I was commissioned by Devon Wildlife Trust through my organization Sustainable Futures to develop the methodology for and to manage and facilitate the process of stakeholder and public participation. This chapter therefore is the result of my own professional involvement in the pilot and has been adapted from a report originally produced for the UK Environment Agency and which I jointly wrote with Philip Moore from Devon Wildlife Trust (Moore and Spencer, 2005). The aim of this chapter is to describe, justify and evaluate methodology used in applying a particular set of participatory techniques to a concrete case study.

12.2 Participants and Participation

12.2.1 Stakeholders

The active involvement of stakeholders through a participatory planning and democratic process is still relatively unusual. However, their involvement is an important means of improving decision-making and creating awareness of environmental issues, and helps to increase acceptance and commitment to the outcomes. Generally, stakeholders are involved in river catchment management through information supply and consultation processes. They are usually members of groups or organizations. They may also be members of the general public who have a stake in the future of the catchment because they are affected by or want to influence the outcome of changes or plans for the future of the catchment.

In the UK there has been some debate among engagement practitioners about the language used to describe participants. The approach used in this chapter uses the word stakeholder to mean local residents, individual members of the general public, people who may be members of a pressure group, and people who may be working on behalf of a government agency or a non-government agency. What all these groups and people have in common is a stake in the future of the river and its catchment.

Stakeholders can participate in statutory processes in a number of ways. These processes can include:

1. Consultation
2. Information gathering
3. Active involvement, including:

- Identification of issues and needs (e.g. identification and characterization of a catchment project)
- Delivery of actions (e.g. identification and implementation of specific catchment management measures)
- Participation in the democratic process.

The general public are usually only involved in the first two levels, with the third being only an ideal. However, increasing the participation of stakeholders in the management of landscapes can positively influence the outcomes and the process. Recognizing that all people living and working in a river catchment or landscape area are stakeholders helps to increase acceptance, ownership, and commitment towards the protection, planning, and management of the landscape.

12.2.2 Why Participation Is Important

Wates (2000) and Porter and McDonagh (1998) highlight the benefits of the active involvement of stakeholders and communities as well as those organizations that are managing projects designed to deliver environmental solutions. The participative themes addressed in the Dart pilot project included the following principles:

- *Empowerment and ownership* – when stakeholders have opportunities to influence decisions, policies, and strategies that affect them, they develop confidence and take ownership of both the problems and the solutions.
- *Building community* – the process of involvement and working together creates a strong sense of community.
- *Collaboration* – when stakeholders participate in planning and decision-making, it helps develop the basis for collaboration. Often complex environmental problems can only be solved through effective collaboration between diverse agencies, non-governmental organizations, stakeholders, and communities who all have a stake in a successful outcome.
- *Inclusiveness* – decisions and plans made by groups that are inclusive help to ensure that the outcomes take account of a wider range of ideas and influences. An inclusive approach also helps to secure the support and sense of ownership of both the problem and solution by those who participate in the planning.
- *Raising awareness* – involving stakeholders in decision-making provides important opportunities to raise their awareness and understanding about the issues and constraints.
- *Better decisions* – by combining the breadth of community and stakeholder knowledge, bringing together laypersons' and expert understandings, new solutions to complex environmental problems can be formulated.

- *Learning through doing and reflecting* – when stakeholders are involved in problem-solving, there are important opportunities for everyone to 'learn together' through doing and reflecting.
- *Building on what local people value* – involving local people as stakeholders in decision-making provides the opportunity to connect into a 'common landscape identity', which builds on what local people value about their environment.
- *The process is at least as important as the product* – any document, be it a strategy, management plan or element of policy, is only as effective as the process from which it derives. A major benefit from involving stakeholders in the development process (through discussion, interaction, recognition of points of view, and agreement) is the increased sense of ownership over the final outcome that those involved can feel. This is essential for successful implementation.

12.2.3 Criteria for Effective Stakeholder Participation

Accepting the different agendas that each participant brings to the process is important. Participants will want to be involved for many different reasons. This reflects the diversity of how individuals relate to an issue – those involved in organizing and facilitating stakeholder events need to be aware of these different agendas. In the Dart pilot project this was used as an opportunity to capture varying perspectives and promote learning and increased understanding between participants.

Clarifying the limitations of the project at an early stage of the process helps to avoid raising participants' expectations that cannot be met. If expectations are not handled effectively, participants can feel frustrated and disillusioned if the project's outcomes do not match up to their original aspirations.

Participative meetings and workshops need to be productive and safe forums in which participants can deliberate on the future protection and planning of the landscape. Setting guidelines and boundaries is particularly important in communities where there can be high levels of contention, or where participants may feel threatened or uncertain about their views being valued. Facilitators need to 'hold the line' on these boundaries to ensure that the meeting and workshop run to the times agreed and that all participants have the opportunity to contribute. Clarifying the aims and objectives of the meetings and workshops and people's roles helps the process to be productive and focused.

A key to creating the partnerships that are needed to deliver an integrated and efficient participative process is to deal with the need to develop the skills and capacity of the participants. Pedlar et al. (2007) underline that it is important that those organizing and facilitating the process are committed to focusing on outcomes that generate change and learning. In the Dart pilot project the sponsoring organizations committed themselves to follow up meetings with an action plan review workshop and newsletters. This helped to support the longer-term involvement and a continuing role of the participants.

With reference to Capra (2002), the systems view is described by Attwood et al. (2003: 23) 'as holistic and organic, whereas conventional thinking is reductionist and blindly mechanical. One sees systems as living, cognitive networks shaped by values and purposes, whereas the other sees a complex system as merely a "click together" collection of components'. Attwood et al. (2003: 114) describe the key principles underpinning a systems approach to participation as: 'Everyone involved in change should be an architect of it. Everyone in the room together – on the pitch playing – as a metaphor and as a reality.' They go on to suggest that the successful involvement of stakeholders can provide opportunities for shared professional, social, and community 'learning'. This, they argue, occurs when people learn together and when changes of perception, understanding, and action take place and are observed by others. Giving participants the opportunity to verbalize what they have learnt during a participative event provides the potential for collective learning. These principles were reflected in the Dart workshops and meetings. This involved a clear commitment to inclusiveness, inviting everyone on to 'the pitch' as an active valued contributor. Meetings were designed to nurture collective learning by encouraging participants to tell each other stories of the past, and to share their perspectives on the present as well as their aspirations or concerns for the future.

Connecting with local knowledge and expertise was a critical factor in the success of this pilot project. Respecting and valuing the wide range of knowledge and perspectives that participants brought to the process revealed a rich tapestry of knowledge and helped to make a strong connection between diverse interest groups and individuals.

The need to allow sufficient time to engage all those who may be interested in a project needs to be balanced against a process that can be over-intensive and can result in 'participation fatigue'. It is also important to recognize that different stakeholders may need different incentives to become involved and to retain an interest in the project. The participatory processes of the Dart pilot project were designed to enable participants to see clearly how their involvement and contributions were acted upon. This helped to build relationships and strengthen trust between all those involved. It is important that those involved are 'given back' the results of their work and have the opportunity to keep in touch with later developments.

The layout of a participative event can also have a significant influence on its effectiveness. At most conferences, people sit in rows – this encourages passive behaviour and may result in the audience listening to a single point of view. The presence of a 'top table' of speakers is designed to signal importance and status. This type of arrangement and layout is ineffective at encouraging equitable communication and dialogue. A more effective and less hierarchical approach, as suggested by Weisbord (2002), was used in the Dart participatory events where participants were seated at tables in groups of six to eight. Each of these groups was intentionally designed to produce a 'maximum mix' of participants. Under this approach each group was made up of participants and stakeholders who represented different interests, priorities, and connections to the river. This helped to foster a greater sense of equality and provided opportunities for active listening, helping the participants engage in conversation, develop working relationships, and gain the opportunity to understand better the opinions of others. Participants were also encouraged to move from one group to another or stand together engaging in conversation and structured reflection. Often considered as 'only' networking, these informal activities offered an important opportunity for learning, gaining insight, and developing empathy.

12.3 The River Dart Catchment Pilot Project – Strengthening the Relationship Between People, Water, and Wildlife

This section examines the participative approach used in the development of the Dart Catchment Project. It reviews the development process and the design of each event held to engage stakeholders and the public in the project and it illustrates how the principles described above were applied in practice.

The project was developed by carrying out the following steps:

(a) An audit of data and information documenting the current state of the catchment area in terms of environment, society, and economics
(b) A review of existing stakeholder projects and their methodologies
(c) An analysis of stakeholders in the catchment area and the creation of a contact database
(d) The identification and review of all the strategies, plans, and policies that relate to activities undertaken in and around the catchment area
(e) The development of partnership opportunities with key organizations
(f) Wider community involvement through public meetings and the identification of key values and threats or issues affecting the catchment
(g) Input from stakeholders, statutory agencies, and specialist groups to prioritize issues and identify solutions (via organizations' stakeholder meetings)
(h) The development of an Action Plan that outlined environmental priorities for the catchment
(i) The evaluation of the project development process.

12.3.1 Designing the Stakeholder Participation Approach

Working with the project team, Sustainable Futures identified a number of key questions that helped to shape the design of the participation process. The questions included:

- What is the most effective 'mix' of stakeholders to represent the catchment system?
- What design principles are relevant to this participation process?
- How can the design ensure the sharing of knowledge and experience?
- How can the design increase ownership of catchment management issues by stakeholders?
- Can the design demonstrate openness and creativity in decision-making?
- Can the design develop a democratic process that is demonstrable to others?

- What can we learn from the experiences of previous participative projects in the catchment?

Following this iterative process the project team and Sustainable Futures agreed a design that included the following participative elements: four public participation meetings; a stakeholder workshop for organizations; a public and organizations' stakeholder review workshop; and a catchment festival.

Another important task in the design process was to create a 'brand' for the participation events and the project as whole. This had to provide an easily understandable goal that embraced the overall aims of the project whilst being meaningful to the target audience. The phrase 'Have your say' with the strap line 'Strengthening the relationship between people, water and wildlife' was developed to encompass the process of identifying values, issues, and solutions in the development of an environmental action plan for the catchment (Fig. 12.3).

Each event was designed to bring together a 'mix' of stakeholder knowledge and perception and provide an opportunity for participants to share knowledge and experiences. The events were also designed to provide a forum for reflection and feedback on the emerging results of the process. In each of the events the facilitators emphasized the importance of implementation, action, and the building of links between the public, organizations, and stakeholders.

A key consideration for all the events was the aim to increase the ownership of catchment management issues through transparency and activities that developed trust between all groups. If stakeholders were to invest their time and to trust that there would be productive and beneficial outcomes, it was important that the workshops and meetings were effective forums in which participants felt their views were valued and acted upon.

12.3.2 Delivery of the Public Participation Meetings

Four public participation meetings were held at geographically diverse locations throughout the catchment, ranging from Princetown, which is high up on Dartmoor, to Ashburton and Totnes located in the middle of the catchment, and finally to Dartmouth, where the tidal estuary meets the sea. Venues for the meetings were selected on the basis of locality, access, facilities, and community significance. The aims of the meetings were to:

- Bring together people who have a stake in the future of the catchment
- Understand and learn together what will best enable the catchment area to thrive in the long term for the benefit of wildlife and people
- Identify what local people value about the river and its catchment
- Identify what issues local people believe need to be addressed
- Pool local knowledge and expertise towards developing the best ways to look after the river's environment and those whose livelihoods depend on the conservation of the catchment.

Fig. 12.3 Poster invitation to public participation meeting concerning the river Dart (Courtesy of Devon Wildlife Trust)

The events were promoted by a written invitation to members of Devon Wildlife Trust in the catchment area. In addition others were invited to attend via media releases generated through newspaper and local radio coverage, parish magazine articles, and posters displayed on village notice boards. Invitations were also sent out to local community representatives and individuals, including elected members of local councils. Information about the project and the participatory meetings was also displayed in local libraries.

The format for each meeting was standardized to allow comparison of results between events. Roles were clearly identified at the beginning of each meeting. The facilitators' role was to set the task and ensure it was completed in the agreed time, manage the large group discussion, and keep the purpose of the task and workshop visible and up-front.

The project team was encouraged to join the small groups with a brief to engage in the discussion without unduly influencing any of the outcomes. Participants were asked to provide information and analysis, interpret the information, manage their own group work, and participate in the discussions.

To create a sense of learning and equality at the events, project information boards were used. These displayed a large-scale map of the area, project literature, a 'Who's Who' directory describing the role of the many environmental organizations within the study area, and a summary of the catchment's environmental, social, and economic characteristics.

Each of the four public meetings generated productive discussions about values and issues. Participants worked in small groups enabling everyone to contribute and think through the task before the whole group worked together to generate a shared picture of the values and issues. Attendance at the public meetings varied: 12 people attended the meeting at Princetown; 29 people attended the meeting at Ashburton; 54 people attended the meeting at Totnes; and 18 people attended the meeting at Dartmouth. Working in 'maximum-mix' small groups, participants were first asked what the essence of the catchment meant for each of them. Some of the participants brought photographs, stories, or mementos to the meetings to illustrate their connection to the catchment. The feedback indicated that for many participants this exercise tapped into a deep level of meaning. Values identified during this exercise are indicated in Table 12.1.

Table 12.1 Values identified in public participation meetings

Farming life	A place of work
The economic value of the river	Historical significance
Diversity of wildlife habitats and species	Recreational opportunities
Cultural and economic institutions	The shellfishery
The river as an educational tool	The water as a vital resource
The social diversity	Community events
Peace and tranquility	The boats and ferries
The towns and villages	Tourism and its value to the local economy
The water quality	Woodlands
The small scale [of the catchment]	Scenic/aesthetic values
Its uniqueness	Cultural significance
A source of inspiration	The Dart as a boundary
The life-cycle of the salmon	It's free
Examples of sustainable tourism and traditional tourism	The river as a water supply
The flora and fauna	

Participants also worked in small groups to consider the long-term trends they believed were affecting the future of the catchment. Emery and Trist (1973) and Weisbord and Janoff (1995) suggest that participants are encouraged to consider future trends and issues that are local, regional, national, or global. Using this approach enabled participants to identify a very wide range of issues. It also provided an important opportunity for creating a shared picture of the future and social and community learning. Trends and issues identified are indicated in Table 12.2.

In order to connect more deeply with local knowledge and 'lived expertise', participants had the opportunity to be interviewed and record their reflections and concerns through a 'video box'. This proved to be very effective, with participants sharing their perspective and knowledge about the river, farming conditions, and cultural and artistic traditions that are associated with the river, its catchment, and its history.

The plenary discussions were recorded through a 'mind mapping' exercise that captured feedback from participants and recorded their responses. This activity provided an engaging process as participants watched a mind map evolve based on their feedback and the diversity and interconnectedness of their values and trends that they identified.

Table 12.2 Trends and issues identified in public participation meetings

Poor co-ordination of data collection especially for species and habitats	A perceived increase in siltation of the river
Alien plant species	The reduction in the 'sponge capacity' of the moorland.
Localized decline in water quality	Low water levels
The use of household chemicals	The erosion of river banks
Acidification	Oil from boats in the estuary
Combined storm overflows	An increase in recreational use of the Dart and tourism
Air pollution from traffic	Population growth
Increase in boat traffic	Limited access to data
Climate change	The use of nitrates on agricultural land
The impact of visitors and 'incomers' to the river	The loss of local traditions and knowledge
A 'disconnection' between nature and humans	The under-use of the Dart as an educational resource
An increasing need for management of visitor and recreation pressure	A top-down approach to management.
A lack of long-term sustainable planning and joined up-thinking	Changes in farming practices and policy
Diffuse pollution	A decline in some key habitats and species
A lack of ownership for managing the catchment area	Demand for and provision of moorings
Increase in litter	

12.3.3 *The Delivery of the Stakeholder Workshop for Organizations*

The stakeholder workshop for organizations was held at Buckfast Abbey, where 75 participants attended. The venue for the meeting was selected because of its central location within the catchment, its transport links, proximity to the river, accessibility, facilities, and cultural significance within the catchment. The aims of this workshop included:

- To bring together organizations which have a stake in the future of the catchment
- To understand and learn together what will allow the catchment to thrive in the long term for the benefit of wildlife and people
- To identify and prioritize key issues that need to be addressed
- To create a vision for the catchment and identify potential solutions
- To pool local knowledge and expertise towards developing the best ways to look after the river's environment and those whose livelihoods depend on the conservation of the catchment.

Following an analysis of the roles and remits from the many organizations (statutory and non-statutory) that exist or operate within the catchment area, the event was promoted by direct invitation to organizational representatives. The initial analysis ensured a representative and appropriate mix of interests. In addition, a media release generated newspaper and local radio coverage.

The workshop began with participants working in small maximum-mix groups identifying long-term trends and issues affecting the future of the catchment (Fig. 12.4). Participants then worked as a large group in plenary mode creating a mind map of the trends and issues identified.

Fig. 12.4 Work in small groups under way at the stakeholder workshop for organizations, 2004 (Photo: Neil Spencer, 2004)

After this analysis participants then considered how organizations, communities, and businesses need to respond to the trends identified. Working in their small groups participants were invited to make recommendations on how project-led work could respond to threats identified and how opportunities for partnership work could be maximized. The group work was followed by a plenary session that enabled the workshop to identify and prioritize key issues that needed to be addressed.

The next step was for participants to identify subject-specific themes that encompassed the issues that were raised in the earlier session. These were used to create self-selecting themed groups in which participants developed a strategic plan for the future of the catchment. Working in their self-selected groups, participants used the following four questions to guide their deliberation:

1. If we build on existing opportunities and develop new initiatives, what are the most important practical projects that need to be in place in the medium term – within 5 years?
2. How do these proposals respond to the trends identified?
3. What do we need to do differently if we are going to succeed?
4. Based on the subject-specific theme, would you please outline a long-term vision for the catchment area?

These questions were designed to enable the groups to formulate a vision for the catchment and identify potential solutions. It was also intended that they should provide further opportunities for learning and the sharing of understanding, as well as emphasizing the need for action and implementation. Participants recorded their work and discussions on large-scale formatted boards. This enabled each group to provide feedback in a format that was broadly similar.

12.3.4 The Public and Organizations' Stakeholder Review Workshop

This workshop was held at Dartbridge with 50 participants attending. The venue for the meeting was selected for its transport links, accessibility, facilities, and its close proximity to the River Dart. Taking place a few weeks after the public meetings and organizational workshop the review workshop aimed to:

- Review the outcomes from the public stakeholder meetings and the organizations' stakeholder workshop
- Evaluate whether the proposed recommendations for action reflect the views of those involved with the catchment
- Identify and prioritize key actions that need to be addressed
- Pool local knowledge and expertise in developing appropriate solutions to issues affecting the catchment
- Clarify the next steps and understand how stakeholders can continue to be involved.

The event was promoted by direct invitation to all stakeholders who had attended or expressed an interest in previous project events. Follow-up phone calls were also made. In addition, a media release generated newspaper and local radio coverage.

Prior to the workshop, participants received a summary of the outline Action Plan to enable them to prepare before the event. The content of the outline plan was based on previous events. The workshop began with participants working in maximum mix groups, formulating feedback around three questions:

1. Did the outline Action Plan capture the values that people regarded as significant?
2. Did it identify the main issues and trends?
3. Did its themes encompass the identified trends?

These questions were also designed to reinforce the role of participants in shaping the solutions that the project would deliver during the implementation phase. An important difference for this workshop was the increased diversity and mix of participants – with members of the public working alongside organizational representatives. This provided an opportunity for participants to understand different viewpoints and was an important step in generating cohesion among the many different interests encompassed by the project's remit.

In the second part of the workshop, participants reviewed and prioritized the proposed recommendations for action, listed under the seven themes generated at the earlier organizations' stakeholder workshop. Working in self-selected groups, participants used large-scale formatted boards to formulate feedback on the following themes:

1. Joined-up thinking and political support
2. Research and understanding
3. Local community, business, and the wider economy
4. Education and awareness
5. Impacts of farming
6. Water quality and habitats
7. Recreation, access, and tourism.

First, the groups were asked to identify how the plan could add value to projects, plans and activities already being undertaken by groups and organizations operating within the area. Second, they were asked to identify gaps in the proposed work in terms of partners, links to other opportunities, and action that needed to be undertaken.

Finally, the participants completed an exercise of using coloured dots to prioritize the issues and actions that individuals thought most important. Groups arranged their boards around the room so that participants could review the progress the project and workshop had achieved. This part of the workshop provided an opportunity for considerable informal discussion in pairs and groups, providing an invaluable opportunity for sharing knowledge and reflecting on what had been collectively achieved.

The results of this workshop provided an important and influential steer for the emerging Action Plan that was released for consultation.

12.3.5 Delivery of the Catchment Festival

The Dart Water Festival was held on the 25 September 2004 in Totnes, which is situated on the tidal limit of the Dart estuary – a significant location in catchment terms. The event embodied the 'systems approach' applied to other participative events. The festival programme included: displays by organizations that have a responsibility or interest in land and water management; trade stands of businesses whose livelihood is associated with and often depends upon the ecological well-being of the river and its catchment; craftspeople, artists and poets who are inspired by the landscape of the catchment; children's activities related to discovering more about the river; and organized boat trips and canoeing.

Against this background, the event provided a high-profile opportunity to launch the project and its draft Action Plan officially for general consultation. The event allowed the project to promote its work programme to a new audience and to the local media whilst also rewarding those involved in the project's development with a fun day out.

The event was a critical step in maintaining the momentum and subsequent interest that had been generated through the series of stakeholder events. Additionally, the event brought members of the public back in touch with the organizations they had worked with at the public meetings, and at the organizational and review workshops.

12.4 Project Evaluation

Achievement can be assessed against both qualitative and quantitative criteria. Those involved can assess the 'value' of the exercise, and how things may or may not be done differently in the future. Evaluation is important, not only from the viewpoint of participants who have invested time and effort, but also from the viewpoint of the organizers and (if different people) those who have funded a process (Cuff, 2001). The results of this participation process demonstrated substantial returns over the cost of investing in this approach.

A specific methodology was developed to evaluate each of the participatory events. Participants who attended the project meetings were asked to complete an evaluation form to rank their impressions on a scale of 1 (ineffective) to 8 (highly effective) of the value of the meetings.

Table 12.3 provides a summary of the principles, application and experience of the participation process, together with samples of participant feedback.

Table 12.3 Principles, application, and experience of the participation process

Principles and guidelines	Application	Participant feedback
Empowerment, ownership, and inclusiveness	The layout of the meeting room, the absence of a 'top table', and the emphasis on transparency were designed to create a sense of empowerment, ownership and inclusiveness. The aims of the events encouraged participants to influence and take ownership of the Action Plan and future of the catchment. As well as the key stakeholders and statutory organizations, invitations to the events were sent to the many diverse social, environmental, and business groups within the catchment.	'This has been truly a "bottom-up" exercise with stakeholders involved right the way through from start to finish and could be a model for other catchment areas to follow.' 'Community ownership, in my opinion, is the key to the long term success of this initiative. The stakeholder involvement so far has been impressive and this must be maintained.'
Building community	The events involved many different communities, providing opportunities to strengthen a shared 'sense of place' and interest. This was in part achieved through participants working together, sharing common values and priorities. The process of decision-making also created the opportunity for new 'temporary communities' to be formed around projects and solutions	'A shared "sense of place" brings people together and it's their commitment which guarantees the sustainability of both the river and the communities which live within its catchment area.' 'Worked in Research Group. Identified need to find funding for small community-based research projects in partnership with universities, research councils and scientific societies.'
Collaboration	The organizations' workshop was designed to develop and strengthen collaboration between key stakeholders and statutory organizations through: maximum-mix groups working together on problem-solving; all participants working together to identify long-term trends and issues affecting the future of the catchment; self-selected groups comprising diverse stakeholders identifying priorities and solutions together.	'Very interesting to hear so many different views on the same subject and to work with so many different people by changing groups halfway through the evening.' 'This will be an excellent opportunity for Cycleau to work in partnership with delivery as well as strategic authorities to tie in with long term sustainability.'
Raising awareness	Participants and members of the project team worked in small groups at events identifying the long-term local, national, and global trends they thought would affect the catchment. Following this, the whole group developed a mind map of the trends and issues.	'Good system of feedback and a lot of material covered effectively in the time.'

Table 12.3 (continued)

Principles and guidelines	Application	Participant feedback
Better decisions	The events aimed to work with local knowledge and expertise together with 'expert' and professional knowledge. This was achieved through maximum-mix small group tasks and combining local knowledge and professional expertise.	'An excellent event which really tapped into local expertise.'
Learning through doing and reflecting	The design of the events provided opportunities for participants to reflect together. Clear framing by the facilitators enabled all groups to learn together how 'we' create solutions to complex interrelated problems. The events were designed to enable participants to engage in dialogue about their perspectives on the present and their aspirations for the future conservation of the catchment for the benefit of both people and wildlife. The design also aimed to create the right conditions for social and community learning.	'. . . I see a dichotomy – there is a need to enthuse, motivate and communicate to ensure a real sharing of a common vision but. . . these responses must be founded on a solid, accurate understanding of the issues. An excellent, well-planned session that certainly successfully raised the profile of the issues/project amongst those present. . . . Ultimately, delivery is all.' 'I think the only way to facilitate all these different voices is to allow people the opportunity to speak and chat to others and discuss ideas/themes/concerns, etc. And I thought that this was achieved by having a more "workshop" feel to the event, rather than simply a succession of different presentations. One of the main concerns I felt about other transnational Cycleau meetings was the lack of time for discussion. Perhaps you should take this forward to Cycleau for future conference events?'
Building on what local people value	Participants at the events were asked to articulate what they valued about the river and its catchment. Working in small groups, participants identified what they valued about the catchment. The results were recorded on mind maps which were displayed at future events.	'I thought the meeting was excellent. I was amazed to see so many different people and organisations attend the event and have something to say about the Dart Catchment.'

12.4.1 Evaluation of the Public Participation Meetings

The evaluation process showed that 95% of participants at the public meetings ranked their overall impression of the effectiveness of the workshop as 6 or above. Feedback from participants included:

> Sessions were very well led. Very effective eliciting of opinion – turned into constructive issues. Nice balance of friendly/business approach.
>
> If the promise of this project is realised, even in part, the benefit for the river, its catchment area and its residents (human & otherwise) is potentially incalculable.

12.4.2 Evaluation of the Organizations' Stakeholder Workshop

The evaluation process showed that 81% of participants at the organizations' workshop ranked their overall impression of the effectiveness of the workshop as 6 or above. Feedback from participants included:

> A very interesting and worthwhile morning. I have learnt from it as well as (I hope) contributed. I look forward to the draft Action Plan and hope for successful use of the current funding for at least some of the projects identified today. Thank you to Devon Wildlife Trust for preparation and presentation of the event and for all that is to come out of it.
>
> Excellent organisation. Excellent breadth & depth of organisations.

12.5 Conclusion

The stakeholder and public participation activities of the Dart Catchment Pilot Project set out to share knowledge and experience of catchment management. It aimed to achieve this by demonstrating how good practice could be developed to involve local communities effectively in the management of a river landscape.

The facilitated meetings and workshops helped to increase stakeholder awareness of catchment management issues and improved water management through the use of local knowledge and experience. The positive feedback from participants and stakeholders indicated that the process increased the feeling of 'ownership' of catchment management issues. The process also made a contribution to community learning by sharing both the know-how of professional decision-makers and the local knowledge and expertise of local people and stakeholders.

In the longer term the participation process developed and connected a network of people with local knowledge and skills. This network has acted as a reservoir of know-how and provided the opportunity for continued learning between members through the sharing of values and beliefs and their involvement in the implementation of the action plan.

The Dart pilot project provided the opportunity to involve stakeholders and members of the public in a catchment-wide project. To deliver this effectively required the provision of adequate resources, not only to plan, manage, and facilitate the meetings and workshops but also to ensure that there were effective mechanisms in place to support follow-up activities. The act of involving communities, organizations, and stakeholders in the future of a landscape or place inevitably builds expectations that need to be resourced. It is not possible to meet all expectations but it is important that the input of participants is translated into visible actions. In the Dart project the Action Plan review workshop provided an opportunity for feedback and refinement of the plan. Following the publication of the action plan a number of activities were organized to continue the engagement process. This included two river festivals that involved over 4,000 local people in celebrating the river and its catchment. The festivals very much built on the values highlighted by the participants who articulated what the essence of the river and its catchment meant for each of them.

A range of longer-term activities were also organized to follow on after the participative decision making activities formally ended. These follow-on activities included volunteers helping to improve the management of 8 km of riverbank to the benefit of water quality and wildlife. The volunteers were trained to help undertake salmon-surveying and habitat restoration, resulting in improved habitats for over 3 km of the river. Advice was given to 33 landowners and 51 farmers who together manage 6,043 ha of land; €30,600 in grant funding was approved for projects and 35 landowners from around the river Dart attended soil-management workshops. Five further workshops were also held for over a hundred farmers on four demonstration farms to show how soil and nutrient management can reduce impacts on wildlife. A number of interpretation points were installed within the catchment and three 'Teacher Resource Packages' for school students using the river as a learning resource were produced. Each of these actions can be tracked back to comments and recommendations made by participants during the course of the meetings and workshops.

The partnership-building that took place through the participation activities provided the basis for addressing the multiple demands on the river and catchment. This has helped to reduce some of the tension that existed between the different interest groups who see the river as their place. The partnerships created have had a positive effect in integrating those multiple demands, enabling all those with an interest to liaise and work more effectively together.

The value of the participative approach and follow-on activities was highlighted by a comment made by an elected councillor for South Hams District Council, who said: 'Overall, I have been very impressed with the importance given to the process of engaging stakeholders throughout the whole project. I like the emphasis in the draft action plan on joined-up thinking and community ownership. I hope to continue my involvement in this initiative as it progresses.'

At a strategic level the experience of the pilot project has been used to influence local and national policy, including Catchment Abstraction Management Plans, a Catchment Sensitive Farming Initiative, and national and regional consultations on the implementation of the Water Framework Directive.

As we develop our understanding and practice of participative decision-making in planning and managing landscapes, it is not always easy for us to anticipate the long-term impact. Attwood et al. (2003) suggest that 'change is not an event; the current state of things will in most cases be the result of a long history, as well as past successes created in different circumstances. Lasting change will be brought about over time'.

The experience of the Dart catchment pilot project indicates that there is an important role for engagement and participative decision-making in the way that we protect, plan, and manage the long-term future of our landscapes and places. The principles and engagement approaches used in this case study provides a source of learning relevant to the implementation of the European Landscape Convention.

References

Attwood M, Pedlar M, Pritchard D, Wilkinson D (2003) Leading change: a guide to whole systems. Policy Press, Bristol

Capra F (2002) The hidden connections: a science for sustainable living. Harper Collins, London

Cuff J (2001) Participatory processes: a tool to assist the wise use of catchments. A guide based on experience, Wise Use of Floodplains, EU Life-Environment Project. Rural Horizons. http://www.fløoodplains.org.uk/technical.htm. Accessed 9 Apr 2010

EC (2000) Directive 2000/60/EC of the European Parliament and of the Council of 23 October 2000 establishing a framework for the community action in the field of water policy. Off J, L 327, 22.12.2000, 1-82. EUR-Lex. http://eur-lex.europa.eu/LexUriServ/LexUriServ.do?uri=CELEX:32000L0060:EN:NOT. Accessed 17 Mar 2010

Emery FR, Trist. EL (1973) Towards a social ecology: contextual appreciation of the future in the present. Plenum, New York, NY

Moore P, Spencer N (2005) Dart catchment project: stakeholder and public participation – a project evaluation. Devon Wildlife Trust & Sustainable Futures, Exeter. http://www.devonwildlifetrust.org/. Accessed 25 Mar 2009

Pedlar M, Warburton D, Wilkinson D (2007) Improving poor environments: the role of learning architectures in developing and spreading good practice. Science Report SC050018/SR3. Environment Agency, Bristol. http://publications.environment-agency.gov.uk/pdf/SCHO0607BMVE-e-e.pdf. Accessed 9 Apr 2010

Porter S, McDonagh C (1998) Toolkit for community involvement and facilitation. Vision 21, Cheltenham

Wates N (2000) The community planning handbook: how people can shape their cities, towns and villages in any part of the world. Earthscan, London

Weisbord M (2002) Discovering common ground: how future search conferences bring people together to achieve breakthrough innovation, empowerment, shared vision, and collaborative action. Berrett–Koehler, San Francisco, CA

Weisbord R, Janoff S (1995) Future search: an action guide to finding common ground in organizations and communities. Berrett–Koehler, San Francisco, CA

Chapter 13
Regional Landscape Strategies and Public Participation: Towards Implementing the European Landscape Convention in Sweden

Anders Larsson, Anna Peterson, Elinor Bjärnborg, Christine Haaland, and Mats Gyllin

Abstract Sweden has recently decided to ratify the European Landscape Convention (ELC). Methods for implementation have been discussed for both the ELC and related national environmental objectives. Thus, the Swedish Government decided that seven County Administrative Boards should undertake pilot studies for Regional Landscape Strategies (RLS) during 2006–2007. The RLS

A. Larsson (✉)
Department of Landscape Architecture, Swedish University of Agricultural Sciences, P.O. Box 58
SE–230 53 Alnarp, Sweden
e-mail: anders.larsson@ltj.slu.se

M. Jones, M. Stenseke (eds.), *The European Landscape Convention*, Landscape Series 13, DOI 10.1007/978-90-481-9932-7_13,

pilot projects aimed at a new and more effective approach for conservation and sustainable land use at regional level. Additionally, RLS should also function as a method for implementing the ELC and other environmental objectives. One specific focus of the pilot studies was on developing methods for public participation. Other important issues were to involve different administrative sectors, municipalities, and regional authorities in the process, as well as attaining a good balance between conservation and profitable land use. This chapter presents some of the results from the RLS case study in Vellinge municipality, Scania (Skåne), Sweden's southernmost province. The focus of the study was on public participation in particular by equestrians and landowners. The first 'bridleway organization' of its kind in Sweden was established as a direct result of the project.

Keywords Regional Landscape Strategies · Public Participation · European Landscape Convention · Stakeholders · Multifunctional Greenways · Greenway Planning · Municipal Planning (Sweden) · Bridleways · Equestrians · Landowners

13.1 Introduction

This chapter presents results and experiences from a pilot study on Regional Landscape Strategies (RLS) in Sweden (Länsstyrelsen, 2007). RLS were suggested by the government as a possible method for implementing both national and international environmental objectives, including the European Landscape Convention (ELC) (Regeringen, 2005a). The aim of the RLS programme was initially to develop methods of operation and planning processes by evaluating a number of pilot studies carried out within the RLS, and then to produce a manual for work with RLS within County Administrative Boards. Important points of departure for the pilot studies were to involve many different administrative sectors, municipalities, and regional authorities in the process, and to attain a positive balance between land conservation and utilization. The activities were to be carried out within a public process in which landowners and other representatives from agriculture and forestry, affected organizations, and local stakeholders were invited to participate (Regeringen, 2005b). The Swedish Environmental Protection Agency (Naturvårdsverket) was given responsibility for the RLS projects and their subsequent evaluation, while the Swedish National Heritage Board (Riksantikvarieämbetet) was given responsibility for implementing the ELC.

The national authorities commissioned out the RLS pilot studies to seven County Administrative Boards. The Board in Scania (Skåne), the southernmost region in Sweden, in its turn commissioned out the work to four municipality working groups, who performed one local pilot study each. The work presented here is the result of one of these local pilot studies, in Vellinge, the southernmost municipality in Scania (Fig. 13.1). The major part of the pilot study was carried out by members of the research project 'Multifunctional Greenways as a tool for strategic landscape planning – proposals for design and implementation in peri-urban landscapes'. This project studies the phenomenon of Greenway Planning

Fig. 13.1 Vellinge, Scania – location of the study area (Map: Mats Gyllin)

from a multitude of perspectives. One of these is the planning perspective, not least aspects concerning public participation. The active participation within the RLS pilot studies gave the researchers an opportunity to experiment and draw preliminary conclusions on one of the aspects covered by the general research project, which was then used as a stepping stone towards a more precise research design. This chapter presents concrete results of the RLS pilot study in Vellinge, especially regarding public participation. This will later be integrated with other findings within the general research project on the formal planning system in Sweden and possibilities for working with strategic Greenway Planning. During the process, the research group was open about their double role as both researchers and coordinators of the RLS pilot study. The involved authorities and individuals accepted the fact that the results of the RLS project would be further analysed within the general research project.

13.2 Brief Policy Background: National Environmental Quality Objectives and Regional Landscape Strategies

In 1999, 15 Environmental Quality Objectives were adopted by the Swedish Parliament with the objective of safeguarding biodiversity and natural environments (Miljömålsrådet, n.d.). The Swedish National Environmental Quality Objectives form an important part of the process of achieving sustainable development, along with the social and economic dimensions involved. These objectives have been designed to promote human health, safeguard biodiversity and the natural environment, preserve the cultural environment and heritage, maintain long-term ecosystem productivity and, finally, ensure prudent management of natural resources. All of this is covered in the now 16 specific Environmental Quality Objectives along with their respective interim targets (Miljömålsrådet, n.d.).

The 16th Environmental Quality Objective was added to the list in 2005 (Miljömålsrådet, n.d.) and aims at maintaining a rich diversity of plant and animal life. It states that:

> Biological diversity must be preserved and used sustainably for the benefit of present and future generations. Species habitats and ecosystems and their functions and processes must be safeguarded. Species must be able to survive in long-term viable populations with sufficient genetic variation. Finally, people must have access to a good natural and cultural environment rich in biological diversity as a basis for health, quality of life and well-being (Miljömålsrådet, n.d.).

The Environmental Quality Objectives are to be considered within all municipal planning and infrastructure planning in Sweden (see Alfredsson and Wiman, 2001 for a description of the Swedish planning system in English).

Also in 2005, the Swedish government proposed that a group of County Administrative Boards should perform pilot case studies on RLS as tools for implementing interim target 3 (sustainable use of biological resources) of the 16th Environmental Quality Objective (Regeringen, 2005b). RLS was also regarded as a possible method for implementing the ELC in Sweden. The pilot studies were completed in 2007 (Naturvårdsverket, 2008).

The ELC (Council of Europe, 2000) was signed by Sweden in 2001. In 2008, the Swedish National Heritage Board presented their suggestions on how to implement the ELC in Sweden. Two of their recommendations were that Sweden should ratify the ELC as soon as possible and introduce Regional Landscape Strategies (Riksantikvarieämbetet, 2008). In November 2010, the Swedish government decided to ratify the ELC. The compatibility and possible synergy effects between the RLS and the ELC are clear.

The Environmental Quality Objectives will influence the way that the ELC will be implemented in Sweden. Even though 'future generations' and the well-being of people living in the landscapes are mentioned, our concern is that the strong Swedish tradition of top-down nature conservation will affect the ELC perspectives on public participation in a negative way.

13.3 The Vellinge Project

The County Administrative Board of Scania delegated the RLS pilot studies to a group of municipalities. The county was responsible for the central coordination and the main communicative link between the municipalities and the central authorities. It also contributed its expertise as appropriate. The County Administrative Board of Scania further specified that the RLS projects, besides involving different sectors and authorities, balancing between conservation and utilization, and engaging in an open process, should also integrate knowledge about cultural heritage and consider the 'everyday landscape' in order to develop more robust methods for public participation (Länsstyrelsen, 2007). Vellinge was one of the municipalities that announced its interest in taking part in a pilot project and in turn commissioned the coordination of the pilot study to our research group. The project group included both the coordinators and participating staff from the municipality and the County Administrative Board.

It is important to notice that there was a certain ambivalence in objectives since both top-down priorities (biodiversity issues) and bottom-up approaches (local participation) were required to be included in the RLS pilot studies.

Vellinge is situated in an intensive farming region south of Malmö. This is also the part of Sweden where one of the most intensive processes of urbanization is taking place, which has resulted in peri-urban structures and islands of urban settlements surrounded by large fields of inaccessible, arable land (Qviström et al., 2007). There is a great need for green structures and recreational opportunities for the increasing urban population. This part of Sweden is also one of the regions with the highest proportion of horses per capita in the country (Jordbruksverket, 2005). Some biological hot-spots exist, especially along the coastline, but these are not connected within a larger green network (Länsstyrelsen, 2007).

It soon became clear that one of the most important planning problems in Vellinge regarding access to the countryside was how to manage the increasing number of horses, particularly within the urban fringes of highly productive agricultural areas with low accessibility and high pressure on the land (Hautbois and Durand, 2004; Ivarsson, 2008). During the process, the planners in the municipality recognized their lack of expertise and experience in considering such issues. The municipality welcomed equestrians to move to the area but had not hitherto handled any associated conflicts and practical problems (Fig. 13.2). A 'horse village' was planned at the same time as our project was running, but no additional plans had been made for bridleways and access to the surrounding landscape for those who wanted to ride outside the village.

Our project was limited to a period of approximately 6 months. In order to plan constructively for a green infrastructure in the 'everyday landscape' during this time frame, planners within the municipality were asked to identify the most critical and pressing issues. One of the main problems identified was the high number of equestrians having difficulties riding in the intensively farmed landscape. Over time, this had led to hostility between landowners and equestrians. Yet there was no local organization embracing these two interest groups which could have mediated in these conflicts. Farmers are mainly organized at national level in the Federation of Swedish Farmers (Lantbrukarnas Riksförbund – LRF), while equestrians are not organized in any general horse organization.

Hence, the project group decided to concentrate on this specific issue in order to test methods for local participation and transforming conflicts into constructive landscape management, while focusing less on the regional perspective. The specific objective was to attain agreement between landowners and equestrians on how and where to locate bridle paths within the intensively farmed landscape. However, we

Fig. 13.2 Narrow roads and little space for equestrians in a typical Vellinge landscape (Photo: Courtesy of Daniel Melchert, God Bostad Kulturmiljökonsult)

did not expect to establish bridle paths in the short period of time available for the project.

To improve biodiversity in the intensively used agricultural landscape by a green network system was seen as a further possible positive outcome of arranging bridleways. Thus, aspects of biodiversity were brought into the discussions, but rather as a possible by-product, which could be reached by a certain design of the bridleways. The involved authorities accepted this as a necessary step to take. First an arena for dialogue regarding collaborative landscape management should be established, and then specific issues, like biodiversity, should be attempted to be solved. Local stakeholders (mainly landowners and equestrians) were chosen as the main focus groups, in order to handle the project within the limited time available.

It has not been the authors' intention to evaluate the complete RLS program, but to show the results of a case study concerning public participation and discuss this especially in relation to the implementation of the ELC and national planning processes.

The specific case study in Vellinge was carried out through a series of meetings, including presentations from the researchers and the authorities, listening to arguments from local participants, and discussions and active participation among all interested parties. The external project meetings consisted of a landowner meeting, a general meeting, and an equestrian meeting. In-depth interviews with different stakeholders were subsequently carried out (Bjärnborg, 2007; Länsstyrelsen, 2007).

The methodology used was inspired by previous results on communicative aspects in countryside planning (e.g. Larsson, 2004) and literature on stakeholder participation in environmental decision-making (e.g. Bierle, 2002). Specific details regarding the process were elaborated in cooperation with the municipal planners. The method might best be described as a modified traditional planning process with active participation by local stakeholders at the beginning of the process instead of a more passive transfer of knowledge and information in the middle and at the end of the process. There were no ready plans for the local participators to react upon. This model has been described by Axelsson (2009), who summarizes the most important aspects to consider as:

1. An area that fits with the main sustainability gaps or tasks at hand (discussed beforehand with municipal planners)
2. Collaboration among actors and stakeholders (main issue discussed at all meetings)
3. A commitment to sustainable development and sustainability profiles as a result of analysis (main objective and outcome of the equestrian meeting)
4. Knowledge production to learn about the area, solve sustainability issues, and improve practices (issues at the landowner and general meetings)
5. A systematic approach to sharing, including networking (all meetings in combination).

The municipality was the formal host of all of the meetings. The project leaders from the County Administrative Board participated but maintained a low profile since they were not a part of the landowner versus equestrian conflict. They acted as observers and experts who could be approached regarding specific questions regarding, for example, biodiversity issues (Länsstyrelsen, 2007).

13.4 Results

All results below, except post-project results, have been extracted from the Vellinge report, Appendix 5 of the RLS assignment (Länsstyrelsen, 2007).

13.4.1 Landowner Meeting

The first meeting was held with landowners, since it was essential to get their support for the rest of the process. Nearly all the landowners in the area were present at this meeting. They were worried about their land and wanted to ascertain that nothing they did not approve of was going to be carried out. The project group was able to clarify that we had not made any concrete plans for bridle paths. The farmers had, as anticipated, a great deal to say about the equestrians. According to the farmers, equestrians went riding everywhere and had little or no respect for private land or growing crops.

The matter of 'responsibility' was a particular concern. The landowners wanted to know the exact rules concerning private land, the right of public access (*allemansrätten*), and who was to be held responsible for any possible accidents. As a result, project members from the County Administrative Board researched some of the legal issues and organized a lecture on the topic of *allemansrätten* for the following meeting, where the equestrians also participated.

Economic considerations were important. Landowners wanted to be compensated for any bridle paths on their land. We validated their concern and told them that both the issue of proper compensation and redirecting the equestrians to more suitable riding places could be positive outcomes of a process of negotiation. If nothing was done about the present situation, landowners would risk having people and horses all over their properties.

The project group witnessed a lot of frustration from the landowners. However our focus was always on objective listening and encouraging further involvement, not arbitrating the situation. Paradoxically, the landowners were especially frustrated by the fact that there were no existing municipal plans to discuss with the authorities. The landowners were not used to the form of procedure adopted, involving participation at an early stage. Typically, they were used to being contacted by the authorities at a much later stage in the planning process.

13.4.2 General Meeting

Second, a general meeting was arranged, where all the local residents were invited. Not unexpectedly, the topic attracted mainly equestrians and landowners, but also members of local heritage organization and persons interested in recreational issues in general took part. Posters concerning biodiversity were presented and short lectures on such topics were held. Experts from the County Administrative Board provided information regarding nature reserves and thereby represented the 'green' perspective. A short lecture on the right of public access was delivered by a representative from the County Administrative Board. A questionnaire was also handed out, in which issues regarding biodiversity gained relatively low scores. Discussions at this meeting, and elsewhere, also clearly indicated this; the local residents did not place 'biodiversity' as high on the list of important issues to solve as did the invited experts from the various authorities.

This meeting resulted in a very fruitful exchange of perspectives between landowners and equestrians (Fig. 13.3). Both parties gained insight into each other's perspectives and discussed possible solutions for the future. The message from the project group was that status quo would not be a positive solution. We also stated that that the landowners and equestrians had equal responsibility for reaching a resolution. At the end of this meeting, some of the initially most aggressive farmers stated that they might actually agree to lease some of their land to the equestrians, especially if this would mean that equestrians would no longer be riding over other parts of their land. The equestrians, in turn, said that it would be possible for them to pay farmers for the benefit of riding on bridle paths, and that they would consider introducing some kind of 'driver's licence' for equestrians before they would be permitted to ride along these paths.

Fig. 13.3 Public participation meeting with landowners and equestrians in Vellinge (Photo: Courtesy of Daniel Melchert, God Bostad Kulturmiljökonsult)

13.4.3 Equestrian Meeting

The project group agreed that it would be necessary for the long-term continuation of the project to be quite frank with the equestrians. We told them that from now on it would be up to them to take responsibility for the process in the future. We found it encouraging that the equestrians created a group for riders where they could meet and discuss issues at hand. Hence, a final meeting where only equestrians were invited was arranged. Only a few equestrians attended this meeting, but within this group there were representatives from both boarding stables and a riding school. As a result of this meeting, the equestrians decided to get better organized and begin the process of negotiating with landowners.

13.4.4 In-Depth Interviews with Farmers and Equestrians

Subsequent interviews were held with two farmers and two equestrians who had participated in the general meeting. The aim of these interviews was to investigate what some of the participants had experienced in the project so far and to ascertain their reasoning about public participation or the process of collaboration in their everyday landscape. These interviews also provided an opportunity for deeper insight into what the participants, relative to their background, perceived as their role in the project, as well as any opportunities and threats they had experienced as a result.

The in-depth interviews revealed that both the farmers and the equestrians were interested in public participation in order to build a common overview of the situation, and to find solutions. They saw public participation as a means of reaching an improved understanding between the various stakeholders. Even so, the farmers were clear that they were not willing to let anyone else make decisions that concerned their private land. The farmers were worried about anything that could threaten their agricultural businesses.

13.4.5 Post-project Results

Although the intentions of the project group did not encompass the establishment of bridle paths, the project did make progress in this direction in the course of its limited duration. A joint organization consisting of both landowners and equestrians was established less than 6 months from the initial landowner meeting. This organization started to discuss how and where to establish bridle paths, and how to solve other practical issues. Farmers agreed to lease land to equestrians, and the equestrians agreed to introduce 'driver's licences' to their members. Some 9–12 months later, the negotiation process seemed to have terminated, allegedly due to lack of discretionary time for residents to organize these efforts and due to the fact that some of the key individuals in the group had moved out of the region. However, the process later on turned out to have gained enough momentum to continue by itself. The first bridleway organization (*ridstigsförening*) of its kind in Sweden has been established (Vellinge Ridstigsförening, n.d.). This is a positive result considering the extremely short time available and the initial conflict, and clearly illustrates how public participation can be productive in many aspects.

13.5 Possibilities

The municipal planners hope that the new joint bridleway organization and the process of starting it, might provide a role model for a consultative body for the municipality regarding future proactive planning projects in the region. If stakeholders are invited to participate at early stages of planning processes, this could result in a more constructive dialogue than within the traditional planning process, where stakeholders chiefly respond to plans already initiated by the experts. This is consistent with results from other participative municipal planning initiatives (Boverket, 2007). Thus, public participation could provide both democratic and economic value.

In Sweden about 85% of horses are used for recreation (Persson, 2003). These horses and their owners are seldom registered in any organization. The equestrians are individualists, who spend discretionary time on their horses and do not recognize any benefits from organizing within groups. In the Vellinge project their understanding and willingness to participate came from the possibility of better access to riding paths (Fig. 13.4). Many potential bridleways had been closed off by landowners, and

Fig. 13.4 Equestrians on a field road in a typical Scanian landscape (Photo: Mats Gyllin)

equestrians would not allow their children to ride on the roads for safety reasons. Now they saw a chance to get old ways re-opened in addition to new and safer riding possibilities. It became clear that up until the start of our project, landowners had deliberately tried to get horses off their land because a few equestrians had behaved inappropriately while riding. Therefore, the equestrians decided to adopt collective management rules and standardized guidelines for how equestrians in the area should behave. Landowners would be able to consult these regulations when problems arose – before a situation turned into personal conflict.

Farmers maintain that farming is the most appropriate and rational land use. One of the equestrians stated that public participation could be a way of improving stakeholders' involvement. This participation could broaden the view of how to use the landscape for various purposes and alternative ways of earning income from the land. Both landowners and equestrians agreed that the process was beneficial for understanding more about the other parties and their situation. Furthermore, the establishment of bridle paths might create opportunities for developing more advanced multifunctional greenways in the future, which could also lead to an improvement of biological diversity and possibilities for other recreation than horse-riding.

13.6 Further Experiences

Other findings indicate that, however desperate the situation might seem at the outset, conflicts should not be avoided but rather dealt with as soon as possible

(Boverket, 2007). A process will be less constructive if conflict issues are withheld from the negotiating table. Consensus is not a prerequisite for reaching positive results (Hagen, 2006; Nordström, 2008). This contradicts some of the communicative planning theories (e.g. Healey, 1997), which are focused on consensus as both an objective and a method. Another positive aspect of the RLS project was the focus on the 'everyday landscape', as opposed to the tradition of focusing on landscapes and objects of especial value. The focus on the everyday landscape is proactive rather than reactive.

It was clear that issues concerning biodiversity were not of concern for the public taking part in the project. This suggests that biodiversity issues are more of interest for experts or for a minority of layman rather than for the general public, at least when it comes to concrete implementation situations. Policies such as the national Environmental Quality Objectives and the ELC might benefit from using a more concrete terminology in order to reach the minds of the public and, thus, be more in line with additional objectives on public participation. However, local residents are foremost concerned about local and personal issues, which affect their immediate surroundings, their economy, etc. Hence, common and shared values, as well as regional considerations, still need to be safeguarded by the authorities. This might seem a contradiction in terms, but conforms to the results of previous studies of agro-environmental policies, which show that the issue is not a question of discussing public participation as something in opposition to interventions from the authorities. All parties involved need to take their part of responsibility in order for the outcome to become as productive as possible and in order to increase dialogue and respect between experts and laymen (Larsson, 2004). This does not indicate a failure of the participatory procedures adopted, or that issues regarding biodiversity cannot also be handled within such a process given another focus or more time, etc., but rather that there is no single method that can solve all issues. Methodological diversity is needed.

Today, the first official RLS are being implemented (2009–2010), and as expected the focus on the ecosystems approach has increased. Therefore, we find it important to emphasize the importance of public participation for improving the quality of the biodiversity in industrial-agricultural landscapes near cities where urban sprawl occurs and where there is strong competition for land as well as lack of recreational areas.

13.7 Conclusions

There are many positive aspects of the RLS, such as the attention to everyday landscapes and the increased focus on local participation. However, the traditional, comprehensive plans also include aspects such as how to improve biodiversity in everyday landscapes, especially where there is a complementary municipal green plan. Dialogue with stakeholders and local interest groups, including landowners, is also part of most Swedish planning processes (road planning, municipal land-use planning, etc.). The new focus of the RLS has been on early and active participation rather than (passive) response later in the decision-making process. However, such aspects of the ELC and the Environmental Quality Objectives could have been implemented by minor administrative and juridical modifications of the current planning system, rather than introducing a completely new package of general policy objectives.

Active public participation proved to be productive in many aspects. Farmers, who were usually hostile towards equestrians, learned a great deal about the equestrian perspective and started to think more constructively how to solve problems for the mutual benefit of both parties. Equestrians, who normally regarded the landowners as stubborn, old farmers, with little knowledge about other issues than farming, soon realized obvious benefits from better communication with the landowners and among themselves. Public participation does not solve every planning problem, but might be very productive in addressing specific, local conflicts where closer cooperation between different parties is essential for moving towards the next step in the planning process. Such conflicts cannot be solved by a top-down approach, and neglecting them will only lead to increasing problems later on in the process.

Democracy comes at a cost. Nonetheless, public participation in early stages of planning processes might be the best way to maximize the democratic outcome, while at the same time optimizing planning from an economic point of view. People feel much more engaged in the process when asked to participate from the start rather than just to comment upon the work of experts, Early participation also leads to fewer appeals within later stages of the planning process and, thus, lower total costs due to smoother processes. This has been presented in case studies at planning conferences (Boverket, 2007), but needs further scientific studies. It is nevertheless our strong belief that, when local stakeholders resolve their own conflicts, they also make good use of the everyday landscape where they are living. Planners might then also have more time to engage themselves in other relevant environmental problems that need more of the planners' specific attention and expert competence.

Acknowledgments The study was carried out with financial support from the Swedish Research Council Formas, the County Administrative Board of Scania, and Vellinge Municipality. We are grateful to administrators in the latter two organizations who have contributed with their time and knowledge.

References

Alfredsson B, Wiman J (2001) Planning in Sweden – fundamentals outlined. In: Christoferson I (ed) Swedish planning – in times of diversity. Swedish Society for Town and County Planning, Gävle, pp 117–120

Axelsson R (2009) Landscape approach for sustainable developmant – from applied research to transdisciplinary knowledge production. Acta universitatis agriculturae sueciae 2009:94. School for Forest Engineers, Swedish University of Agricultural Sciences, Skinnskatteberg

Bierle TC (2002) The quality of stakeholder-based decisions. Risk Anal 22(4):739–749

Bjärnborg E (2007) Samverkan och dialog i landsbygdens landskap. Examensarbeten inom landskapsarkitektprogrammet 2007:29. Sveriges lantbruksuniversitet, Alnarp

Boverket (2007) Hur bygger vi staden? Dokumentation från Boverkets Plan- och byggdagar 3–5 september 2007 (DVD). Boverket, Karlskrona

Council of Europe (2000) The European Landscape Convention, Florence, 20.X.2000. ETS No. 176. http://conventions.coe.int/Treaty/en/Treaties/Html/176.htm. Accessed 30 Jan 2010

Hagen A (2006) Practice seeking understanding: a study of county planning in oppland 1974–1996 – Praksis søker forståelse: Studie av fylkesplanlegging i Oppland 1974–1996. Doctor Scientiarum (Dr.Scient.) Thesis, 2006:20. Norwegian University of Life Sciences, Department of Landscape Architecture and Spatial Planning, Ås

Hautbois C, Durand C (2004) Public strategies for local development: the effectiveness of an outdoor activities model. Manag Leis 9(4):212–226

Healey P (1997) Collaborative planning – shaping places in fragmented societies. Macmillian, London

Ivarsson M (2008) Framtidsspaning om hästsektorn. Report. Jordbruksverket. http://www.margaretaivarsson.se/margareta/documents/framtidsspan.pdf. Accessed 30 Jan 2010

Jordbruksverket (2005) Kartläggning och analys av hästverksamheten i Sverige. Jönköping

Länsstyrelsen i Skåne län (2007) Det regionala landskapsstrategiuppdraget, Länsstyrelsens missiv 2007-12-19, inklusive bilagor (CD). Malmö

Larsson A (2004) Landskapsplanering genom jordbrukspolitik: En kritisk granskning av EU:s agrara miljöstödspolitik ur ett planeringsperspektiv, Acta universitatis agriculturae sueciae 442. Department of Landscape Planning, Swedish University of Agricultural Sciences, Alnarp

Miljömålsrådet (n.d.) Environmental objectives portal. http://www.miljomal.nu/Environmental-Objectives-Portal/. Accessed 11 Nov 2010

Naturvårdsverket (2008) Landskapsstrategier i sju pilotlän 2006–2007: En utvärdering av pilotprojektens organisation och arbetssätt. Rapport 5906. Naturvårdsverket, Stockholm. http://www.naturvardsverket.se/Documents/publikationer/978-91-620-5906-4.pdf. Accessed 14 Apr 2010

Nordström S (2008) Illusionernas harmoni – Samhällsplanerandets tankestil och dess kraftfullaste topos: Diskrepansförnekandet. Doktorsavhandling. Karlstad University Studies 2008:55. Fakulteten för samhälls- och livsvetenskaper, Kulturgeografi, Karlstads universitet, Karlstad

Persson H (2003) Planering för tätortsnära ridning – mer än bara ridhus. Specialmeddelande 244. Institutionen för jordbrukets biosystem och teknologi (JBT), Sveriges lantbruksuniversitet, Alnarp

Qviström M, Larsson A, Ode Å (2007) Peri-urban landscape analysis as a strategic instrument in environmental planning. In: 18th international annual ECLAS conference, Belgrade. Department of Landscape, Architecture and Horticulture, Faculty of Forestry, Belgrade, pp 43–50

Regeringen (2005a) Svenska miljömål – ett gemensamt uppdrag, Prop. 2004/05:150. Miljö- och samhällsbyggnadsdepartementet, Stockholm

Regeringen (2005b) Regleringsbrev för budgetåret 2006 avseende Länsstyrelserna. Finansdepartementet, Stockholm

Riksantikvarieämbetet (2008) Förslag till genomförande av den europeiska landskapskonventionen i Sverige: Slutredovisning av regeringsuppdrag. Stockholm

Vellinge Ridstigsförening (n.d.) http://www.vellingeridstig.se/ Accessed 14 Apr 2010

Chapter 14
The Role of Information, Knowledge, and Acceptance During Landowner Participation in the Natura 2000 Designations: The Cases of Otepää and Kõnnumaa, Estonia

Monika Suškevičs and Mart Külvik

Abstract Across the European Union, the Natura 2000 network is among the most important measures for preventing the biological degradation of landscapes. However, land-use conflicts in several member states show that the designation of Natura 2000 areas has not been an effective process, foremost due to insufficient public and stakeholder involvement. This chapter presents an investigation of landowner involvement during the Natura 2000 designations in Estonia, focusing on two aspects: first, the role of information and knowledge in the participatory process; and second, the acceptance of Natura 2000 among landowners. Insights gained from two case studies in northern Estonia (Kõnnumaa) and southern Estonia (Otepää) indicate that despite extensive communication processes during designations, many landowners lacked basic knowledge on Natura 2000 issues and on consultation procedure at the beginning of involvement processes and afterwards. Our results additionally suggest that addressing the needs, expectations and knowledge claims of different stakeholders within participatory processes is a necessary precondition for gained acceptance in biodiversity-related landscape planning.

Keywords Natura 2000 · Site selections and designations · Knowledge · Acceptance · Landowner participation

---

M. Suškevičs (✉)
Institute of Agricultural and Environmental Sciences, Estonian University of Life Sciences, EE-51014 Tartu, Estonia
e-mail: monika.suskevics@emu.ee

M. Jones, M. Stenseke (eds.), *The European Landscape Convention*,
Landscape Series 13, DOI 10.1007/978-90-481-9932-7_14,

Photo: Courtesy of Voldemar Rannap

14.1 Introduction

14.1.1 Participation in Landscape and Nature Conservation Issues

The European Landscape Convention (ELC) aims to bring together natural and cultural approaches in landscape protection, management, and planning (Council of Europe, 2000). Participatory approaches are promising ways to bridge the gap between different disciplines in landscape and environmental research, and are intended to tackle several inherent deficiencies of hierarchical top-down decision-making, for example the democratic legitimacy crisis (Luz, 2000; Biermann et al., 2007; Reed, 2008). Broadly defined, participation denotes those processes that enable non-elected citizens to incorporate their concerns into political decision-making (see e.g. Creighton, 2005) but also cooperation between academia and lay people in applied research (Tress et al., 2006). The ELC acknowledges the principles of the Aarhus Convention on Access to Information, Public Participation in Decision-making and Access to Justice in Environmental Matters (UNECE, 1998) and calls for the active involvement of the public and interested parties in defining and implementing landscape policies.

Estonia is currently among the few European countries that have not signed the ELC, mainly because of a mismatch between the interpretations of the term 'landscape' in the ELC and in the Estonian language, and due to the vague definition of responsibilities between the governmental stakeholders about who should take the lead in implementing the Convention (Palang, 2009). However, several trends and processes that could lead to signing the Convention are already on the way. During the past two decades, Estonia has developed a democratic decision-making system and adopted several international regulations (e.g. ratification of the Aarhus Convention in 2001) that require introducing principles of public participation into national legislation. Thus, participatory approaches have become important elements of environmental decision-making in Estonia.

A nation-wide participatory delineation of valuable landscapes was carried out from the late 1990s to 2007 as part of a larger spatial planning initiative. Public discussions, surveys, and interviews with different stakeholders were undertaken in order to gain insights into local people's landscape preferences and to motivate them to participate in discussions about landscape, especially concerning its cultural-historical, recreational, aesthetic, and identity aspects (Alumäe et al., 2003).

In addition to social dimensions of landscapes, biodiversity conservation is also an essential aspect of sustainable landscape policies (Naveh, 2000; Antrop, 2006). Selecting and designating Natura 2000 areas in Estonia, based on the Birds Directive (EEC, 1979) and the Habitats Directive (EEC, 1992) of the European Union (EU), was accompanied by a set of informational activities for the general public and formal consultations for certain stakeholders (landowners, and regional and local governments). As the Natura 2000 network is among the main measures to preserve and enhance the ecological qualities of landscapes at the EU scale, it is relevant to explore participatory approaches within Natura 2000 in the wider context of Estonia's possible accession to the ELC.

In this chapter, the term 'participation' refers to processes organized by nature conservation authorities for members of the public and certain stakeholders to allow for their contribution to the selection and designation of Natura 2000 areas. Although it was not required in the Habitats Directive, the practice of site selections included consultations with the public and stakeholders in several EU member states (EC, 2004b) – although participation remained controversial in several countries, e.g. France (Alphandery and Fortier, 2001; Pinton, 2001), Finland (Hiedanpää, 2002, 2005), and Germany (Stoll-Kleemann, 2001b; Eben, 2006). Among other issues, debates during the site selections and designations have gathered around two questions: how to ensure adequate information dissemination and effective knowledge management; and how to gain public acceptance for the designation processes.

14.1.2 Information and Knowledge in Participatory Processes During Natura 2000 Designations

Although the provision of information on the issue in question and on the participatory procedure does not in itself empower the public or other stakeholders, it allows people to make informed judgements when their opinions are sought by the authorities (Konisky and Beierle, 2001; Diduck and Sinclair, 2002; Hartley and Wood, 2005; Kujinga and Jonker, 2006). During the Natura 2000 designations, the stakeholders have not always been provided with advice and information early and sufficiently enough (Eben, 2006). However, not only is adequate information flow from experts to lay people needed – other stakeholders might also have relevant knowledge to contribute to decision-making (Soini and Aakkula, 2007; Soliva et al., 2008; Collier and Scott, 2009). We understand knowledge here as cognitive factual information (e.g. scientific knowledge), as well as knowledge based on personal experiences (e.g. local knowledge) (Glicken, 1999). The French experience of Natura 2000 designations suggests that not acknowledging some knowledge-holders, for example the local people, can result in their strong resistance towards designations (Alphandery and Fortier, 2001; Pinton, 2001). Further, participatory approaches can help to create common awareness among participants on the issue under discussion (e.g. Cote and Bouthillier, 2002; Sipilä and Tyrväinen, 2005). This awareness has the potential to build a mutual understanding and shared language among different parties, which is a necessary precondition for successful participation in further steps in the management of Natura 2000 areas.

14.1.3 The Role of Acceptance in Natura 2000 Designations

According to Sattler and Nagel (2010), acceptance in relation to nature conservation measures (in agriculture) has three components: *object* of acceptance, *subject* of acceptance, and *context*. The designation of protected areas is an example of an acceptance object. The subject of acceptance can be farmers and their personal attitudes or, in a more general sense, other stakeholders who are affected by nature conservation measures (like landowners). The attitudes of people show how they perceive and evaluate some kind of environmental management measure (Seeland et al., 2002). In the Natura 2000 site selections and designations, fundamental differences in stakeholders' worldviews and values triggered opposing attitudes to the designations. For example, some stakeholders felt their personal freedom to decide on land-use issues to be threatened by the designations (Stoll-Kleemann, 2001a; Hiedanpää, 2005). Certain attributes of participatory processes, for example the quality of deliberation (Schenk et al., 2007), can be the most significant contextual factors influencing peoples' attitudes towards nature conservation policies. Lack of deliberation during the consultations over site designations and insufficient empowerment of stakeholders caused the decision-processes to be regarded as unfair by many stakeholders (Stoll-Kleemann, 2001a; Paavola, 2004; Hiedanpää, 2005; Eben, 2006).

14.1.4 Aims of the Study

This chapter takes a retrospective look at participatory processes during the Natura 2000 site selections and designations in Estonia. The focus is on landowners, as they have been among the largest and most diverse stakeholder groups within the Natura 2000 designations across several countries in the EU. On the basis of two cases, the study aims to:

- explore the role of information and knowledge within participatory processes during the Natura 2000 designations
- examine acceptance of Natura 2000 designations among landowners.

We explore participatory processes targeted at landowners within: (1) selection of potential Natura 2000 areas for submission to the European Commission (EC) from the start of preparatory work for site identifications in Estonia in the mid-1990s up to the spring of 2004; and (2) designation of these areas as under national protection, starting from summer 2004.

14.2 Participation Within Natura 2000 Site Selections and Designations in Estonia: Providing Information and Consulting the Landowners

The first draft list of potential Natura 2000 areas in Estonia was compiled by a set of experts representing the Estonian Ministry of Environment (MoE) and its regional departments, universities and research centres, and non-governmental organizations (NGOs) in conservation (e.g. the Estonian Ornithological Society and Estonian Fund for Nature). According to the national strategy and action plan (2000–2007) for implementing the Natura 2000 network in Estonia (Riikliku programmi, 2000), one task for the environmental authorities was to introduce the concept of Natura 2000 to the public and certain stakeholders, including landowners. Accordingly, the MoE as the main actor responsible for the designations initiated a general information campaign in 2002. The campaign included the launching of a national Natura 2000 webpage (Eesti Vabariigi Keskkonnaministeerium, 2009), production of several posters, booklets and leaflets, and some radio and television broadcasts. Information days, mainly targeted at landowners, were arranged by county environmental departments and protected areas' administrations. These information events also served as the main means for distributing the booklets and leaflets on Natura 2000. The information campaign and the following consultation periods were accompanied by printed media coverage of Natura 2000 issues.

Two formal consultations were organized in 2004, mostly based on the Law on Protected Natural Objects (in force 1994) and the Nature Conservation Act, developed on the basis of the previous Act and entering into force in spring 2004. The core aim of these consultations was to negotiate the boundaries of the selected areas with the landowners, who were expected to comment on the lists of potential Natura 2000 areas. The landowners were invited to express their consent for the pre-selections and designations, or give their reasoning in case they did not agree with them. Landowners could also propose additional areas to the pre-selection list. Only those landowners whose land did not have a conservation status of any kind by the beginning of the Natura 2000 process in Estonia were consulted. This was because their interests were expected to be those most infringed upon by the designations (e.g. through the introduction of new land-use restrictions).

In the first official consultation period (spring 2004), landowners were invited to submit written comments on the preliminary list of Natura 2000 areas, and on the temporary land-use restrictions on those areas. This period is not included in the case descriptions and analyses below since the relevant documents were unavailable to the authors. The temporary land-use restrictions were the same for all new Natura 2000 areas in Estonia and were enforced for a maximum of 1 year, i.e. until the final protection status of each new area was clarified. Landowners were notified about the opportunity to make submissions via national printed media because the circle of the landowners to be consulted was considered (in accordance with the Administrative Procedure Act of 2001) to be too wide to contact them personally. However, some county environmental departments and protected areas' administrations also sent personal notifications to landowners in addition to the newspaper announcement.

The second consultation round (starting from summer 2004) concerned designating the initially selected areas as under national conservation. Administrative acts outlining the planned land use conditions and paper-based maps of potential Natura 2000 sites were made publicly available in county environmental departments, municipalities, and protected areas' administrations. In addition, starting from 2002, maps of potential Natura 2000 areas were permanently available on the national Natura 2000 website, though it was not a legal requirement. All concerned landowners received an official letter from nature conservation authorities with basic information about Natura 2000 and an invitation to comment on the issue. The results of this commenting period were discussed at public meetings arranged by protected areas' administrations or county environmental departments. The meetings aimed at introducing the basic information on Natura 2000 to the landowners, answering their questions, and clarifying misunderstandings.

Prior to the official consultations in 2004 and separately from the national information campaign, informal negotiations and several information events took place in the framework of different projects and which contributed to the selection list for the Estonian Natura 2000 areas.

14.3 Materials and Methods: The Cases of Otepää and Kõnnumaa

The case study approach (Gerring, 2007) was selected in order to study the role of information, knowledge and acceptance in landowner participation during selection

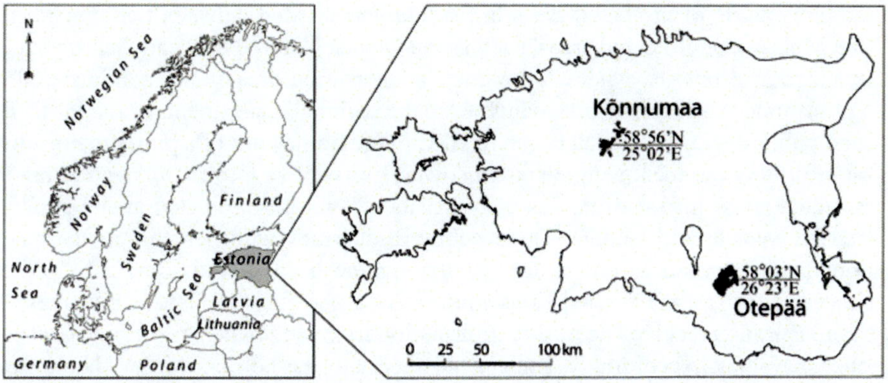

Fig. 14.1 Location of the Otepää and Kõnnumaa case study areas in Estonia

and designation of the Otepää Special Conservation Area, southern Estonia, and Kõnnumaa and Kastna Special Conservation Areas, northern Estonia (Fig. 14.1).

The Special Conservation Area is a new type of protected area that was first introduced into the Estonian legal system through the Nature Conservation Act of 2004, which transposed the principles of the Habitats and Birds Directives into national law. Special Conservation Areas do not have concrete land use restrictions; however, all private owners are required to concert their land management decisions with the nature conservation authorities, for example when changing land use, conducting land readjustment, building, or undertaking forestry actions. Each Special Conservation Area has a specific purpose of protection, which serves as the basis for the nature conservation authorities to decide whether the planned actions interfere with the purpose of protection or not.

14.3.1 Participation in the Designation of Otepää Special Conservation Area

Otepää Special Conservation Area (3.64 km^2), situated in Valga County, was designated in 2005 to secure certain types of semi-natural communities and old-growth forests as habitats for particular bird species listed in the Habitats and Birds Directives. The area is situated next to the Otepää Nature Park, which due to its interesting landscape and good accessibility is a popular tourist destination and subject to recreational development. At the time of designation, most of the land in the Special Conservation Area was in private ownership, divided into c.30 land parcels which were owned by approximately the same number of landowners. Roughly half of the landowners were not local people, and several parcels were owned by real-estate development firms and small-size forestry, agriculture, or tourism enterprises.

The decision process regarding designation was coordinated by the local administration of the Otepää Nature Park. At the time of Natura 2000 designations, the administration of protected areas in Estonia was divided between two divisions of the MoE: some protected areas, such as the Otepää Nature Park, had their own local management boards, whereas the other conservation areas were administered by county environmental boards, which were the regional departments of the MoE. The first public events in Otepää began in 2002 in the framework of a pilot project aimed at informing the local people and landowners as well as informally negotiating the selected sites. Information was also distributed via the protected area's administration website and leaflets. Three public discussions with personal invitations to landowners served as arenas for negotiating the borders and some management issues of the Otepää Special Conservation Area. As a result of these meetings, the administration received necessary information from landowners regarding local ecological values. Accordingly, some adjustments to the borders of the Natura 2000 area were made. The negotiations functioned as a preparatory phase for the official consultation period starting from 2004.

14.3.2 Participation in the Kõnnumaa Case Study

The Kõnnumaa case study included both the Kõnnumaa (5.96 km^2) and Kastna (8.37 km^2) Special Conservation Areas (both in Rapla County), which were first designated in 2006, mainly for the conservation of certain mire habitats and forest types listed in the Habitats Directive. At the time of the designations, there were 13 private landowners altogether, but only two or three were local in the true sense, i.e. living there the whole year. Less than 20% of the land in the two areas was privately owned. Several parcels in both areas were owned by peat-extraction and forestry companies, as well as real-estate development firms. As neither of the areas had a local administration, the designation process (including participation) was coordinated by the county environmental board, which operated at a regional level for the management of all protected areas in the region. Since there were no participatory activities arranged specifically for the areas addressed in this study, participatory approaches were organized on a county-wide basis.

The informal communication process regarding Natura 2000 designations in the Kõnnumaa case began in 2000 during an international cooperation project between the Estonian MoE and the Danish Environmental Protection Agency. The aim was to prepare a list of Natura 2000 areas in Rapla and Lääne Counties. However, promoting public and stakeholder awareness about Natura 2000 was among the main goals of the project. A detailed investigation of stakeholders in the area was carried out, with the result that landowners were identified as one of the key stakeholder groups. The Natura 2000 concept was communicated mainly through information days at local municipalities, where posters and pamphlets were distributed and a video film on Natura 2000 was shown. Information was also distributed via a Natura 2000 homepage.

Following the requirements in the Nature Conservation Law, the design of the consultation processes in the summer of 2004 was in principle the same in both the Otepää and Kõnnumaa cases. Landowners were notified by an official letter from the nature conservation authorities about the basics of the Natura 2000 concept and the opportunities to be involved in the designation process. During a 2-week public display of maps of the areas and the draft of the Nature Conservation Act, and in the course of the following public discussions, the borders of the Natura 2000 areas were negotiated. In both cases, approximately half of the landowners made submissions to the nature conservation authorities. Most of the submissions were negative towards designation. The Special Conservation Areas in both cases were finally designated with a slightly reduced size of area compared to the initial selection.

14.3.3 Interviews and Document Analysis

The study relied mainly on semi-structured face-to-face or telephone interviews with landowners from both case study areas and on document analysis. Eighteen landowners from Otepää Special Conservation Area were interviewed in 2006 and 13 interviews were made in 2007 with landowners from Kõnnumaa and Kastna Special Conservation Areas. The interview partners were chosen randomly, although the choice of participants depended on the availability of respondents. The main topics covered during the interviews (see Box 14.1) included landowners' perceptions of Natura 2000 as a concept as well as of the designation process and their experiences with it. For the analysis, interview protocols were written.

Box 14.1 Discussion Guide for Interviews with Landowners from Special Conservation Areas

- Have you heard about the concept of Natura 2000?
- How would you explain the concept? What does it mean?
- What is the purpose of protection on your land and near surroundings?
- What were the main information channels for you regarding Natura 2000? How do you evaluate your knowledge base on Natura 2000? Would you like to receive more information on Natura 2000?
- How would you describe your experience with the designation process? Did you know about the public in-volvement events? Did you take part of them? Why (not)?
- To date, has the designation process had a more positive influence on your activities, a more negative influence, or no influence at all?

Additionally, in order to create a systematic overview of the design of the landowner participation processes, the nature conservation authorities who had been directly responsible for organising the participatory events in the case study areas were consulted. The authorities were asked about the principles of designing the events and about general responses from the landowners to those events.

All available documents concerning participatory process at the case level were accessed, e.g. minutes of public meetings, written submissions from landowners to the nature conservation authorities regarding designation, as well as relevant national documentation regarding participation. Interview protocols and other documentation were content-analysed. The texts were screened in order to detect: statements regarding landowners' perceptions of the whole process and of key decisions that were taken within it; landowners' knowledge of the Natura 2000 topic and consultation procedures; and how information and knowledge were treated in the process. The main statements found were categorized according to the core research questions, along with key issues that emerged from the data.

14.4 The Role of Information and Knowledge Concerning Designations

14.4.1 Landowners' Perceptions of Information Provision

In general, landowners were aware of the information sources on Natura 2000 and the consultations that had taken place during the designations. Still, they were often uncertain about the exact rules of the consultation procedures, e.g. what the aims of public discussions were, how to make written submissions, and what responsibilities the authorities had to respond to the submissions. Many landowners were unsatisfied with their main information source (printed media), claiming it was not specific enough and too biased. In contrast, targeted and personal ways of communication, such as direct contact with the nature conservation authorities or the official letters to landowners, were much more appreciated.

Most of the respondents in both cases were interested in receiving more information about the content of Natura 2000, especially its socio-economic aspects (concrete land-use restrictions, financial compensation mechanisms, subsidies, etc.). The socio-economic implications of designations turned out to be the main concern of landowners during the consultation process as well. However, the following excerpts from a public meeting in the Otepää case indicate that a great deal of uncertainty and ambiguity surrounded the discussions on those issues, and that the nature conservation authorities were far from providing clear answers to landowners' questions:

> Will there be some kind of restrictions in the planned Special Conservation Area? Could you just name the conditions of land use? And how will the state compensate the reduction of economic revenue for the landowners? I think we should find some kind of a compromise here (Landowner, male, tourism entrepreneur).
>
> Well, concerning the land around the river, our aim is to manage and restore the meadows. In other parts of the Special Conservation Area, the purpose is to protect valuable forest habitats which are necessary for several rare bird species (Nature conservation manager).
>
> Maybe we should discuss what the exact land use restrictions are? (Landowner, female, local government employee).

For us it is important that important habitats will be preserved. It means managing the meadows in case of semi-natural communities (Nature conservation manager).

Does this constrain the activities of landowners? (Landowner, female, local government employee).

In general, the purpose of land use shouldn't be changed. There are also some restrictions to building. Forest management conditions are probably the strictest (Nature conservation manager).

Interviewees also requested more site-specific information about the justifications for why their land had been selected, for instance what the specific biological value of their land was. This is illustrated in the following excerpt from a written submission from a landowner addressed to the county environmental department:

In the letter I received from the county environmental board it was noted that my land was incorporated into the European network of protected areas, Natura 2000. But the explanations why the land had been selected were missing. During the public display of Natura 2000 areas in the Rapla County the nature conservation authorities couldn't explain to me which habitats, plant or animal species need protection on my land, or which parts of the land would be included into the network. So I have the impression that my land has been incorporated into this network for 'just in case'. I regard this as unwarranted restriction of my owner rights and therefore I don't approve the designation (Landowner, female, Kõnnumaa case study).

In both cases, the question of inadequate information provision was repeatedly raised by the landowners at public meetings and in written submissions to the nature conservation authorities. In the Kõnnumaa case, several landowners did not know by the time of the consultations that their land had been selected to be included into the Natura 2000 network, or what the exact boundaries of the selected areas were.

14.4.2 Landowners' Knowledge and Information Management in the Consultations

Interview partners were asked to describe their familiarity with and understandings of the Natura 2000 concept. In general, a great deal of confusion was associated with the concept. Even when the respondents had heard of Natura 2000, they admitted that the content of the concept had remained rather vague for them. Thus, many of our respondents could not give specific explanations about the meaning of the concept (Fig. 14.2). When elaborating on the issue, keywords often used in the communication campaigns, such as 'European Union-wide network of protected areas' or 'protected areas based on European Union directives', were known to few respondents.

Although many landowners claimed to be unaware of the exact conservation purposes of the Natura 2000 area in question (Fig. 14.2), most of them, especially local landowners, nevertheless had multi-faceted ideas of the local biodiversity in their mind. Some publicly well-known species characteristic of the case regions, e.g.

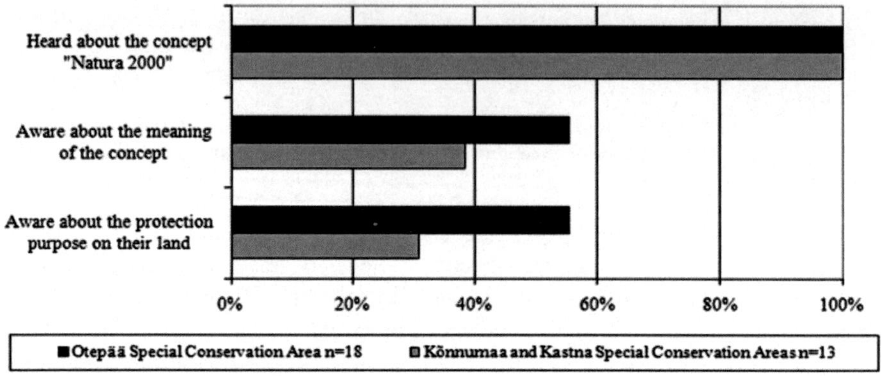

Fig. 14.2 Landowners' general knowledge about the concept of Natura 2000

hawk species in the Otepää case and ant species in the Kõnnumaa case, as well as certain types of ecosystems (semi-natural grasslands, forests) were highlighted by the interviewees as significant components of the local biodiversity. Perceived landscape values were highly appreciated and some respondents revealed their concerns about activities such as intensive tourism, logging, and building, which they considered as threats to these values. However, several landowners took a critical view towards the scientific inventories (the underlying basis for designations), relying on their own expertise about the local biodiversity. An interview quote from a local landowner in the Otepää case shows that the judgements of the nature conservation authorities were hardly trusted:

> The question of what is really the purpose of protection on my land came up several times during the public meeting. They said that it is the corncrake but I don't believe it, this bird just does not live here! I have seen several other species here, like moose, lynx, or hazel grouses, but not the corncrake (Landowner from Otepää case, male, retired farmer).

One of the aims of consultation with landowners during the designations was to gain information from them regarding ecological values on their land. Our cases showed that in practice this goal was barely achieved – landowner submissions concerned mainly socio-economic aspects rather than information on local biodiversity. When reviewing and responding to the submissions, the nature conservation authorities relied on the scientific information gained from ecological inventories. However, in the case of Otepää, several landowners suggested various management options for the Special Conservation Area. In the final designation document, it was specified that their opinions were to be taken into account during management planning of the areas concerned.

14.5 Acceptance: Landowners' Attitudes Towards the Designation Process and the Final Decision

The overall impression gained from the interviews was that landowners rather unenthusiastically responded to the Natura 2000 as a general notion. When we asked them to describe their mental associations to designations in general, their first reactions were mostly negative connotations, e.g. 'restrictions', 'constraints', and in more extreme cases 'nothing can be done on designated areas' or 'scandals', even when this was not the case in reality. While discussing their experiences with the designation process more specifically, many landowners from both cases perceived it as imposing the EU laws while paying little attention to local conditions (e.g. imprecise inventories, unclear compensation measures). The nature conservation authorities were blamed for not listening to landowners' views, and decisions about designations were believed already to have been made. In the case of Kõnnumaa, landowners continuously stressed during public meetings that their land had been designated without asking their opinions. Thus, the process was in general regarded as a top-down initiative:

> The principle of Natura 2000 is right but the way it is implemented is wrong. Designation of Natura 2000 areas should be negotiated with landowners and followed by mutually beneficial agreements between landowners and nature conservation authorities (Landowner from Otepää case, male, retired farmer).
>
> I had the impression from the public meeting that Natura 2000 areas will be designated regardless of what we think of it (Landowner from Otepää case, male, employee in glass industry).

Some landowners had more extreme notions in mind, comparing the designation process with certain characteristics of decision-making processes during the Soviet period (e.g. land expropriation). The following excerpt from a written submission by a landowner addressed to the county environmental department illustrates this:

> Natura 2000 equals a new expropriation. Therefore we categorically reject the decision to designate our land as a Natura 2000 area. For me, the Natura 2000 network does not exist, there's only our farmland! (Landowner, male, Kõnnumaa case study).

However, when discussing the final designation decision, about half of our interview partners from both cases held a quite indifferent position regarding the designation of their land (Fig. 14.3), either because the designations had neither significant negative nor significant positive implications on their land management decisions, or their land was not their main source of income:

> Natura 2000 may be problematic for those who intend to divide their land into parcels, build houses, or do something else. For me it is not a problem, I just have the land and that's all. I haven't got any economic plans for it (Landowner from Otepää, male, local government administration).
>
> No, I haven't had any problems and probably won't have them in the future either because I don't plan to cut forest there, build something, or construct roads, so everything will remain there as it is now (Landowner from Otepää, female, higher education administration).

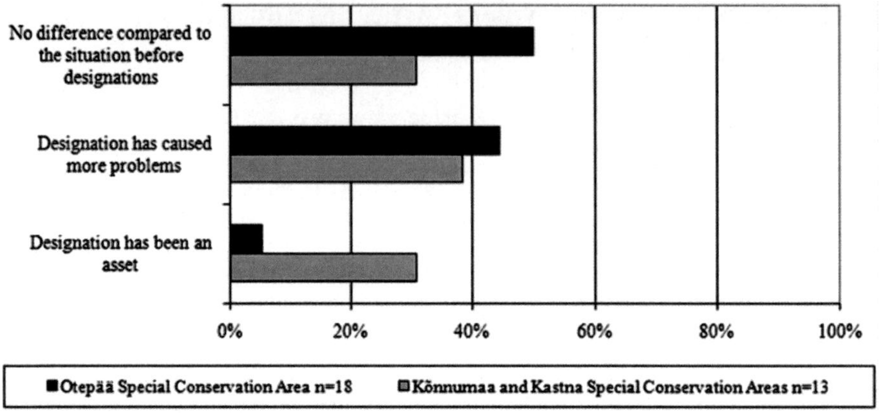

Fig. 14.3 Landowner perceptions of the current experience with Natura 2000 designations

However, most of the landowners did not perceive the future as very promising. Interviewees referred to their right to manage the land independently and were concerned that this right would be constrained without their being consulted. The designations were perceived as taking away the landowners' right-to-decide:

> I can't decide anything entirely on my own; I will have to concert everything with the conservation authorities. But in this way I will no longer be a master of my actions (Landowner from Otepää case, male, employee in glass industry).

Natura 2000 as an asset was mentioned a few times, with compensation being among the most important reasons, as well as the preservation of beautiful landscape.

14.6 Discussion

Initiating the implementation of the EU Birds and Habitats Directives was one of the widest nature conservation projects in recent decade in Estonia, affecting many social groups, especially landowners. The following section discusses the main lessons learned from our two cases regarding information dissemination and knowledge management, and the factors that appear to have influenced acceptance by landowners. The results will be compared to similar studies, including results gained from a project on the delimitation of valuable landscapes in Estonia.

14.6.1 Lessons Learned from Knowledge and Information Management in the Natura 2000 Designations

Despite the relatively extensive communication campaign during the Natura 2000 site selections and designations, information dissemination was still perceived as insufficient by most landowners in our case areas. One reason for this might be that the information was mostly disseminated in an untargeted and impersonal manner (e.g. through media or leaflets in public meetings). As Schenk et al. (2007) note, direct information channels can far more effectively convey messages. Many landowners in the cases of Kõnnumaa and Otepää appreciated the personal sources of communication (for example, public meetings or directly addressed letters to landowners) in terms of concreteness and clarity.

In order to contribute effectively in participatory processes, people need adequate information about the opportunities to participate (Hartley and Wood, 2005), as well as sufficient information about the content of the issue in question (Kujinga and Jonker, 2006). Landowners' low awareness of the specific meaning of the designation of Natura 2000 areas might have been one barrier for them to formulate informed judgements regarding designations. Our cases have shown that by the time landowners were expected to submit their final opinion towards designation (written claims in summer 2004), many landowners did not have at their disposal enough information about some basic issues regarding designations, for example, the socio-economic implications of designations, how to participate in the consultations, or even whether their land has been selected for designation or not. In addition to the information deficiencies, an earlier analysis of the Otepää case study (Suškevičs and Külvik, 2007) and some other studies (e.g. Diduck and Sinclair, 2002) suggest that expectations of having limited impact on the ultimate decision can also prevent people from participating.

However, local people and other stakeholders can also potentially give relevant input to decision-making with their experiential knowledge (Soini and Aakkula, 2007; Soliva et al., 2008). The lay people–expert interface (Palang and Fry, 2003) was evident in our cases, especially highlighted by the question of who were holders of legitimate knowledge. Many landowners were opposed to expert judgements, questioning the validity and relevance of scientific expertise. In contrast – although one of the aims of landowner involvement was to complement the scientific inventories with their knowledge of local biodiversity – the nature conservation authorities implicitly regarded scientific ecological expertise as the only true knowledge (Collier and Scott, 2009). An exception was in some instances in the Otepää case, where landowners' propositions concerning biodiversity management were acknowledged by the authorities (although in the future and not in the designation phase). The discussions held in conjunction with the selection and designation of Natura 2000 areas in Estonia, similarly to some other EU countries (Alphandery and Fortier, 2001; Pinton, 2001), were subtly designed as scientific talks, which made it difficult for the landowners to contribute with their knowledge, and for the authorities to accept these knowledge claims as legitimate for the designations. However, considering that stakeholder participation, for example in the format of public-private partnerships, is encouraged by the European Commission in the management of Natura 2000 areas (EC, 2004a), we find that the perspectives of different actors and the knowledge management issues deserve further attention (both academic and in practice) in the actions towards ensuring that the Natura 2000 areas are received favourably.

14.6.2 Factors Influencing Landowners' Acceptance

The cases show that expectations regarding participatory processes play a leading role in determining attitudes towards the whole decision process. As Sauer (2006) notes, conflicts over Natura 2000 designations in Germany have partially resulted from the fact that nature conservation authorities and affected actors, such as foresters or farmers, had different understandings about what to expect from the process. The information provision and consultation processes in the cases of Otepää and Kõnnumaa were designed to inform the landowners about the importance of Natura 2000 (mainly in ecological terms), to provide them an opportunity to express their opinions towards the designations, and to gather information from them about habitat types and other conservation values of the land. In contrast, landowners mainly regarded the consultations as an arena in which to discuss socio-economic issues. Since the involvement opportunities had been created for other purposes, misunderstandings regarding decision processes between landowners and nature conservation authorities occurred, resulting in a mainly negative perception by the landowners about the participatory process.

One reason why these differing expectations could not be met lies in the contextual constraints to free deliberation in the participatory processes of Natura 2000 designations. As the Habitats Directive requires, the topics of discussion in our two cases were mainly limited to ecological issues. Yet, the socio-economic aspects were the main concerns for landowners, who brought them continuously on to the consultations' agenda. Nevertheless, due to the ambiguity of land-use restrictions in the case of Special Conservation Areas and the unclear financial compensation mechanisms and subsidies, the discussions on socio-economic issues remained rather abstract, increasing uncertainty about the exact implications of the designations for the livelihoods of landowners. The delineation of valuable landscapes in Estonia, for example, had a different nature, leaving much more room for the participants to elaborate on their personal experiences and views about human-nature-culture relationships. The valuable landscapes project had a wider scope, where natural values of landscapes (rare species and communities) were only one aspect among cultural-historical, aesthetical, recreational, and identity values of landscapes (Alumäe et al., 2003). One can suppose that this difference in the process design – Natura 2000 being restricted to ecological issues only and the valuable landscapes project having a wider thematic scope – could be one reason why no considerable conflict situations have been registered in the case of the latter project. It can be further suggested that adopting the landscape approach to nature conservation, integrating community involvement, spatial planning, and biodiversity management (Mitchell et al., 2004), could be useful as a means of making the management of Natura 2000 areas more flexible.

Despite the fact that the designation process in the Kõnnumaa and Otepää cases was not well-accepted by most landowners, quite a remarkable proportion of landowners did not strongly criticize the final decision to designate their land. Several explanations can apply. According to Wallner et al. (2007), landowners' perceptions of protected areas are mostly determined by individual interests and aims. Many landowners in our cases claimed that they lacked direct personal interest in the issue since using their properties for economic purposes was currently not their primary interest. Thus, they were indifferent about the designation at that time, although their perceptions of the future reflected rather negative attitudes towards Natura 2000. However, we did not systematically examine the exact role of the economic situation of the landowners in determining their attitudes towards conservation. Although economic considerations are not suggested as the primary determinants of stakeholders' attitudes towards nature conservation measures (Schenk et al., 2007; Wallner et al., 2007), further investigation is needed to find out in what respects and to what extent economic factors influence the attitudes of the landowners towards Natura 2000 issues.

14.7 Conclusions

The study revealed two main aspects that play a crucial role in the participatory processes concerning the Natura 2000 designations. First, the cases suggest that the rules of the participation process as well as expectations of different stakeholders regarding the process should be made more explicit. Our results also indicate that the specific information regarding Natura 2000, which is relevant for stakeholders in order to formulate their own opinion, should have been communicated in due time, in a targeted manner, and in a context-specific format to the landowners. This would have helped clarify misunderstandings between participants and allowed stakeholders to contribute more effectively in the consultations.

Second, the results of the Otepää and Kõnnumaa cases show that room for deliberation and decision alternatives in participatory processes are critical factors for acceptance among landowners. The Natura 2000 designations were exclusively based on scientific knowledge which left little leeway for discussing the issues that landowners regarded as important. However, the next steps in implementing the Birds and Habitats Directives, i.e. managing the Natura 2000 areas, tend to take a more flexible approach towards stakeholder partnerships and sustainable use of natural resources on those areas. This trend seems to acknowledge several principles outlined in the ELC, for example multiple uses of landscapes, the ability of landscapes to enhance peoples' quality of life, and encouraging cooperation between stakeholders, and can thus be an important step towards setting the EU's nature conservation policy in the wider landscape and participatory context.

Acknowledgments We wish to thank Ms Annika Puhkan for her meaningful contribution to data collection for the case study in Kõnnumaa and Kastna Special Conservation Areas. We also thank Mr Augustin Berghöfer and Dr Dimitrios Zikos from Helmholtz Centre for Environmental Research (UFZ) and two anonymous reviewers for their valuable comments that helped us to refine this work. In addition, thanks go to Ms Tia Ruutopõld and Dr Helen Alumäe for checking and improving the language, and Dr Ain Kull and Mr Peeter Vassiljev for technical assistance with the illustrations.

References

Administrative Procedure Act (2001) Administrative Procedure Act passed 6 June 2001 (RT I 2001, 58, 354), entered into force 1 January 2002. Estonian Legal Language Centre. http://www.legaltext.ee/et/andmebaas/ava.asp?m=022. Accessed 30 Nov 2009

Alphandery P, Fortier A (2001) Can a territorial policy be based on science alone? The system for creating the Natura 2000 network in France. Sociol Rural 41(3):311–328

Alumäe H, Printsmann A, Palang H (2003) Cultural and historical values in landscape planning: locals' perception. In: Palang H, Fry G (eds) Landscape interfaces: cultural heritage in changing landscapes. Landscape Series, vol 1. Kluwer, Dordrecht, pp 125–145

Antrop M (2006) From holistic landscape synthesis to transdisciplinary landscape management. In: Tress B, Tress G, Fry G, Opdam P (eds) From landscape research to landscape planning: aspects of integration, education and application. Wageningen UR Frontis Series, vol 12. Springer, Dordrecht, pp 27–50

Biermann F, Chan M, Mert A, Pattberg P (2007) Multi-stakeholder partnerships for sustainable development: does the promise hold?. In: Glasbergen P, Biermann F, Mol A (eds) Partnerships, governance and sustainable development. Edward Elgar, Cheltenham, pp 239–260

Collier MJ, Scott M (2009) Conflicting rationalities, knowledge and values in scarred landscapes. J Rural Stud 25(3):267–277

Cote M-A, Bouthillier L (2002) Assessing the effect of public involvement processes in forest management in Quebec. Forest Policy Econ 4(3):213–225

Council of Europe (2000) European Landscape Convention, Florence, 20.X.2000. ETS No. 176. http://conventions.coe.int/Treaty/en/Treaties/Html/176.htm. Accessed 19 Jan 2009

Creighton JL (2005) The public participation handbook: making better decisions through citizen involvement. Jossey-Bass, San Francisco, CA

Diduck A, Sinclair J (2002) Public involvement in environmental assessment: the case of the nonparticipant. Environ Manage 29(4):578–588

Eben M (2006) Public participation during site selections for Natura 2000 in Germany: the Bavarian Case. In: Stoll-Kleemann S, Welp M (eds) Stakeholder dialogues in natural resources management: theory and practice. Springer, Berlin, pp 261–278

EC (2004a) LIFE-nature: communicating with stakeholders and the general public – best practice examples for Natura 2000. European Commission. http://ec.europa.eu/environment/life/publications/lifepublictions/lifefocus/documents/natcommunicat_lr.pdf. Accessed 8 Mar 2009

EC (2004b) Report from the commission of 5 January 2004 on the implementation of the directive 92/43/EEC on the conservation of natural habitats and of wild fauna and flora [COM(2003) 845]. European Commission. http://eur-lex.europa.eu/LexUriServ/LexUriServ.do?uri=COM:2003:0845:FIN:EN:PDF. Accessed 7 Mar 2009

EEC (1979) Council directive 79/409/EEC of 2 April 1979 on the conservation of wild birds. Off J L 103:7–18, 25/04/1979. EUR-Lex. http://eur-lex.europa.eu/LexUriServ/LexUriServ.do?uri=CELEX:31979L0409:EN:HTML. Accessed 15 Apr 2010

EEC (1992) Council directive 92/43/EEC of 21 May 1992 on the conservation of natural habitats and of wild fauna and flora. Off J L 206:7–50, 22/7/1992. EUR-Lex. http://eur-lex.europa.eu/LexUriServ/LexUriServ.do?uri=CELEX:31992L0043:EN:html. Accessed 15 Apr 2010

Eesti Vabariigi Keskkonnaministeerium (2009) Natura 2000: linnu- ja loodushoiualade võrgustik. http://www.envir.ee/natura2000/. Accessed 20 Oct 2009

Gerring J (2007) Case study research: principles and practices. Cambridge University Press, Cambridge

Glicken J (1999) Effective public involvement in public decisions. Sci Commun 20(3):298–327

Hartley N, Wood C (2005) Public participation in environmental impact assessment – implementing the Åarhus convention. Environ Impact Assess Rev 25(4):319–340

Hiedanpää J (2002) European-wide conservation versus local well-being: the reception of the Natura 2000 reserve network in Karvia, SW-Finland. Landsc Urban Plan 61(2–4): 113–123

Hiedanpää J (2005) The edges of conflict and consensus: a case for creativity in regional forest policy in Southwest Finland. Ecol Econ 55(4):485–498

Konisky DM, Beierle TC (2001) Innovations in public participation and environmental decision-making: examples from the Great Lakes Region. Soc Nat Resour 14(9):815–826

Kujinga K, Jonker L (2006) An analysis of stakeholder knowledge about water governance transformation in Zimbabwe. Phys Chem Earth 31(15–16):690–698

Law on Protected Natural Objects approved by the Riigikogu on 1.06.1994 (RT I 1994, 46, 773), took effect on 9.07.94 (1994). Estonian Legal Language Centre. http://www.legaltext.ee/et/andmebaas/ava.asp?m=022. Accessed 27 Oct 2009

Luz F (2000) Participatory landscape ecology – a basis for acceptance and implementation. Landsc Urban Plan 50(1–3):157–166

Mitchell N, Brown J, Beresford M (2004) Conclusions – the protected landscape approach: conservation for a sustainable future. In: Brown J, Mitchell N, Beresford M (eds) The protected landscape approach: linking nature, culture and community. IUCN, Gland, Switzerland and Cambridge, pp 233–246

Nature Conservation Act Passed 21 April 2004 (RT I 2004, 38, 258), entered into force 1 May 2004 (2004) Estonian Legal Language Centre. http://www.legaltext.ee/et/andmebaas/ava.asp?m=022. Accessed 27 Oct 2009

Naveh Z (2000) What is holistic landscape ecology? A conceptual introduction. Landsc Urban Plan 50(1–3):7–26

Paavola J (2004) Protected areas governance and justice: theory and the European Union's habitats directive. Environ Sci 1(1):59–77

Palang H (2009) Landscape and rural heritage. In: Landscape and rural heritage, Sibiu, Romania 20–21 September 2007: sixth meeting of the workshops of the Council of Europe for the implementation of the European Landscape Convention, proceedings. European spatial planning and landscape, No 88, pp 89–96. http://www.coe.int/t/dg4/cultureheritage/heritage/Landscape/Publications/ATEP-88_bil.pdf. Accessed 8 Oct 2009

Palang H, Fry G (2003) Landscape interfaces: Introduction. In: Palang H, Fry G (eds) Landscape interfaces: cultural heritage in changing landscapes. Landscape series, vol 1. Kluwer, Dordrecht, pp 1–13

Pinton F (2001) Conservation of biodiversity as a European directive: the challenge for France. Sociol Rural 3(41):329–342

Reed SM (2008) Stakeholder participation for environmental management: a literature review. Biol Conserv 141(10):2417–2431

Riikliku programmi (2000) Riikliku programmi «Eesti NATURA 2000» kinnitamine aastateks 2000–2007. Elektrooniline Riigi Teataja – eRT. http://www.riigiteataja.ee/ert/act.jsp?id=83109. Accessed 27 Oct 2009

Sattler C, Nagel UJ (2010) Factors affecting farmers' acceptance of conservation measures – a case study from north-eastern Germany. Land Use Policy 27(1):70–77

Sauer A (2006) European nature conservation policy: challenges for local implementation in Germany. In: Tress B, Tress G, Fry G, Opdam P (eds) From landscape research to landscape planning: aspects of integration, education and application. Wageningen UR Frontis Series, vol 12. Springer, Dordrecht, pp 173–189

Schenk A, Hunziker M, Kienast F (2007) Factors influencing the acceptance of nature conservation measures – a qualitative study in Switzerland. J Environ Manage 83(1):66–79

Seeland K, Moser K, Scheuthle H, Kaiser FG (2002) Public acceptance of restrictions imposed on recreational activities in the peri-urban nature reserve Sihlwald, Switzerland. Urban Forestry Urban Green 1(1):49–57

Sipilä M, Tyrväinen L (2005) Evaluation of collaborative urban forest planning in Helsinki, Finland. Urban Forestry Urban Green 4(1):1–12

Soini K, Aakkula J (2007) Framing the biodiversity of agricultural landscape: the essence of local conceptions and constructions. Land Use Policy 24(2):311–321

Soliva R, Rønningen K, Bella I, Bezak P, Cooper T, Flø BE, Marty P, Potter C (2008) Envisioning upland futures: stakeholder responses to scenarios for Europe's mountain landscapes. J Rural Stud 24(1):56–71

Stoll-Kleemann S (2001a) Barriers to nature conservation in Germany: a model explaining opposition to protected areas. J Environ Psychol 21(4):369–385

Stoll-Kleemann S (2001b) Reconciling opposition to protected areas management in Europe: the German experience. Environment 43(5):3–32

Suškevičs M, Külvik M (2007) Assessing the effects of public participation during the designation of Natura 2000 areas in Otepää Nature Park area, Estonia. In: Chmielewski TJ (ed) Nature conservation management: from idea to practical results. a long-term biodiversity, ecosystem and awareness research network. Lublin University of Agriculture, Lublin, pp 220–235

Tress B, Tress G, Fry G (2006) Defining concepts and the process of knowledge production in integrative research. In: Tress B, Tress G, Fry G, Opdam P (eds) From landscape research to landscape planning: aspects of integration, education and application. Wageningen UR Frontis Series, vol 12. Springer, Dordrecht, pp 27–50

UNECE (1998) Convention on Access to Information, Public Participation in Decision-making and Access to Justice in Environmental Matters done at Aarhus, Denmark on 25 June 1998. United Nations Economic Commission for Europe. http://www.unece.org/env/pp/documents/cep43e.pdf. Accessed 8 Oct 2009

Wallner A, Bauer N, Hunziker M (2007) Perceptions and evaluations of biosphere reserves by local residents in Switzerland and Ukraine. Landsc Urban Plan 83:104–114

Chapter 15
Conclusion: Benefits, Difficulties, and Challenges of Participation Under the European Landscape Convention

Marie Stenseke and Michael Jones

Abstract The ambition of this volume has been to evaluate and discuss the state of public participation in landscape issues a decade after the establishment of the European Landscape Convention. This concluding chapter summarizes the insights from various countries and discusses vital issues for future landscape research. While the merits of the ELC and public participation are acknowledged in the

M. Stenseke (✉)
Department of Human and Economic Geography, University of Gothenburg, SE–405 30 Göteborg, Sweden
e-mail: Marie.Stenseke@geography.gu.se

M. Jones, M. Stenseke (eds.), *The European Landscape Convention*, Landscape Series 13, DOI 10.1007/978-90-481-9932-7_15,

cases presented, a number of weaknesses and difficulties are also recognized. The main challenges to participation identified include public indifference, political and administrative power structures, scepticism regarding participatory approaches in government quarters, diverging perspectives between experts and stakeholders, and how to ensure democratic involvement. Positive lessons and cases of good practice show, nonetheless, that there are democratic gains to be made from participation. Methods may vary in detail, but techniques to ensure effective two-way communication are essential. The spectrum of participatory methods and communicative concepts examined indicates a need for mediation and arbitration. This is particularly so as the number of conflicts over the role of participation in environmental and landscape issues is likely to increase as the participatory approach spreads. Finally the chapter discusses the ELC in relation to European Union (EU) Directives, the future role of science in participatory approaches, and new issues emerging. There is a need for further knowledge concerning landscape perceptions, the interface between the ELC and other societal goals concerning landscape and land use, and policy strategy discourses. Since participatory approaches challenge the role of experts, questions are raised about how this field is to be researched. New questions also arise regarding options for participation in the face of contemporary trends and issues such as tourism, climatic change, biodiversity loss, and multiculturalism.

Keywords Challenges to participation · Positive lessons · EU Directives · The role of science · Contemporary issues

15.1 Introduction

Public participation has been an emerging issue for more than a decade in landscape management and planning. The European Landscape Convention (ELC) has made a significant contribution to this development due to its participatory approach. While the ELC can be understood as a political framework, it is not yet self-evident how its general aims concerning participatory engagement with landscape are reflected in practical management and planning (Jones, 2007). Therefore, the ambition of this volume has been to make a contribution towards evaluating and discussing the state of public participation in European landscapes and the implementation of the Convention a decade after its establishment. Empirical studies from a number of countries shed light on benefits as well as difficulties and limits of the participatory approach. Furthermore, the various cases give insights into how public participation, as it is implemented and used today, relates to democratic ambitions and implications of participatory approaches in different types of area. We also get indications of how the implementation of the ELC interacts with other societal objectives related to landscape and land use.

In this concluding chapter we summarize, on the basis of the selected examples presented in the previous chapters, insights from different parts of Europe, and identify key challenges to increased public participation. We also condense positive lessons and good examples to indicate potential paths and measures for fulfilling the Convention. The various contributions in the book raise a number of vital issues for future landscape research. Furthermore, the lessons learned pose new questions regarding options for participation in the face of major contemporary trends and issues such as tourism, climatic change, biodiversity loss, and multiculturalism.

15.2 Challenges to Participation

The establishment of the ELC stems in part from aspirations supported by scientific results on positive effects of increased public involvement in landscape management and planning. Arguments in favour of participatory approaches to landscape are well acknowledged, though not fully agreed on, but problems and limitations have received less attention thus far. Hence we pay particular attention to difficulties and challenges when summarizing the lessons from the case studies in this volume. While the contributing authors acknowledge the ELC and the merits of public participation as a complement to official decision-making, a number of weaknesses and difficulties are recognized when applying participatory approaches and when the European Landscape Convention meets practice. Challenges identified in the case studies range from indifference or scepticism regarding public participation in government quarters to problems of implementing ideas produced through participatory exercises. However, it is important not to mix inevitable difficulties related to participatory approaches in general with negative results arising in concrete cases where there is seemingly community involvement – cases labelled participatory in rhetoric – but where in reality the degree of influence from the public is low (Zachrisson, 2004; Jones, 2007).

15.2.1 Challenges Related to Indifference

A fundamental prerequisite for implementing the ELC is the existence of concern for landscape in both official and civic circles. In the Polish case, lacking insights were identified. This can be related to lack of landscape research and low political ambitions as the ongoing socio-economic transformation in the country makes landscape subordinate to economic pursuits. There was found to be no national landscape policy, and lack of commitment at ministerial level. Responsibility for landscape matters was split among different sectors, which lacked coordination. The focus was on designating exceptional areas and objects, while no mechanism was in place for organising stakeholder involvement in landscape issues. The socio-economic transition in Poland appears to be accompanied by general indifference towards landscape in society, with consideration of landscape viewed as an obstacle to development.

The question about how to implement the ELC when landscape is not considered an issue is also raised in the Greek example. Greece signed the Convention in 2000 but did not ratify it until 2010. Here the environment with its many historic landscapes attracts many foreign visitors. Tourism promotes romanticism and a pressure to create the views and surroundings that tourists prefer, rather than an interest in the landscape itself. Thus, we can note lack of a sense of landscape as a common good, and lack of a 'landscape conscience'. There is little public involvement in landscape issues, local interests are frequently marginalized, and development decisions are left to public and private interests.

15.2.2 Challenges for Political and Administrative Structures

For the implementation of the ELC, the various European examples show a number of difficulties related to established power structures and institutional arrangements. Likewise, there is often an incoherency between various sector policies regarding the concept of 'landscape'. Furthermore, 'landscape' is commonly regarded as a specific sector interest, instead of as an area where everything takes place. The Spanish example indicates that proposals for public participation might fall through the cracks in a political structure built up with another logic. In Spain, where the Convention was only ratified at the end of 2007, its implementation is the responsibility of the autonomous regions. The approach so far appears to be limited to 'raising awareness' of landscape issues through the establishment of 'landscape observatories' and other similar institutions.

As for the implementation of community involvement, relatively recently democratized countries in the east might, like Poland, have a relative weakness in civic organization as a result of the past period of central planning. A number of problems are identified: old habits in the political and administrative system of decision-making; citizens with low trust in authorities; lack of a participatory tradition; and feeble non-governmental organizations, all of which make public participation appear a long-term project.

In countries with a well-established structure of local authorities, demands for new routines for participation in the political and administrative system are raised. When the local level gains an increasing degree of influence, there is need for inter-local collaboration, as revealed by the study of national parks and other large conservation areas in Norway. This means that local authorities, where they have developed policies that differ somewhat from authority to authority, have to find areas of agreement. It also means that it is necessary to consider what kind of representation should be instituted in inter-local platforms and discussions. In the French study, it was found that participatory exercises might be negated if there is a change in political strategy after elections. In both the French and Swedish cases, it proved to be difficult to continue participation after the completion of the research exercise, or it was deemed problematic to translate the results of participation into policies.

15.2.3 Challenges in Diverging Perspectives

Differing views between experts and stakeholders emerged from the studies in several countries. A major challenge is how to combine Landscape Character Assessment, undertaken by experts, and a 'sense of place' approach, which brings out the values of residents and tourists. As was shown in the Norwegian study of landscape planning for recreation, experts and the general public often did not agree on what sort of knowledge is relevant. For the experts, knowledge should be 'objective', quantifiable, and instrumental; functional values dominated, while the users favoured experiential values. The study of French and Belgian 'rurban' areas showed that one dimension of the observed difference between experts and farmers was that planners and officials tend to view farmland from an urban perspective rather than from an agricultural one. The danger in this case was the non-involvement of farmers. In Poland, experts have prepared typologies, provided biophysical knowledge, and delineated cultural landscapes, but there is a discrepancy between management concerns and citizen preferences. Similar differing views of experts and stakeholders are reported from Portugal and France.

Diverging perspectives not only exist between experts and the general public, but also between different groups of experts, as illustrated in the French and one of the Norwegian cases. In the latter, disagreements were found to occur in a number of dimensions: there were also disagreements between national and local level, and the character of disagreement varied from area to area. In France, experts who conceived of landscape as the product of biophysical processes made policies related to these processes, while experts who conceived of landscape as a social construction made policies related to cultural heritage; yet in neither case were these policies necessarily accepted locally, where people viewed landscape more in terms of personal experience and values. The study of recreation in Norway also indicated strong sector- and discipline-oriented approaches. There was lack of coordination between the different sectors, as well as varying approaches to landscape and differing priorities according to which academic discipline dominated the planning apparatus in different local administrations.

Related to the issue of diverging perspectives are challenges emanating from the heritage of top-down planning. In the Netherlands, where the landscape planning profession is well established, there was initially some opposition within central agencies to giving municipalities responsibility for landscape issues. The ability of the latter to make 'sound' decisions was questioned, and potential conflict between local economic targets and improvement of landscape quality was feared. A similar fear, specifically concerning the danger of reducing the biological quality of conservation, was found where the ELC meets strong nature conservation legislation. The study of delegated management of large nature conservation areas in Norway demonstrated that the role and power of experts set limitations on local management. Furthermore, the Swedish and the Estonian cases showed that attempts to introduce participatory approaches in nature conservation have not been fully successful as landowners are reluctant to limit discussion to the narrow approach of biological diversity and nature protection, but want to integrate other local aspects. This could be regarded as a limit for public participation from a sector policy point of view. However, it might also be read as a demand for policies related to landscape to be more integrated and coherent. The Swedish case study indicated that biodiversity was of little interest to local people, and there was an obvious challenge in negotiating biological concerns such as species loss with the utilitarian and anthropocentric perspectives of place-based stakeholders. Usually, we find different expectations regarding participation between public authorities and stakeholders, and one feature recognized is a patronising belief that local resistance to nature conservation could be overcome by education or economic compensation (Stoll-Kleeman, 2001; Selman, 2004). It remains to be seen what will happen regarding the application of the ELC in countries with well-established institutions and regulations for nature protection. Will nature conservation dominate over the ELC, with the Convention therefore being more applied in areas without significant biological qualities?

15.2.4 Challenges of Democratic Settings

Almost all the cases presented in this volume concern local participation. Even though there are good reasons to pay attention to how people living in specific landscapes can have a say in plans and decisions concerning their physical surroundings, there is, as pointed out in several chapters, need to develop strategies and methods on how to involve non-locals. Further challenges related to the issue of democracy are who should participate, and the relationship between deliberative democracy based on broad participation and representative democracy based on elected officials. The Swedish and the Portuguese cases show that local participants often have a narrow local approach, as they tend to be 'trapped in their common landscape', making it hard to envision alternative pathways for the future. One conclusion drawn in the Swedish case is that wider common values still have to be safeguarded by authorities when they are not recognized by local people. A complementary strategy would be to expand the possibilities for non-local stakeholders to participate.

There is diversity within any local population and, even though participation can help marginal groups to have a say in planning processes, there is also a risk of exclusion from participatory processes. These issues raise questions of how real participation is. In the Norwegian study of large conservation areas, women turned out to be almost absent in delegated management; further, the role of interested parties from outside the local community was not clarified, and a difficulty identified was how to select who should represent the local. Where local management was based on cooperation between several local authorities, the broader representation of local interests beyond politicians was lacking.

A specific issue is how to arrange community involvement in marginal areas that are being abandoned by people. The Portuguese case identified a challenge in depicting a desired future through a participatory process in places where it is difficult to foster visionary thinking among the local population. In such areas, the prospect for many of the present inhabitants is to move somewhere else. The landscape might be in a process of transformation from a place for production to a place for leisure and recreation, and potential stakeholders in envisioned scenarios may not yet be known.

An even greater challenge is to enhance community involvement in urban areas. Here the number of competing demands on land is greater; official decision-making often leaves less room for public negotiations; the representativeness of potential participants is difficult to distinguish and there is an even greater risk that marginal groups will not be heard; and non-engagement and indifference due to less perceived or articulated relationships to the physical landscape are more common.

15.3 Positive Lessons and Cases of Good Practice

Although participation meets many challenges, there are also positive lessons. Results from the empirical studies presented in this volume give support to a number of benefits identified in earlier studies (e.g. O'Riordan and Stoll-Kleeman, 2002; Zachrisson, 2004; Jones, 2007), and also give new examples that might be instructive.

15.3.1 Democratic Gains

Not surprisingly, the cases show that public participation has strengthened the relationship between people and their physical surroundings. It has meant an increased legitimacy for landscape-related policies, as the viewpoints and knowledge of various actors have been included in landscape planning, implying an improved and more effective management. Furthermore, many of the cases presented give evidence of the usefulness of public participation for awareness-raising, providing participants with better abilities to construct their own understanding and opinions concerning landscape issues.

The Swedish case illustrates how a participatory approach can help solve local conflicts and thereby facilitate the work of planners. Despite limited follow-up, there were some interesting indications of the potential of initiating participation at an early stage in planning, and a method for turning conflicts into constructive landscape management was explored. In the growing conflict between landowners and horse riders in Scania (Skåne), meetings with the opposing groups individually and together helped bring a mutual understanding of the problems and led to a certain willingness to find solutions. The need for local authorities to develop expertise in equestrian matters was also identified.

Furthermore, where there is a lack of coherent and effective strategies and measures concerning landscape qualities, as shown in the Spanish case, the implementation of the ELC can stimulate the development of more powerful landscape management policies. The study of public participation experiments in Norwegian nature conservation areas indicated that local involvement is one factor that can ease the commonly conceived scepticism towards protection. Likewise, in the Estonian case, it was found that information about the Natura 2000 designations was better among participating landowners than among those who did not participate, and despite criticisms this led to greater acceptance of the outcomes.

15.3.2 Methods

In implementing participatory approaches in accordance with the ELC, new or modified methods are frequently sought. A challenge for researchers is therefore to investigate and test the feasibility and the quality of methods. Public participation is to a significant degree about relations. A major intention of many of the methods developed is to improve the ability of people to express their opinions and perspectives on landscapes, to be able to construct their own understanding, and to communicate this to others. The aim is improve the ability to understand each other. At the same time, experts are likely to contribute with an outsider view, complementing a possibly narrow local perspective. Another potential function for professionals is to provide insights into more general forces that lead to transformation of the landscape. On the basis of participatory approaches in a number of cases, the French study presents a typology of techniques for use in discussions of landscape preferences, based on the aspects visual–literary, descriptive–analytic–synthetic, and views from inside–views from outside.

Various forms of meetings to discuss landscape issues are presented in the case studies. They include discussions indoors in workshops as well as outdoors in the landscape. Some discussions embrace the broad public while others are for specific groups. A participatory exercise in the Dart River Catchment in England even involved a festival, paving the way for broader awareness-raising and consultation.

Scenarios have provided a means for comparing the perceptions of different types of experts and stakeholders and are commonly used as a method of envisioning future landscapes, discussing advantages and disadvantages of alternatives, presenting preferred futures, and assessing different outcomes in order to develop a common vision (Emmelin, 1982; Emmelin et al., 1990; Schoute et al., 1995; Emmelin, 1996; Tress and Tress, 2003). In remote areas, as in the Portuguese case, scenario techniques proved to be successful in accommodating the high level of uncertainty concerning the future of the area. One method, tested in some of the cases, with instructive results, was the 'worst case scenario'. A participatory method for mediating landscape knowledge is 'Prospective Visions', tested in Belgium and France. This method is designed to facilitate the integration of farmers' perspectives in planning processes. It involves the construction of visual representations, with workshops being held on both negative and positive Prospective Visions, and stimulates a collaborative learning process in which the visions are ultimately less important than the questions asked around the visions.

Dialogues and exchanges bring out areas of disagreement and, therefore, another methodological concern is about mediation and arbitration. The aim of mediation, as it is for example practised in dealing with landscape issues in New Zealand, is to help to produce areas of agreement rather than attempting to arriving at full consensus, which might not be necessary (Menzies, 2007, 2010). An important lesson learnt is that by involving local interests at an early stage, time and costs are saved at a later stage of the process. In the case of the Netherlands, it was found that local landscape coordinators who are appointed to stimulate the implementation of plans greatly contributed to the success of landscape development plans.

15.3.3 Communicating Concepts

The various European cases present a number of concepts that have been developed in order to facilitate the integration of important aspects and stimulate discussions. Two concepts will be recapitulated here, addressing particular aspects: landscape resource analysis, stressing a resource perspective, and landscape biography, focusing on the historical dimension of a piece of landscape.

Landscape resource analysis is a concept presented in one of the Norwegian studies. It is aimed at assessing qualities in the landscape, complementing expert-based Landscape Character Assessments with the public's 'sense of place'. A landscape resource analysis thus attempts to unite a description of the general characteristics of a landscape with its importance for people. This is in line with the ECOVAST (European Council for the Village and Small Town) method (ECOVAST, 2006), designed to help local residents assess the landscape, which is also applied in the Dutch case.

In the Netherlands, landscape biographies have provided inputs for formulating future visions while facilitating the inclusion of historical landscape aspects. Experts and local stakeholders cooperate in the work of making landscape development plans and village plans. The wishes and expertise of the local population are incorporated through the landscape biographies, which combine expert scientific knowledge and the knowledge and perceptions of the local people. By this means, the gap between 'official' heritage and 'local' heritage can be closed. An important stimulus for this approach was the Belvedere Manifesto of 1999, promoting the idea of 'conservation for development'. In the rapidly developing region of the eastern Netherlands, for example, ideas were developed for taking account of the past in planning for a changing environment.

15.4 The ELC Compared with EU Directives

Although 22 of the 27 member states of the European Union (EU) have ratified the European Landscape Convention, and a further two members have signed but not yet ratified it (although Sweden announced its decision to do so in November 2010), the EU is not in itself a Party to the Convention. Alongside the Council of Europe's Landscape Convention, there are a number of directives related to landscape management emanating from the EU. When it comes to implementation of the ELC in specific areas, we may experience conflicts between two sets of aspirations set out in international conventions, one concerned about nature conservation, as expressed in the Convention on Biological Diversity (1992) and in Natura 2000

(EEC 1992), and the other concerned with local involvement, as expressed in the Rio Declaration (UN 1992: Principle 10), the Aarhus Convention (UNECE 1998), the EU's Water Framework Directive (WFD) (EC 2000: §46), and the ELC (Council of Europe 2000: Article 5c). As the participatory approach spreads, the number of conflicts over the role of participation in environmental and landscape issues is likely to increase. It is hard to see any general solution, but transparency in the arguments put forward is essential. Inescapably, the answer to how priorities should be made has in the end to be political (Stenseke 2009).

One EU policy related to landscape management is Natura 2000, a network of conservation areas in Europe initiated in accordance with the Habitats Directive of 1992 (EEC, 1992). Natura 2000 is the EU's principal contribution to fulfilling the Convention on Biological Diversity (CBD) of 1992, and has been labelled the 'centrepiece of EU nature & biodiversity policy' (European Commission, 2010). The main purpose of Natura 2000 is to prevent further extinction of species and destruction of habitats. Natura 2000 specifically targets biological diversity, and a number of researchers, doing studies in various member countries, reveal problems related to the lack of consideration of societal objectives (Rekola et al., 2000; Alphandéry and Fortier, 2001; Young et al., 2005). In Estonia, although it has not signed the European Landscape Convention, participatory exercises have been undertaken with landowners in connection with the European Union's Natura 2000 designations. Nature conservation authorities expected participation to provide a means of informing of the benefits of designation, whereas landowners were more concerned with socio-economic issues. Differing expectations led to misunderstandings, which contributed to negative acceptance of the designations. Many landowners felt that the information they had received was insufficient, or that they had little influence on the final decisions, with consequent lack of interest and motivation. Many felt that the selection and designation of Natura 2000 areas was imposed from above, and there was a belief that local interests were not listened to. The restrictions involved in the designations were felt to be taking away landowners' rights to decide. The landowners were also critical to the scientific inventories.

Another EU directive is the Water Framework Directive (WFD) of 2000. This directive is a legal framework to protect and manage water resources within the union. Spatially, it is based on river basins, and public participation in planning is among the core requirements (EC, 2000: §46). Thus, the WFD seems to be well in accordance with the intentions of the ELC. The British case, concerning the implementation of the directive in the River Dart Catchment in the south-west of England, presents a good example of community involvement.

15.5 The Role of Science

Over time much research effort has been spent on illuminating the prospects of public participation, and this to some extent paved the way for the establishment of the ELC. Now there is a need to nuance and deepen our insights into methods, contextual aspects, discursive studies of policy strategies, and the interface between the

ELC and other societal goals concerning landscape and land use. One task in this respect is to expose taken-for-granted hierarchies and contribute to a better understanding of underlying structures and power relations. Throughout the present book the role of science in the participatory approach is implicitly addressed. It could be noted that while in some countries (such as Sweden and the Netherlands) researchers have contributed to the development of participatory techniques at the request of or in collaboration with the authorities, in other countries (such as Poland) there is very little connection between landscape research and policy-making.

Undoubtedly, the participatory approach raises questions about how this relatively new field is to be researched. Science faces the demand to develop further methods for participation and communication between various actors. Related to this is a request for more knowledge about landscape perceptions, as the landscape becomes increasingly the subject of public discussion. This also necessitates investigating the meaning of landscape to people. What are people really talking about when they talk about landscape? We need to understand better underlying symbols and implicit perspectives.

Another task for researchers is to examine critically participatory approaches. A central issue is what it means for the role of researcher that the European Landscape Convention is explicitly designed to promote public participation. The question to address is what new demands the Convention puts on research and researchers. What is the future role of 'landscape experts'? Should they act as agents for awareness-rising? Should their principal role be instrumental in designing procedures for participation? Or should they keep a critical distance in examining the role of participation and how it works in practice? There is a need for both applied research and critical research. Important issues are participatory writing, transdisciplinarity, and different modes of knowledge. Vital issues for landscape research concern the role of local people and their relationship to landscapes specialists and other experts, in particular when landscape quality is to be assessed and determined, for example in the formulation of Landscape Quality Objectives.

15.6 The Participatory Approach and New Issues Emerging

Communicative planning is challenged by new public management, implying entrepreneurialism and depoliticization (Sager, 2009). In landscape management there are emerging examples of green partnership, implying 'trade' of landscape values and management based on economic agreements. Conservation covenants, i.e. voluntary agreements between individual landowners and authorities for protecting and managing landscape qualities, are found in the environmental schemes within the EU's Common Agriculture Policy (CAP); similar ones have been introduced in Norway (Jones, 1993; Rønningen, 1998; Statens landbruksforvaltning et al., 2007). There are aims of economic efficiency behind this kind of management. When landscape management is made a market issue, possibilities of public involvement and collaborative planning become easily weakened.

Another phenomenon, driven by the market, is the ever increasing tourist industry. As indicated by the example of Greece, there is a temptation to create landscapes that are deemed attractive for tourists. This risks promoting a trend towards homogeneity, which very well could be enforced by public participation, as local communities and regions perceive the possibility to compete for more tourists by creating what is believed to be appealing to the tourist gaze.

The issue of biodiversity and public participation is raised in some chapters of this book. The declaration of 2010 as the International Year of Biodiversity by the United Nations indicates that this issue is of continuing concern. However, there is still a widespread understanding that nature conservation is only about biology – even though it has been widely acknowledged that stakeholder involvement, communication, and collaboration are of vital importance in nature resource management (Folke et al., 2003; Robertson and Hull, 2003), and that local knowledge, awareness, and involvement as well as awareness among the public at large play important roles in keeping and enhancing biological qualities, at least in rural areas (Stenseke, 2006). One likely reason for the perception that nature conservation is primarily a matter of biology is that executives working with biodiversity issues have mainly been trained as conservation biologists, with deficient insights into social aspects. On a general note, a complicating factor is that in biology and ecology the term 'landscape' has a well established definition that differs fundamentally from that of 'landscape' expressed in the ELC. From an ecological point of view, a landscape often functions as a scale measure related to the movement of different organisms, usually within a set area and including components such as heterogeneity and habitat mosaics. However, during the last decade, what appears to be a more holistic 'landscape perspective' has been launched in ecological and environmental research and practice. This perspective is increasingly seen as an alternative in conservation processes, where efforts traditionally have mostly targeted relatively small single objects (Lindborg et al. 2008). However, social and immaterial features are still excluded, as what are taken account of in the 'landscape perspective' are species, habitats, and ecosystems (Naturvårdsverket 2010). It can be noted that this concept of landscape and the one linked to the agenda of the ELC have developed in separate scientific fields and have seldom confronted one another. In order to produce more effective and consistent policy-making and management concerning landscape, there is need to clarify how these two concepts relate to each other, both analytically and in practice.

A further matter is what happens as climate change gains increasingly urgent attention. The emergence of climate change as an issue is to a large degree dependent on expert investigations and analyses, while it is only to a limited degree immediately apparent – at least in the short term – for people in their lived landscapes. It is likely that the climate change issue has more of a top-down flavour than, for example, biodiversity among the public at large, since professional biologists and amateurs with a great interest in birds and plants are often active in local societies, while climate experts are rarer in such forums. Another feature of the climate change issue is that measures taken at the local level are rarely directly observable as far as their possible effects on mitigating harmful consequences of climate warming are concerned. Even global agreements will take time before their results show. Nonetheless, there is potential for public participation in deciding how to act upon the signals from climate change predictions. While experts can inform the public of what changes might be expected in the landscape as a result of climate warming, the public needs to be involved in discussions of the landscape changes that might be the result of mitigatory measures such as the promotion of renewable energy.

A final issue concerns the growing multicultural character of European society and landscapes. Minorities include not only old-established regional and ethnic minorities, but increasingly also recent immigrants, often living together in particular areas of towns, and leaving their mark on the landscape in ways that are perceived both positively and negatively by the old-established majority. The building of mosques and other unfamiliar religious structures, and the development of ethnically distinctive urban districts, are frequently matters of heated debate. More contentious still are temporary migrants such as guest workers, asylum seekers, and refugees, who have restricted rights and entitlements. Even more contentious are illegal immigrants, often working as labourers in the construction industry or in agriculture, or as hotel workers and domestic helpers, who are without formal rights. A participatory democracy without discrimination means that such groups should not be dealt with summarily but also be heard in matters concerning the physical environment in which they find themselves (Jones, 2007: 622–623). In the same way as deprived groups are often in practice excluded from real participation, immigrants are often forgotten when participation in landscape issues is discussed. Landscapes that physically reflect neglect, social inequality, discrimination and exploitation raises questions of justice that the Aarhus Convention explicitly deals with and the European Landscape Convention implicitly through its democratic ambitions.

References

Alphandéry P, Fortier A (2001) Can a territorial policy be based on science alone? The system for creating the Natura 2000 network in France. Sociol Rural 41:311–328

Convention on Biological Diversity (1992) Text of the Convention on Biological Diversity. United Nations Environment Programme (UNEP). http://www.cbd.int/convention /convention.shtml. Accessed 17 Mar 2010

Council of Europe (2000) European Landscape Convention, Florence, 20.X.2000. ETS No. 176. http://conventions.coe.int/Treaty/en/Treaties/Html/176.htm. Accessed 6 Mar 2007

EC (2000) Directive 2000/60/EC of the European Parliament and of the Council of 23 October 2000 establishing a framework for the Community action in the field of water policy. Off J L 327:1–82, 22.12.2000, EUR-Lex. http://eur-lex.europa.eu/LexUriServ/LexUriServ.do?uri=CELEX:32000L0060:EN:NOT. Accessed 17 Mar 2010

ECOVAST (2006) Landscape identification: a guide to good practice. http://ecovast.org/papers/good_guid_e.pdf. Accessed 30 Nov 2009

EEC (1992) Council Directive 92/43/EEC of 21 May 1992 on the conservation of natural habitats and of wild fauna and flora. Off J L 206:7–50, 22/7/1992, EUR-Lex. http://eur-lex.europa.eu/LexUriServ/LexUriServ.do?uri=CELEX:31992L0043:EN:NOT. Accessed 17 Mar 2010

Emmelin L (1982) Painting the future: visual impact analysis of changes in the Swedish landscape. Forskningsrådsnämnden, Delegationen för långsiktsmotiverad forskning, Stockholm

Emmelin L (1996) Landscape impact analysis: a systematic approach to landscape impacts of policy. Landsc Res 21:1–13
Emmelin L, Jones M, Ree H, Forshed N (1990) Landskap på Jæren og Romerike: Utvikling og framtidsbilder. In: Sødal DO, Vatn A, Emmelin L, Jones M, Forshed N (eds) Jordbrukspolitikk og miljø, del II Kulturlandskapet. Landbruksforlaget, Oslo, pp 1–32
European Commission (2010) Natura 2000 network. http://ec.europa.eu/environment/nature/natura2000/. Accessed 25 Mar 2010
Folke C, Colding J, Berkes F (2003) Synthesis: building resilience and adaptive capacity in social-ecological systems. In: Berkes F, Colding J, Folke C (eds) Navigating social–ecological systems: building resilience for complexity and change. Cambridge University Press, Cambridge, pp 352–387
Jones M (1993) Economy versus ecology – challenges for agriculture in Norway in the light of some West European experiences. Bebyggelsehistorisk tidskrift 26:43–64
Jones (2007) The European landscape convention and the question of public participation. Landsc Res 32:613–633
Lindborg R, Bengtsson J, Berg Å, Cousins SAO, Eriksson O, Gustafsson T, Hasund KP, Lenoir L, Pihlgren A, Sjödin E, Stenseke M (2008) A landscape perspective on conservation of semi-natural grasslands. Agric Ecosyst Environ 125:213–222
Menzies D (2007) Mediating conflicts over landscape issues. Paper presented at the conference Management of Landscapes in Conflict Regions, Birzeit University (Ramallah, Palestine), 29 Nov to 1 Dec 2007
Menzies D (2010) Effective mediation. Paper presented at effective RMA participation, Environmental Defence Society Workshop, 27 Feb 2010
Naturvårdsverket (2010) Arbetssätt för biologisk mångfald och andra värden i ett landskapsperspektiv: En handledning. Rapport 6342. Naturvårdsverket, Stockholm. http://www.naturvardsverket.se/Documents/publikationer/978-91-620-6342-9.pdf. Accessed 1 Apr 2010
O'Riordan T, Stoll-Kleeman S (2002) Deliberative democracy and participatory biodiversity. In: O'Riordan T, Stoll-Kleeman S (eds) Biodiversity, sustainability and human communities: protecting beyond the protected. Cambridge University Press, Cambridge, pp 87–112
Rekola M, Pouta E, Kuuluvainen J, Tahvonen O, Li C-Z (2000) Incommensurable preferences in contingent valuation: the case of Natura 2000 Network Finland. Environ Conserv 27: 260–268
Robertson DP, Hull RB (2003) Public ecology: an environmental science and policy for global society. Environ Sci Policy 6:399–410
Rønningen K (1998) Agricultural policies and countryside management: a comparative european study. Dr.polit. thesis, Department of Geography, Norwegian University of Science and Technology, Trondheim
Sager T (2009) Planners' role: torn between dialogical ideals and neo-liberal realities. Eur Plan Stud 17:65–84
Schoute JFT, Finke P, Veeneklaas FR, Wolfert HP (eds) (1995) Scenario studies for the rural environment: selected and edited proceedings of the symposium Scenario Studies for the Rural Environment, Wageningen, The Netherlands, 12–15 September 1994. Kluwer, Dordrecht
Selman P (2004) Community participation in the planning and management of cultural landscapes. J Environ Plan Manage 47:365–392
Statens landbruksforvaltning, Direktoratet for naturforvaltning, & Riksantikvaren (2007) Utvalgte kulturlandskap i jordbruket. Oslo
Stenseke M (2006) Biodiversity and the local context: linking seminatural grasslands and their future use to social aspects. Environ Sci Policy 9:350–359
Stenseke M (2009) Local participation in cultural landscape maintenance: lessons from Sweden. Land Use Policy 26:214–223
Stoll-Kleeman S (2001) Opposition to the designation of protected areas in Germany. J Environ Plan Manage 44:109–128

Tress B, Tress G (2003) Scenario visualisation for participatory landscape planning – a study from Denmark. Landsc Urban Plan 64:161–178

UN (1992) Rio Declaration on Environment and Development. United Nations Environment Programme (UNEP). http://www.unep.org/Documents.Multilingual/Default.asp?DocumentID=78&ArticleID=1163. Accessed 17 Mar 2010

UNECE (1998) Convention on Access to Information, Public Participation in Decision-making and Access to Justice in Environmental Matters done at Aarhus, Denmark on 25 June 1998. United Nations Economic Commission for Europe. http://www.unece.org/env/pp/treatytext.htm Accessed 11 Mar 2010

Young J, Watt A, Nowicki P, Alard D, Clitherow J, Henle K, Johnson R, Laczko E, McCracken D, Matouch S, Niemela J, Richards C (2005) Towards sustainable land use: identifying and managing the conflicts between human activities and biodiversity conservation in Europe. Biodivers Conserv 14:1641–1661

Zachrisson A (2004) Co-management of natural resources: paradigm shifts, key concepts and cases. Mountain Mistra Programme report no. 1. Mountain Mistra Programme, Umeå

Index

Note: The index covers chapters 1 to 15, but not abstracts or references. Relevant Tables and Figures are indicated by t or f after the page number. Referenced authors are indexed only where their work is specifically discussed in the text.

M. Jones, M. Stenseke (eds.), *The European Landscape Convention*, Landscape Series 13, DOI 10.1007/978-90-481-9932-7,

O

P

T